AF477130

Industry 4.0 – The Global Industrial Revolution: Achievements, Obstacles and Research Needs for the Digital Transformation of Industry

Industry 4.0 – The Global Industrial Revolution: Achievements, Obstacles and Research Needs for the Digital Transformation of Industry

Editor

Johannes Winter

Basel • Beijing • Wuhan • Barcelona • Belgrade • Novi Sad • Cluj • Manchester

Editor
Johannes Winter
L3S Research Center
Hannover
Germany

Editorial Office
MDPI
St. Alban-Anlage 66
4052 Basel, Switzerland

This is a reprint of articles from the Special Issue published online in the open access journal *Sci* (ISSN 2413-4155) (available at: www.mdpi.com/journal/sci/special_issues/Industrie_The_Global_Industrial_Revolution_Achievements_Obstacles_Research_Needs_Digital_Transformation_Industry).

For citation purposes, cite each article independently as indicated on the article page online and as indicated below:

Lastname, A.A.; Lastname, B.B. Article Title. *Journal Name* **Year**, *Volume Number*, Page Range.

ISBN 978-3-0365-9663-1 (Hbk)
ISBN 978-3-0365-9662-4 (PDF)
doi.org/10.3390/books978-3-0365-9662-4

Cover image courtesy of Pugun & Photo Studio

Contents

About the Editor

Johannes Winter

Dr. Winter is a multi-faceted professional renowned for his expertise in digital transformation, innovation research and management, and strategic leadership. He assumed the role of Chief Strategy Officer and Member of the Board of Directors at the esteemed L3S Research Center in 2022. This institution is internationally recognized for pioneering research in data science and Artificial Intelligence. Simultaneously, he holds the position of Managing Director at the Lower Saxony Center for AI and Causal Methods in Medicine, highlighting his commitment to advancing AI-driven methodologies in healthcare.

Previously, Dr. Winter served as the founding Managing Director of the National German Platform for Artificial Intelligence, under the leadership of the Federal German Minister of Education and Research and the president of the National Academy of Science and Engineering (acatech). His extensive career spans leadership roles at the Technology Department of the National Academy of Science and Engineering in Munich and as an executive assistant of Professor Henning Kagermann, the former CEO of SAP SE and one of the conceptual fathers of Industrie 4.0.

He also holds advisory positions in tech startups and research institutions, along with participating in judging panels for prestigious tech and business awards such as Best of Consulting by Wirtschaftswoche. At the same time, Dr. Winter continues to teach innovation management and technology management at the University of Applied Sciences for Economics and Management, Munich, and the Berlin Professional School.

Dr. Winter holds a Ph.D. in Regional Economics from the University of Cologne. His research interests include the strategic management of advanced technologies, digital business model innovation and servitization, and implementing and using Artificial Intelligence. His previous work experience includes positions in the automotive industry, management consulting, and academic research.

Preface

The genesis of industrial revolutions, sparked by mechanization, electricity, and IT, set the stage for the fourth revolution, termed Industrie 4.0 (Industry 4.0). This paradigm shift embraces hyperconnected, smart, decentralized, and autonomous systems, heralding increased complexity yet promising individualized products and services with unparalleled value and user experiences. This is achieved through the fusion of mass production's cost efficiency with innovative advancements.

Every revolution contends with the established norms of the past. Industrie 4.0 transcends the boundaries of the analog and physical realms, extending them into a digital sphere. This expansion entails not just physical growth but virtual growth as well. Smart factories, driven by myriad sensors, operate seamlessly through high automation and self-organization. These intelligent systems perpetually strive for enhanced productivity and top-notch quality. Achieving this relies on cyber–physical systems and the astute interconnection of machines, products, and workforces. Products themselves relay necessary production data to smart factories, guiding each step towards the desired outcome.

The integration and operation of Industrie 4.0 solutions unveil unprecedented opportunities alongside fresh challenges in the digital transformation journey for both organizations and value networks.

This Special Issue delves into the strides made, challenges encountered, and research imperatives within the realm of Industrie 4.0 from both a scientific and practical standpoint. This publication features the voices of Industrie 4.0 pioneers Henning Kagermann and Wolfgang Wahlster, as well as leaders in research and industrial application of smart manufacturing concepts.

Johannes Winter
Editor

Editorial

Implementing Smart Services in Small- and Medium-Sized Manufacturing Companies: On the Progress of Servitization in the Era of Industry 4.0

Johannes Winter 

L3S Research Center, Leibniz University of Hannover, 30167 Hannover, Germany; winter@l3s.de

Citation: Winter, J. Implementing Smart Services in Small- and Medium-Sized Manufacturing Companies: On the Progress of Servitization in the Era of Industry 4.0. *Sci* **2023**, *5*, 29. https://doi.org/10.3390/sci5030029

Received: 7 July 2023
Accepted: 10 July 2023
Published: 12 July 2023

1. Introduction

For a long time, the challenge has been to provide products and services that precisely match the preferences, habits, and needs of users. This is easier to accomplish with custom manufacturing and small batch sizes than in a rigid production environment (mass production) [1,2]. Mass production is a hierarchically organized system in which largely uniform and standardized products are produced in repetitive steps based on the division of labor. Mass production is characterized by low unit costs and high economies of scale. The low prices of mass products encourage the concentration of demand on uniform products—a self-reinforcing effect in favor of the producers of mass products. The diversity of variants is correspondingly low. It is difficult for mass producers to fulfill the individual wishes of individual customers—particularly in the early digital age—at the cost of a mass product. Individualization is more likely to be found where consumers are willing to pay a higher price or to forgo certain functionalities.

When, in an individualized society [3], the demand for higher quality, extended functionalities, and stronger personalization of a product [4,5] increases, classical product development and production processes reach their limits. The way out does not lie in customized products or small series; both the manufacturing process and the product are too customer specific, which means that economies of scale do not apply, and that costs and product prices are higher than those of mass-produced products.

However, when mass production approaches are combined with customized product development and manufacturing, additional value can be created [6]. Mass customization takes advantage of industrial organizational processes and highly automated, flexible production systems, and it combines them with digital innovations in the area of customer co-design and personalized product development. The integration of cognitive technologies and methods, combined with the exponential growth of storage capacity, computing power, and networks, and the simultaneous availability of vast quantities of data, now makes it possible to offer highly personalized product–service systems. Smart products are being refined with digital services (smart services) and connected to industrial metaverses [7]. This has far-reaching consequences for the production and use of devices, as they should be adaptable, reconfigurable, customizable, flexible, and able to interact with their environment at any time via user-friendly interfaces. In addition, products are now highly adaptable to the needs of the user. As a result, mass markets for uniformly equipped products are expected to become less important in the medium term.

This paper presents three empirical case studies for the implementation of smart services in medium-sized manufacturing companies and discusses the progress of the concept of servitization in industry.

2. Methods

This editorial follows a qualitative research design. The chosen method was to conduct explorative, semi-structured interviews that took place in a virtual environment due to the COVID-19 pandemic.

A total of 30 qualitative interviews were conducted between fall 2020 and summer 2021 by members of the "Platform Learning Systems" project, which is part of acatech—the German Academy of Science and Engineering. The "Platform Learning Systems" brings together key players in the field of artificial intelligence from industry, the service sector, trade, science, civil society, and politics, and accompany the introduction and use of AI in Germany and Europe with studies, use cases, conferences, and public information services such as the Map on AI [8,9].

For each company example, at least one appropriate expert was interviewed. The in-depth interviews will be analyzed using qualitative content analysis methods and presented in textual and graphical form. Qualitative content analysis is "a research method for the subjective interpretation of the content of text data through the systematic classification process of coding and identifying themes or patterns" [10,11]. In addition, the relevant literature and statistics were reviewed and processed.

3. Servitization in Industry: From Selling Products to Smart Product–Service Systems

Smart products are intelligent everyday objects, machines, systems, or means of transportation that are equipped with sensors, controlled by embedded software, and connected to the Internet worldwide [12,13]. Smart services are digital services that complement and enhance physical, increasingly intelligent products by enabling flexible and personalized adaptation to specific customer expectations based on processed data. The combination of smart products and services is unique in its disruptive potential. Smart product–service systems enable better user experiences and altered value propositions [14]. In the case of a manufacturing company, this can mean that networking and data-driven intelligence turn the machine tool into a smart product. Add a digital dashboard that provides information about the status of the machine or the value-added step, and the smart machine tool is complemented by a smart service. When smart products and smart services are bundled into a digital business model with a billing model based on usage or machine hours, it is called a smart product–service system [15]. If a digital marketplace with an APP store is created in relation to the machine tool so that additional APPs can be used and booked, this becomes a smart service platform, or an innovation ecosystem related to the tool. This step-by-step model is emblematic of the digital transformation of machine tool manufacturers and users.

Many companies have already connected their smart products to the Internet; they are in the process of collecting and analyzing relevant data. The speed and radicalness with which current business models need to change is often underestimated. Figure 1 illustrates the process of moving from optimized production to data-driven business model innovation. Connectivity and real-time responses pertaining to the original product or service are followed by optimization and efficiency at the product and process levels, including new after-sales services. Extending the business model to products-as-services and value-added services transforms the company into a service organization. Through the new digital business, the company ultimately becomes a platform company or a participant in a digital ecosystem. While the best networks today have latency periods of ten to fifteen milliseconds, the upcoming 5G/6G mobile communications standard will provide near real-time mobile Internet; data latency, the time between data request and data delivery, will be reduced to just one millisecond; hence, 5G/6G will be ultra-fast, latency-free, energy-efficient, and reliable—a fundamental requirement for the next generation of products and services [16].

	Connect & operate live	Optimise & supply efficiently	Expand & boost sales	Innovate & develop ecosystem
Business model	Products & support services	Product services & after-sales services	Product-as-a-service & value-added service	Data-driven digital business model
Business driver	Product sales	Process optimisation	Service growth	Expanded ecosystems
IoT capacities	Embedded systems, augmented reality	Analytics, machine learning, optimisation	Service management (portfolio, product management)	Ecosystem business development
Integration & technology	Vertical integration (OT-IT), machine connectivity	Horizontal integration (planning to delivery)	Services platform, SLA management	Open data platforms, business networks
Standards	Connectivity (e.g. OPC-UA)	Semantic standards	Service interoperability	Cross-sectoral standards
	Optimised production		**Smart services**	**Innovation business**

Figure 1. From optimized production to innovation ecosystems (source: own illustration, 2023, based on [6]).

3.1. Usage Models of Smart Offerings

The evolution of physical products and traditional services using data and analytics is still a relatively new trend. Until now, the sale of a product has been at the end of the value chain, even though the pay-per-use models shown in Figure 1, as well as the sharing economy, have been on the rise for about two decades. Traditionally, raw materials and components are purchased, supply chains are orchestrated, and labor and machinery are used in the manufacturing process. The finished product is then sold with a profit margin and delivered to the customer. With the sale of the product, the ownership and availability of the product changes from the manufacturer to the user. This tried-and-true practice is losing its dominance in the digital age, as the change in question impacts both the development and production process of products and services and the way they are marketed.

The business model describes the product or service that a market participant offers and the characteristics of how it interacts with customers and suppliers [17]. At its core are (1) a unique value proposition; (2) a revenue model or revenue mechanism that describes how the value of the offering can be converted into revenue; and (3) the resources and processes used to deliver the value proposition [18,19]. The key resources are brand, people, technology, partnerships, and data. In the digital age, data amount to a key resource for implementing a company's value proposition in the form of intelligent, networked products and services, as well as digitized organizational, development, production, and logistics processes.

It should be emphasized that a product or service delivered to the customer is no less interesting from the supplier's point of view. This is because smart products generate operational data throughout their lifecycle, which opens up new opportunities to drive innovation, provide services, and engage with customers on an ongoing basis. An example from the mobility sector illustrates this: small-scale weather events, such as regional precipitation or fog, can be recorded in the intelligent, connected vehicle via the on-board camera, windshield wiper sensors, or other connected objects, and sent via the vehicle backend to the manufacturer, who, in turn, can provide this safety-related data to weather services, traffic radio, or other connected vehicles and fleets. This has the potential to increase road safety, and this vehicle and mobility data can also be monetized [20].

This development is being driven by rapid technological progress and building technologies have reached the level of technological maturity appropriate for use in the field (and at affordable prices).

3.2. Smart Services: New Value Propositions for the Digital Age

Industry 4.0, in the sense of networked, highly automated, and adaptable production, enables the manufacture of smart products at marketable costs. By analogy, smart services are individualized services at the price of standard services [21]. Disruptive business models are based on extending smart products with a bundle of smart services to offer users new user experiences and new value propositions. Such an enhanced user experience might include a recommendation via the multimodal mobility APP to switch to the metro as a mode of transport because the e-scooter booked is less safe on rainy roads or in dense fog. Weather data were collected in real time and used for the APP travel recommendation. A new service promise could be the guaranteed arrival time for a long-distance trip because the weather and traffic data are processed and made available in the navigation APP in real time, making dynamic route guidance much more precise and reliable than we have become used to.

The lubricant for smart services is therefore smart data, i.e., large quantities of processed data that provide information about the user's preferences, the optimal selection decision, or the environmental conditions in which the smart service is used.

Current challenges include the monetization of smart services in addition to the company's traditional product range, as well as their economical operation. There is potential for revenue generation, for example, in the area of flexible pricing models. In this way, opportunities to generate additional revenue on the basis of data can be exploited throughout the entire product lifecycle. An example of this is services that can be booked "over the air", such as the above-mentioned activation of a higher performance of the engine or the battery in the vehicle. The machine tool shown in Figure 1 can also be priced according to consumption (pay-per-use), but it can also be priced according to the number of units sold or proportionally to the revenue generated. Payment based on the number of records exchanged is also conceivable for any application domain. Accordingly, metadata hubs and data marketplaces are currently being created, such as the Mobility Data Space, in which leading mobility providers, namely, cities and municipalities, research institutions, the German National Academy of Sciences and Engineering, and the German Federal Ministry of Transport, are participating [22]. The goal is to enable innovative and sustainable mobility services by sharing and using a wide range of data.

3.3. Outcome-Oriented Smart Services

In the digital age, the value of the smart product–service package lies more in the result achieved than in its mere material existence, including specific product characteristics. The product is no longer a means to an end; the smart services developed around the product enhance it, making it unique and valuable. A driverless vehicle can navigate passengers safely and efficiently through traffic, giving them time to spend as they wish instead of sitting behind the wheel in a traffic jam. Measurable results of the intelligent product can also include lower costs, higher revenues, or improved environmental sustainability [23].

The everything-as-a-service concept (see Figure 2) stands for products and services that focus on their outcomes. In as-a-service or outcome-oriented business models, companies move from selling a product through a single transaction to providing a service with a guaranteed outcome, often offered on a usage basis or via a profit/risk-sharing model. Selling the outcome rather than the product shifts ownership, risk, and responsibility for maintenance back to the provider. This is the example of the machine manufacturer who rents the machine and charges on a usage basis instead of selling the machine. This amortizes over a longer period of time, rather than immediately upon sale.

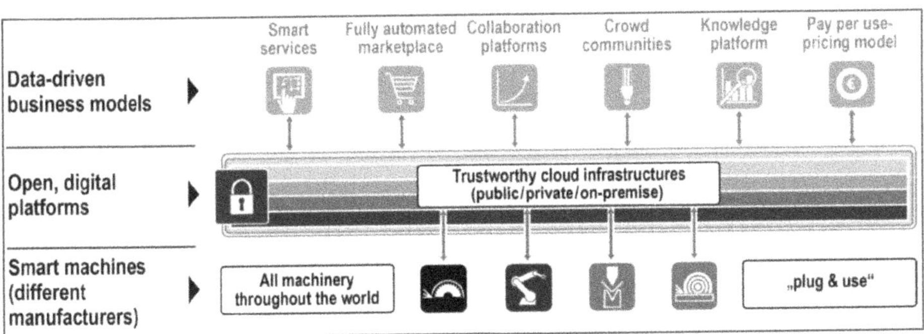

Figure 2. Value architecture of data-driven business models (source: [24]).

Another example is the provision of features that can be activated on demand and paid for as they are used. In this way, consumers can rent additional navigation and infotainment services in order to arrive at their destination faster or more relaxed. Such as-a-service models are therefore outcome-based and often combined with flexible pricing. This monetization approach is based on the in-app purchase model of the smartphone industry, and it enables an ongoing customer relationship and continuous upselling potential. Upselling refers to the sales method of offering customers a higher-value—and usually more expensive—product than they originally wanted [18].

The trend towards "everything-as-a-service" (Xaas) is not new; it was already marketed in the dot.com era with performance promises such as "holes instead of drilling" and "temperature instead of air conditioning" [25]. However, the hardware and software performance parameters and data are now much more comprehensive and cheaper to obtain, allowing the more mature technology to break through on a broad scale. Software updates and the intelligent use of data generated during operation will make the product adaptable and intelligent in the future. This requires a high level of competence from the product provider and the partners involved in the innovation ecosystem. After all, the product is not rigid and interchangeable but dynamic and changeable; therefore, it is not completely predictable in its product behavior. This creates significant challenges for customization development, maintenance, and customer service. Finally, software updates must ensure that the product profile and feature set are maintained. To manage this complexity, product manufacturers need to build an entirely new, innovative ecosystem around their products, or to collaborate with leading ecosystems through various platforms.

4. On the Progress of Servitization in Industry: Three Empirical Case Studies from Small- and Medium-Sized Manufacturing Companies

Large amounts of data are available virtually free of charge from sensors, the Internet, and other data sources. Collecting, structuring, evaluating, and interpreting these data presents an immense potential that can already be tapped into today via artificial intelligence platforms. These data are used to improve physical products or services. The following examples show how specific added value can be created by intelligent services in different areas. The examples were collected and prepared in the context of a qualitative company survey of the German AI platform [8].

4.1. Case Study #1: The Refined Machine Tool—Smart Services in Plastics Processing

The Berlin-based manufacturing company India-Dreusicke uses about 70 machines for plastics processing. Several times a week, the injection molds must be completely disassembled, and the precision parts relubricated. If maintenance is not carried out as planned, there is a risk of damage to machines, molds, or the product itself, resulting in high costs, waiting times, and production downtime. This makes maintenance essential but also costly and time-consuming. In addition, the exact timing of maintenance is unpredictable, so an intelligent service could add real value.

With the help of predictive maintenance based on artificial intelligence methods, the possible time window for maintenance of the systems should be maximized, and potential faults in the systems should be detected at an early stage. Injection molds are large steel elements into whose cavity the product is injected under very high pressure. Ejectors then push the finished parts out of the mold. The necessary lubricating film in the molds wears off gradually during operation and must therefore be applied permanently.

The smart service is based on data pertaining to acoustic signals that indicate the normal condition of a system or a possible need for maintenance. Together with a start-up company, the company collected the data over several months and then used them to train an AI system. Today, the AI system is able to detect acoustic impulses that are inaudible to employees and to provide information about the optimal maintenance period for the system. This allows for the early detection of damage or restrictions to the machines, preventing unplanned downtime or loss of production. At the same time, necessary maintenance can be scheduled and performed in a timely manner without jeopardizing the production goals. Microphones, software, and hardware products are used around the machine fleet for this digital service.

The added value of this intelligent maintenance approach lies in the more efficient execution of maintenance work, as well as in cost and time savings achieved by minimizing equipment downtime and production losses (see Figure 3). The data-based value proposition can be linked to guaranteed asset availability, which outperforms existing offerings that require machine maintenance at fixed points in time.

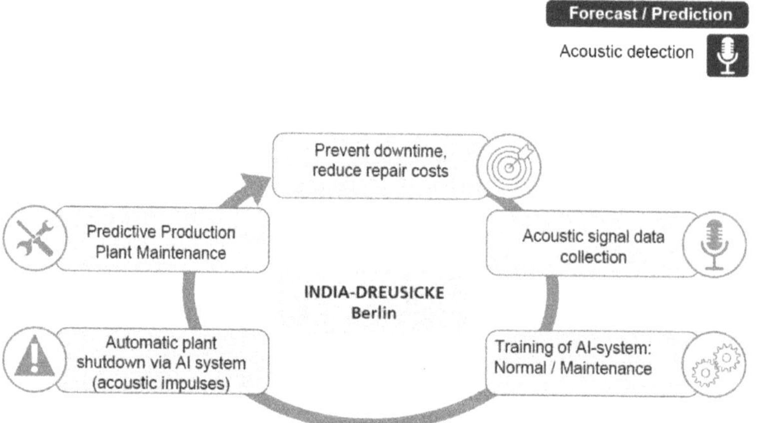

Figure 3. Acoustic analysis of maintenance needs in plastics processing (source: own illustration, 2023, based on based on [8]).

4.2. Case Study #2: Intelligent Planning Assistance—Smart Services in Metal Processing

META-Regalbau, a metalworking SME based in Arnsberg, Germany, develops and produces shelving systems for commercial and private use. These include shelving and pallet racking, storage platforms, and multi-level shelving systems for industrial warehouses, workshops, offices, and private rooms. The smart service aims to optimize internal logistics processes via sensor and AI-based data analysis processes.

In the course of incoming shelving orders, logistics employees have to assemble many individual components at different locations in the warehouse then pack them and prepare them for outgoing goods. The picking process had a lot of potential for optimization, as there were long waiting times for tools and the overall layout could be improved.

To realize this potential, the picking process was analyzed together with a software service provider. By automatically analyzing the manual work processes (motion mining), important measurements could be taken in the warehouse. Small transmitters (Bluetooth beacons) were attached to the walls. Employees also wore sensors as they walked around the warehouse to record walking distances and longer waiting times at shelves or machines, as well as to analyze the existing picking process with the help of artificial intelligence. This allowed the company to identify long tool queues and employee coordination needs and to redesign the process accordingly. The AI-based analysis enabled the company to design an intelligent layout with the best possible arrangement for a new warehouse and to make internal logistics processes more efficient. According to the company, the annual savings amount to more than EUR 20,000, which means that the cooperation with the start-up will have already paid for itself in under two years.

The logistics management actively involved the works council and employees in the process optimization (see Figure 4), informing them of the goal of the AI-based optimization and how the technologies worked, which contributed to the success of the project. In this case, the data-based value proposition can refer to an optimal picking process that can deliver faster shipment of goods and higher customer satisfaction.

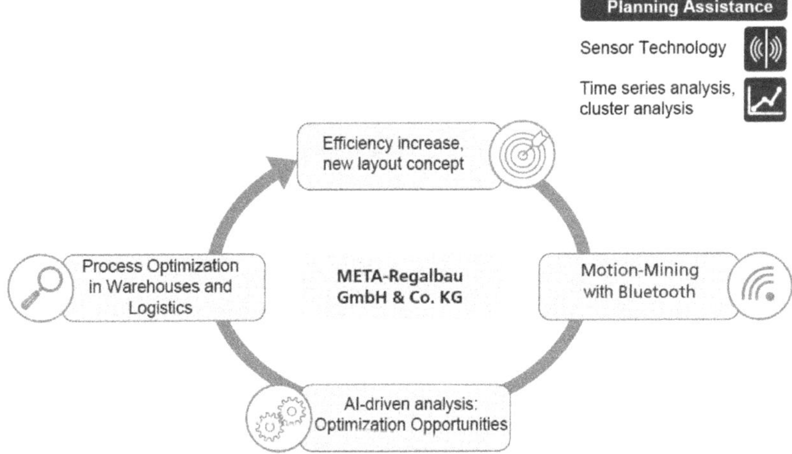

Figure 4. Smart optimization of the picking process in metal processing (source: own illustration, 2023, based on based on [8]).

4.3. Case Study #3: Autonomous Palletizing—Smart Services in Wood Processing

Eifelbrennholz, a small company based in Monschau, North Rhine-Westphalia, Germany, has been a producer and supplier of firewood for 25 years. The company covers all stages of the firewood production process, from harvesting to processing to shipping. So far, only private households are customers.

Business-to-business marketing to DIY stores or retailers has not yet been carried out, as large customers only buy firewood on pallets for efficiency of storage and delivery logistics. Manual palletizing is not economically viable in high-wage Germany. Therefore, the company decided to automate the firewood handling process using a 6-axis kinematic system. On the input side of the system, disordered firewood logs are identified using computer vision technology. The software can recognize and classify objects such as logs in digital still and moving images. The logs are then placed in a fixed arrangement on a pallet by an automated gripping tool. Because each log has a unique, natural surface, traditional bin-picking approaches (automated reaching into the box) cannot be used.

The intelligent service includes a cost-effective, autonomous pick-and-place application for firewood handling which optimizes the overall process and can serve large customers at competitive prices. This AI-based process for automating the palletizing of firewood, developed with RWTH Aachen University and Digital in NRW, is based on appropriate camera technology (sensors) and gripping technology (actuators). A 3D camera captures the disordered logs in lattice boxes. With the help of point clouds, a digital image (digital twin) of each log is created. An algorithmic segmentation can identify the geometry of the nearest trunk. The geometry of the trunks, the attack points, the deposition structures, and the movement paths of the objects must be "learned" from a new database and then algorithmically clustered.

Based on these data, a robot and a custom gripper can reliably pick up the uneven logs. AI supports the robot in the autonomous pick-and-place application (see Figure 5).

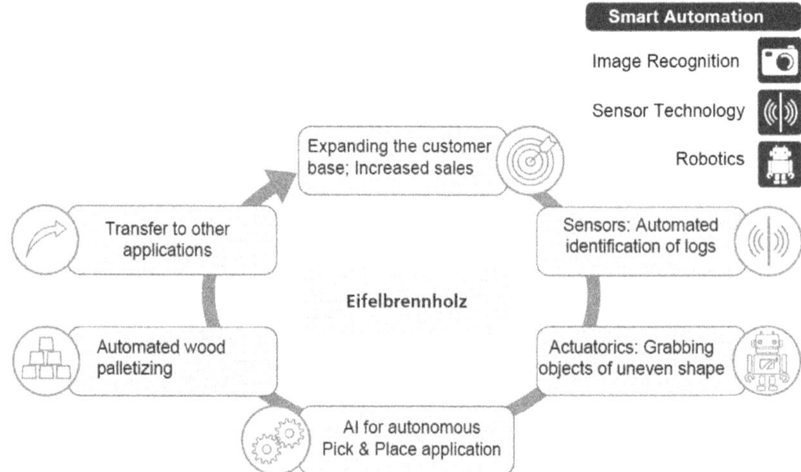

Figure 5. Autonomous stacking of logs in firewood production (source: own illustration, 2023, based on [8]).

The value proposition of this smart service is that autonomous log stacking can guarantee measurable savings (labor costs) and revenue increases by expanding the customer segment (key accounts). In this case, the investment in technology also pays off for firewood producers in high-wage countries. In addition, regional fuel trade eliminates the need for transportation, saving costs and reducing the carbon footprint.

5. Discussion: Data-Driven Product and Service Revolution

Smart data, smart products, and smart services are on the agenda of many innovative companies. Driven by real-time 5G networks, exponential growth in compute and storage performance metrics, the near-infinite availability of sensor and Internet data, and advances in machine and deep learning, business strategies are increasingly based on data-driven value propositions and enhanced user experiences around products and services. Data are being used not only to optimize processes and functions but also to create entirely new business models. Data enables disruption, and data-driven innovation is revolutionizing the one-sided-markets for products and services [26].

Data can contribute to a new kind of value proposition in a variety of ways, as examples from manufacturing, online retail, and automotive industries have shown. At its core, there are two thrusts: data can add value to a company's most important resource, or it can become the company's most important resource. Companies that focus on the latter—that is, using data as their primary key resource—are often startups; that is, they are without extensive product and service offerings, traditional processes and organizational structures, and a large number of demanding existing customers. Startups start "from scratch" and can radically put data at the center of their strategy. This is easier for startups because they are not burdened by the structural inertia or legacy of an established large company—an influencing factor often referred to as "the innovator's dilemma" [27,28].

6. Conclusions

As large companies move toward data-driven value propositions and the delivery of intelligent products and services, such as self-learning, predictive, personalized, and speech processing, they will need to break down existing silos in order to fully leverage data for the customer. Some companies are already doing this and experimenting with data as a key resource; they are harnessing the power of innovation ecosystems, wherein they form new value-creating partnerships with young and established players [29]. This is a promising path, as the practical examples in this article and other practical studies from industry and services have shown.

This requires a clear digital strategy within the company. After all, data are only valuable if they are used to reorganize business processes or create new revenue models. The memorable application examples from industry, services, and retail are intended to inspire market participants with concrete learning paths and best-practice views to resolutely implement their previously defined digital strategy and seize the opportunities of the data and platform economy. The product of the future is intelligent and enhanced by digital services. The benchmark has been set by digital pioneers; now, it is time to implement it across the economy. This is important in order to remain globally competitive with China, the U.S., and other leading innovation regions, and at the same time to enable users to enjoy convincing performance promises and new experiences arising from the digital product–service system.

Funding: This research was funded by the Federal Ministry of Research and Education, project title: Plattform Lernende Systeme (2017–2022; funding number: 01IS17097).

Acknowledgments: The author wishes to thank Ursula Ohliger and the other project members of the Learning Systems Platform for their support in data collection and data preparation.

Conflicts of Interest: The author declares no conflict of interest.

References

1. Gu, X.; Koren, Y. Mass-individualisation: The twenty first century manufacturing paradigm. *Int. J. Prod. Res.* **2022**, *60*, 7572–7587. [CrossRef]
2. Hu, S.J. Evolving paradigms of manufacturing: From mass production to mass customization and personalization. *Procedia Cirp.* **2013**, *7*, 3–8. [CrossRef]
3. Reckwitz, A. The Society of Singularities: Reply to Four Critics. *Anal. Krit.* **2023**, *45*, 177–187. [CrossRef]
4. Wan, J.; Li, X.; Dai, H.N.; Kusiak, A.; Martinez-Garcia, M.; Li, D. Artificial-intelligence-driven customized manufacturing factory: Key technologies, applications, and challenges. *Proc. IEEE* **2020**, *109*, 377–398. [CrossRef]
5. Korneeva, E.; Hönigsberg, S.; Piller, F.T. Mass customization capabilities in practice–introducing the mass into customized tech-textiles in an SME network. *Int. J. Ind. Eng. Manag.* **2021**, *12*, 115. [CrossRef]
6. Kagermann, H.; Winter, J. The second wave of digitalization. In *Germany and the World 2030: What Will Change. How We Must Act*; Mayr, S., Messner, D., Meyer, L., Eds.; Econ Verlag: Berlin, Germany, 2018.
7. Mourtzis, D.; Panopoulos, N.; Angelopoulos, J.; Wang, B.; Wang, L. Human centric platforms for personalized value creation in metaverse. *J. Manuf. Syst.* **2022**, *65*, 653–659. [CrossRef]
8. Platform Learning Systems. *AI for SME (KI im Mittelstand)*; Federal Ministry of Education and Research/Acatech—National Academy of Science and Engineering: Munich, Germany, 2021.
9. Platform Learning Systems. Map on AI. 2021. Available online: www.ki-landkarte.de (accessed on 9 July 2023).
10. Mayring, P. Qualitative content analysis: A step-by-step guide. *Qual. Content Anal.* **2021**, *1*, 1–100.
11. Hsieh, H.-F.; Shannon, S.E. Three approaches to qualitative content analysis. *Qual. Health Res.* **2005**, *15*, 1277–1288. [CrossRef]
12. Khan, S.; Tomar, S.; Fatima, M.; Khan, M.Z. Impact of artificial intelligent and industry 4.0 based products on consumer behaviour characteristics: A meta-analysis-based review. *Sustain. Oper. Comput.* **2022**, *3*, 218–225. [CrossRef]
13. Zheng, P.; Wang, Z.; Chen, C.H.; Khoo, L.P. A survey of smart product-service systems: Key aspects, challenges, and future perspectives. *Adv. Eng. Inform.* **2019**, *42*, 100973. [CrossRef]
14. Koldewey, C.; Hemminger, A.; Reinhold, J.; Gausemeier, J.; Dumitrescu, R.; Chohan, N.; Frank, M. Aligning strategic position, behavior, and structure for smart service businesses in manufacturing. *Technol. Forecast. Soc. Chang.* **2022**, *175*, 121329. [CrossRef]

15. Zhou, T.; Chen, Z.; Cao, Y.; Miao, R.; Ming, X. An integrated framework of user experience-oriented smart service requirement analysis for smart product service system development. *Adv. Eng. Inform.* **2022**, *51*, 101458. [CrossRef]

16. Firyaguna, F.; John, J.; Khyam, M.O.; Pesch, D.; Armstrong, E.; Claussen, H.; Poor, H.V. Towards industry 5.0: Intelligent reflecting surface (irs) in smart manufacturing. *arXiv* **2022**, arXiv:2201.02214.

17. Seiberth, G.; Gründinger, W. Data-driven business models in connected cars, mobility services and beyond. *BVDW Res.* **2018**, *1*, 18.

18. Ibarra, D.; Ganzarain, J.; Igartua, J.I. Business model innovation through Industry 4.0: A review. *Procedia Manuf.* **2018**, *22*, 4–10. [CrossRef]

19. Johnson, M.W.; Christensen, C.M.; Kagermann, H. Reinventing your Business Model. *Harv. Bus. Rev.* **2008**, *86*, 50–59.

20. Werne, J.; Winter, J. Point of no return: Turning data into value. *J. AI Robot. Workplace Autom.* **2021**, *1*, 43–57.

21. Winter, J. Business Model Innovation in the German Industry: Case Studies from the Railway, Manufacturing and Construction Sectors. *J. Innov. Manag.* **2023**, *11*, 1–17. [CrossRef]

22. Mobility Data Space. Available online: www.mobility-dataspace.eu (accessed on 9 July 2023).

23. Aas, T.H.; Breunig, K.J.; Hellström, M.M.; Hydle, K.M. Service-oriented Business Models in Manufacturing in the Digital Era. Towards a new Taxonomy. *Int. J. Innov. Manag.* **2020**, *24*, 1–15. [CrossRef]

24. Acatech—National Academy of Science and Engineering. *Digital Service Platforms (Digitale Serviceplattformen)*; Acatech: Munich, Germany, 2016.

25. Winter, J. Digital Business Model Innovation: Empirical insights into the drivers and value of Artificial Intelligence. *Int. J. Comput. Technol.* **2021**, *21*, 63–75. [CrossRef]

26. Kagermann, H.; Oesterle, H.; Jordan, J.M. *IT-Driven Business Models: Global Case Studies*; Wiley: Hoboken, NJ, USA, 2010.

27. O'Reilly, C.A., III; Tushman, M.L. *Lead and Disrupt: How to Solve the Innovator's Dilemma*; Stanford University Press: Redwood City, CA, USA, 2021.

28. Christensen, C.M.; Raynor, M.; McDonald, R. What is disruptive innovation? *Harv. Bus. Rev.* **2015**, *93*, 44–53.

29. Cusumano, M.; Gawer, A.; Yoffie, D.B. *The Business of Platforms: Strategy in the Age of Digital Competition, Innovation, and Power*; Harper Business: New York, NY, USA, 2019.

 **Sci**

Review

Ten Years of Industrie 4.0

Henning Kagermann [1,*] and Wolfgang Wahlster [2,*]

[1]   acatech—National Academy of Science and Engineering, 80333 Munich, Germany
[2]   German Research Center for Artificial Intelligence (DFKI), 10559 Berlin, Germany
*   Correspondence: kagermann@acatech.de (H.K.); wahlster@dfki.de (W.W.)

Abstract: A decade after its introduction, Industrie 4.0 has been established globally as the dominant paradigm for the digital transformation of the manufacturing industry. Amalgamating research-based results and practical experience from the German industry, this contribution reviews the progress made in implementing Industrie 4.0 and identifies future fields of action from a technological and application-oriented perspective. Putting the human in the center, Industrie 4.0 is the basis for data-based value creation, innovative business models, and agile forms of organization. Today, in the German manufacturing industry, the Internet of Things and cyber–physical production systems are a reality in newly built factories, and the connectivity of machinery has been significantly increased in existing factories. Now, the trends of industrial AI, edge computing up to the edge cloud, 5G in the factory, team robotics, autonomous intralogistics systems, and trustworthy data infrastructures must be leveraged to strengthen resilience, sovereignty, semantic interoperability, and sustainability. This enables the creation of digital innovation ecosystems that ensure long-term adaptability in a volatile economic and geopolitical environment. In sum, this review represents a comprehensive assessment of the status quo and identifies what is needed in the future to reap the rewards of the groundwork done in the first ten years of Industrie 4.0.

Keywords: Industrie 4.0; intelligent manufacturing; smart factories; industrial artificial intelligence; digital twins; zero-defect manufacturing; digital ecosystems

Citation: Kagermann, H.; Wahlster, W. Ten Years of Industrie 4.0. *Sci* **2022**, *1*, 26 https://doi.org/10.3390/sci4030026

Academic Editors: Sharifu Ura and Paolo Bellavista

Received: 18 May 2022
Accepted: 21 June 2022
Published: 28 June 2022

Publisher's Note: MDPI stays neutral with regard to jurisdictional claims in published maps and institutional affiliations.

1. Introduction

Our initial article, *Industrie 4.0: With the Internet of Things Towards the 4th Industrial Revolution*, was published in German on 1 April 2011 in cooperation with Wolf-Dieter Lukas, shortly before the opening of the Hanover Fair took place [1]. At this time, under the impact of the global financial crisis, we aimed to make the German economy more resilient and competitive by strengthening adaptability and resource efficiency.

This review discusses, from a conceptual and technological perspective, which elements of Industrie 4.0 have been fully implemented ten years after it had been drafted by us and which technological trends are now required for deepening the digital transformation of the manufacturing sectors.

2. Industrie 4.0: From a Conceptual Framework to an International Brand

Our main idea was to merge real and virtual spaces in so-called cyber–physical production systems, building on progress that German industry had already made with the lighthouse projects on the Internet of Things (IoT) and the Internet of Services (IoS) [2]. This was technologically interesting but would only have had an impact in specialist circles, not in practical implementation. Our term 'Industrie 4.0' got to the heart of the subject and attracted significant attention.

We received strong political support. As early as 3 April 2011, German Chancellor Angela Merkel spontaneously picked up on the new brand 'Industrie 4.0' in her opening speech at the Hanover Fair. However, also the business community, trade unions, and, very importantly, representatives of other industrialized countries recognized the magnitude of this concept. Our initial focus on the manufacturing sector was of considerable importance. It was widely accepted that economies with a strong industrial backbone such as Germany recovered faster and better from the global financial and economic crisis.

The term 'Industrie 4.0' has spread virally and is now associated with Germany all over the world, similar to 'kindergarten' and 'autobahn'. Industrie 4.0 is an export hit that has received attention and recognition in business, science, and politics around the globe. For the first time in the high-tech world, we have once again been able to establish an innovative concept from Germany internationally, after they had mostly come from North America or Asia for many years. Industrie 4.0 has made Europe the most innovative factory supplier of the world. There does not exist any 'smart factory' anywhere in the world where a large number of software and hardware components does not come from European companies. However, for the next decade of Industrie 4.0, the continuing support of stakeholders and international cooperation are required to reap the rewards of the groundwork done in the first ten years of Industrie 4.0. This also encapsulates leveraging the six key trends: industrial AI, edge computing up to the edge cloud, 5G in the factory, team robotics, autonomous intralogistics systems, and trustworthy data infrastructures.

3. Basic Prerequisite and Success Factor: Putting the Human at the Center

The networking and connectivity of people, intelligent objects and machines, the use of service-oriented architectures, and the composition of services and data from different sources to form new business processes is opening opportunities. Industrie 4.0 does not lead to factories empty of people. On the contrary, employees are supported by physical and cognitive assistance systems realized by collaborative robots (Cobots) and software agents (Softbots), which support the humans in complex manufacturing tasks (see Figure 1).

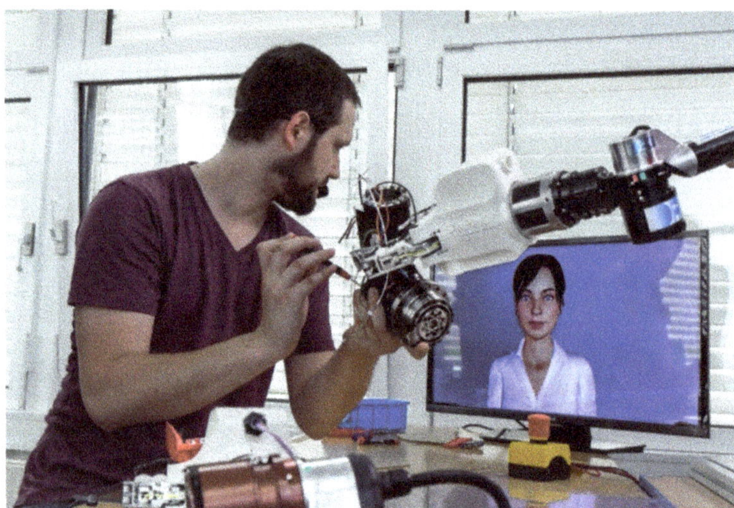

Figure 1. A Cobot and a Softbot helping a human worker (Source: DFKI).

Industrie 4.0 is the basis for data-based value creation, innovative business models, and agile forms of organization, but also for new solutions in areas such as energy, health, and mobility.

This vision is compelling because it puts people in the center, promising significant progress for the economy and society at large. In economic terms, it initially involved a shift from traditional automation with predetermined outcomes to learning and self-adapting machines and environments that respond in real time to changes in customer demand, as well as to unexpected disruptions. This is accompanied by a move from mass production to mass customization, i.e., the competitively priced production of individualized, tailor-made artefacts [3].

In social terms, the focus was set on implementing social partnerships for Industrie 4.0. Therefore, trade unions were closely involved in the entire process and contributed constructively. Focus points were set on the promise of better and more meaningful human–machine cooperation without the fear of losing control, the creation of jobs through 'nearshoring', and the inclusion of older and disabled people, supported by physical and cognitive worker assistance systems.

Ecologically, resource and energy efficiency has been a central goal from the outset: Industrie 4.0 has the potential to establish a circular economy that decouples economic growth from resource consumption. Sustainability through upcycling and the resilient factory have been two of the use cases proposed in our recommendations [3].

4. Key Challenge: Managing the Digital Transformation of the Manufacturing Industry

The success of Industrie 4.0 is closely interrelated with the broad support of the mainstays of society. The wide-scale roll-out of Industrie 4.0 during the last ten years was based on the effective cooperation of trade unions, industry, politics, and academia, institutionalizing their collaboration via an appropriate digital and organizational platform. Industrie 4.0 has set standards for how quickly a concept that initially emerged in cutting-edge research can develop out of companies and industry associations and, with the active accompaniment and support of the trade unions, can lead Germany to success as a location for business and innovation. Today, Industrie 4.0 is at the top of the agenda for federal policy—in the past ten years, more than 1000 project consortia, 10,000 conferences, and 100,000 publications have dealt with its technical and scientific implementation (see wiso-net.de in www.genios.de, accessed on 4 June 2022).

The Internet of Things (IoT) and cyber–physical systems are now a reality in newly built factories [4]. At the same time, in existing factories, the connectivity between machines, tools, workpieces, and skilled workers was improved, relying on various migration and bridging technologies for Industrie 4.0 [4]. Retrofitting—the digital upgrade with new low-cost sensors and their wireless connectivity—is steadily advancing. More and more production steps can be monitored in real time through multi-sensor fusion—for example, for quality control. The emerging product controls its own production via its digital twin. As in a marketplace, it selects the production services that match the customer's requirements, relying on the digital twins of the networked production facilities.

Today, there are a number of 'smart factories' that implement the basic principles of Industrie 4.0 [5,6], including 'Plug & Produce' and the virtual commissioning of new plant components relying on various types of digital twins (e.g., product twins, process twins, or machining twins), as well as cycle-independent matrix production architectures or multi agent architectures, with heterarchical and modular holonic control regimes, with configurable production cells and short set-up and changeover times even for the smallest batch sizes, and with a high degree of product individualization. This also holds for variable intralogistics combined with real-time production planning, as well as for location-based services for all workers, operating resources, and the products being created. Factory floor positioning has been greatly improved for mobile systems such as autonomous forklifts using AI-based visual SLAM (Simultaneous Localization and Mapping) techniques. GPU computing for the massively parallel execution of neural networks on very powerful graphics cards has significantly improved the necessary recognition of landmarks to enable the free and precise navigation of mobile robots.

After the experience of the COVID-19 pandemic, we need to develop solutions to avoid disruptions in supply chains or production stoppage due to short-term staff shortages [7]. Home-office technologies are hardly helpful in this regard. So-called 'home workbenches' that enable the mobile control, maintenance, and repair of factory equipment as software solutions with remote access to cyber–physical systems through tele-operation with physical avatars are needed instead.

5. What Is Next? New Megatrends for the Next Decade of Industrie 4.0

What is next? We must continue to drive semantic interoperability and international collaboration in open ecosystems. Six new megatrends (see illustration in Figure 2) will decisively influence the development of the next 10 years: industrial AI, edge computing up to the edge cloud, 5G in the factory, team robotics, autonomous intralogistics systems, and trustworthy data infrastructures.

Industrial AI will enable a second wave of digitalization of production. The first level, making all production and supply chain data available digitally and mobile via cloud systems, is largely achieved. These data can now be analyzed by AI systems in real time and interpreted in context even on the edge (e.g., signal-based machine learning with time delay on sensors [8]) so that they can be actively used for new value chains and business models.

With digital training data for machine learning systems, AI systems can be used not only for predictive maintenance, which is already widespread, but increasingly for incremental quality control, mostly via video sensors. Thus, the next phase of Industrie 4.0 will aim for AI-based zero-defect production (see Section 6). Self-learning capability and modular long-term autonomy rather than simple automation will characterize the new generation of 'smart factories' and, in addition to extreme flexibility, guarantee extremely robust production, high occupational safety, energy efficiency, and a high degree of resource conservation. A capability-oriented production architecture ensures expandability and mutability at the next level of Industrie 4.0 to respond quickly to volatility in the markets.

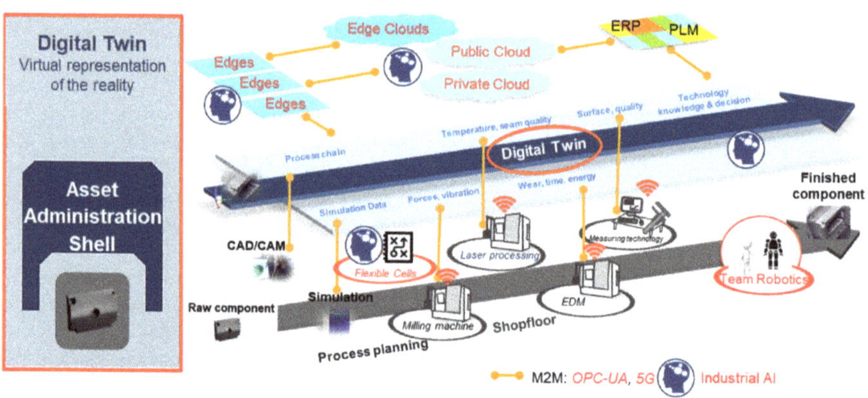

Figure 2. Megatrends for the next level of Industrie 4.0 (own illustration).

In 5G campus networks, edge devices can exploit the high bandwidth and low latency guaranteed with 5G to build a local edge cloud that can then meet real-time requirements on the factory floor. Mobile and real-time teleoperation, combined with multimodal sensor fusion, will also enable remote maintenance, repair, and installation.

In 'smart factories', intra-logistic planning and production planning are coordinated in real time, highly flexibly: mobile robots, factory drones, and driverless transport systems ensure that the parts and tools needed for the next planned production step are available just in time, at the right production island (see Figure 3).

Figure 3. Industrie 4.0 in a smart factory (Source: SmartFactory [KL] and DFKI IFS).

Production planning is revolutionized by a new service-oriented production architecture: the specification of the digital twin of the emerging product tries to find production capabilities that will transform the semi-finished product into its final state. Thus, digital twins become active agents in a multi-agent architecture, where the required skills of workers and machines are coordinated in real time. This enables the specification of products by semantic matchmaking.

Hybrid teams of workers and collaborative robots with different skill sets enable a new form of team robotics that focuses on human–machine interaction led by skilled human personnel. To solve complex manufacturing tasks, they are working hand-in-hand with robots as a team.

Data infrastructures must integrate industry requirements for data sovereignty, decentralization in heterogeneous multi cloud systems, and edge support. After the first decade of Industrie 4.0, factories digitally record, transmit, and store all production and machine data as sensors capture all relevant process data on edge devices. This is a first step towards higher productivity and more transparency of manufacturing processes. However, the interpretation of these data sources still requires manual data analysis by human experts using various digital data visualization and data analysis tools. Due to the massive amount of data provided in real time in an Industrie 4.0 factory, human data analysts will soon reach their limits.

An important goal for the next decade of Industrie 4.0 is therefore the automatic interpretation of industrial data based on artificial intelligence (AI). It is an enabler, e.g., for zero-defect production, and it is the decisive innovation to ensure that the superior quality of our products remains a unique selling point compared to similar products from the US or China. This requires the implementation of the entire cognition cycle from perceiving over understanding to acting, with all phases supported by various forms of machine learning relying on digital mass data from cloud and edge platforms [9] (p. 68). In addition, we must enable industrial AI systems to learn new knowledge not only autonomously from empirical data but also from being taught by human experts in interactive human–machine conversations, or from machine understanding of relevant technical documents.

6. Strategic Field of Action I: Towards Zero-Defect Manufacturing Based on Industrial AI

Zero-defect manufacturing can create a competitive advantage over 'low-wage and low-tech countries', since most consumers prefer high-quality, reliable, and sustainable products, even if they come with a somewhat higher price. Detecting anomalies and defects in the production process too late causes immense costs and has a negative impact on sustainability and productivity, as it leads to an enormous waste of time, energy, and material. It is therefore of the utmost importance to detect, explain, and eliminate such errors as early as possible—ideally immediately when they occur—by taking appropriate measures.

Typical sources of errors are the incorrect actions of a worker or a robot, or the incorrect interaction of workers and robots in the process. AI-based plan recognition, intention, and interaction recognition modules use video streams, wearable sensors, and IoT devices for incremental error detection. Thus, instead of one big loop for error correction after the traditional final quality check of the product, the next generation of systems consists of many small quality management loops. This eliminates the need for final inspection and partial disassembly of the already finished but faulty product for repair (see Figure 4).

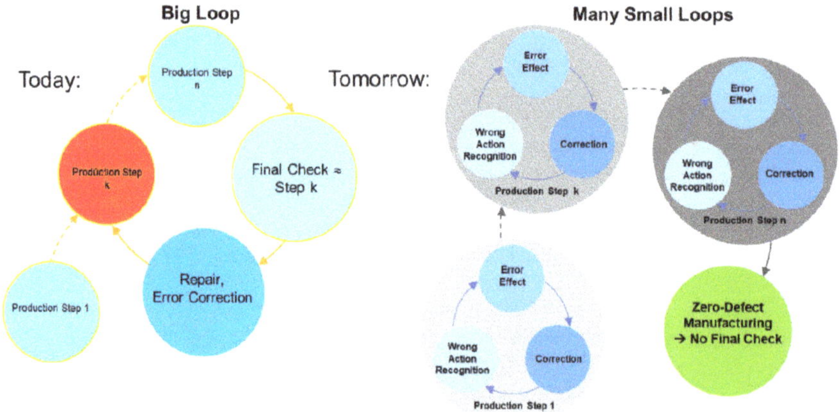

Figure 4. AI-based quality management with incremental error recognition (own illustration).

Of course, the earliest time to reliably detect an incorrect action, interaction, or out-of-control state is the moment it emerges. For this purpose, relevant quality-, task-, and interaction-specific parameters and constraints are continuously evaluated by AI-based methods during incremental real-time checks. These AI-based methods combine statistical deep learning methods on diverse sensor data streams with semantic models encoded in digital twins and symbolic reasoning. As real recorded data for training models are often not available, synthetic data must be generated, e.g., through accurate simulations involving digital twins.

Detected errors must be immediately reported to the responsible workers with a comprehensive explanation and a checklist to avoid them in the future and on how to proceed further if the error occurs. This requires intelligent user interfaces with massively multimodal explanation capabilities for human workers and production experts.

Thus, not only deep learning but also deep understanding by AI systems need to be strengthened in the next decade of Industrie 4.0 to allow for the implementation of explainable, more robust, and trustworthy systems (see [10]). We must include novel architectures beyond current deep learning, capturing causality and meta-learning to enable more powerful forms of compositional generalization. On the one hand, current machine learning systems lack the ability to leverage the invariances included in causal relations, which would be needed to boost their generalizability, robustness, and explainability [11].

Current causal inference methods, on the other hand, lack the ability to scale up to higher-dimensional settings, where current machine learning systems excel. Recently, a shift in research direction and new tools are opening the door to the development of novel architectures for addressing more sophisticated tasks, capturing causality and systematic generalization in error diagnosis, repair planning, and recovery. We predict that meta-learning, compositional generalization, and representation learning are needed for the next generation of industrial AI systems during the next decade of Industrie 4.0.

7. Strategic Field of Action II: Shaping Digital Ecosystems

In 2019, experts from the 'Platform Industrie 4.0' updated the vision of Industrie 4.0 for 2030 with the headline 'Shaping digital ecosystems globally' [12]. We must continue to drive semantic interoperability and international collaboration in open ecosystems, which permits plurality, diversity, flexibility, and a corporate culture of sharing success with business partners. We strive for a sustainable economy where economic growth is decoupled from resource consumption. We also strive for sovereignty—self-determination—at all levels. In a networked economy, self-determination means, above all, the freedom to select the technology of choice, the business partner of choice, the location of choice—especially the place where data are stored and processed in accordance with the legal system in force there. Against the background of recent developments and geopolitical challenges and the resulting shortages and bottlenecks in supply chains, with significant effects on industrial value creation, in particular, rethinking the security and resilience of supply becomes more important. Diversified supply chains and the ability to redesign value chains on demand seamlessly are fundamental in this regard. In the next phase of Industrie 4.0, companies must therefore exploit the advantages of digital factories and distributed modular production architectures to build trustworthy and reliable industrial digital ecosystems [13,14].

An additional challenge is business model innovation: understanding the customers' processes and extracting enterprise value from customer value. The value proposition in a digital economy is smart services [15]: individualized product–service bundles on demand, with superior user experience and low effort in switching to alternative business partners. The supporting value-creating architecture is illustrated in Figure 5, demonstrating the need to rethink and reengineer business processes as well as workplaces exploiting the power of AI [16] and replacing manual or cognitive routine tasks by autonomous systems. For all activities of the value chain, dynamic business networks must be established with dedicated orchestration models and governance. Obviously, a secure and trustworthy data supply chain and frictionless interoperability in technological and business terms are fundamental for success (see Figure 5).

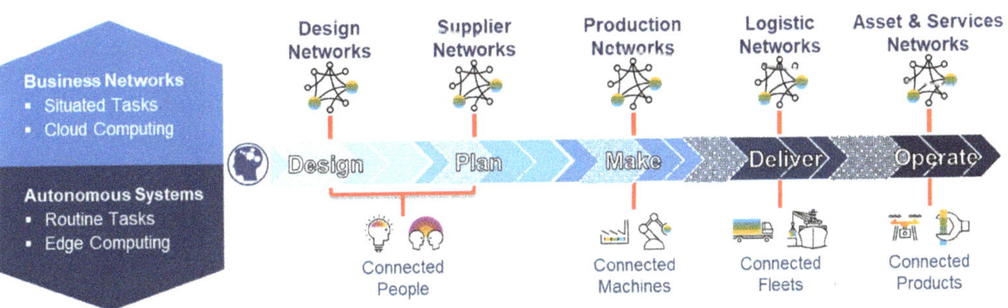

Figure 5. The digital enterprise (own illustration).

Digital enterprises have higher capabilities to operate in digital value creation network models. Within these decentralized networks of firms, governed by reciprocity and shared success, collaborative and coordinated elements for joint value creation are balanced to pursue the joint development of platform-oriented business models [17].

Digital ecosystems are the base layer and are dependent on a digital economy because many of the ecosystem members operate in different countries under different regulations and legal conditions [18,19]. This is why international cooperation on standards and secure data exchange across borders are of the utmost importance, particularly to guarantee sovereignty in an interconnected digital economy.

Many efforts have been undertaken in European initiatives such as Gaia-X in building a data infrastructure allowing for the sovereign exchange of data supported by an architecture for data spaces comprising technological standards, guidelines, and rules [20].

8. Outlook: Industrie 4.0 Has Still a Long Way to Go

For the next decade of Industrie 4.0, we even need to go beyond today's cloud and multi-cloud systems, since advanced distributed production systems need sky computing as a cloud of clouds [21]. Today's cloud market is fragmented, with many proprietary services running on proprietary hardware accelerators (e.g., TPUs, GPUs) and offering incompatible APIs. Based on the compatibility and intercloud layers of the emerging sky computing platforms, APIs can be used without changes, allowing applications to run on multiple clouds transparently. Such platforms are urgently needed if we want to realize the vision of full circular economy loops in distributed solutions for Industrie 4.0 with thousands of data providers and data consumers.

Many of the challenges of Industrie 4.0 are transnational and require continued international cooperation. We must simultaneously preserve our digital sovereignty while sharing our knowledge, experience, and best practices internationally. Other countries will favor different solutions in some cases, due to different political systems or culturally different approaches to problem solving. Nevertheless, our answer can only be self-determination and open collaboration based on our own values. For example, we presented the first comprehensive AI standardization roadmap in December 2020 [22].

We must not reduce our efforts in research and innovation for the next phase of this fourth industrial revolution. For the second wave of industrial digitalization, a major investment in industrial AI is required. Digital twins, which are already of the utmost importance in almost all sectors of industry, will become even more decisive. The semantic interoperability of software and hardware components plays a crucial role, especially to ensure international market access for German SMEs and startups, but also to safeguard Europe's technological sovereignty.

Standards, norms, and certificates are decisive drivers for interoperable solutions. The Asset Administration Shell (AAS), developed by the Platform Industrie 4.0 [23], is a promising attempt in this regard. Semantic interoperability also contributes to strengthening ecological sustainability, e.g., an AAS-based demonstrator, developed by Platform Industrie 4.0 and CESMI, creates transparency regarding greenhouse gas emissions across the value chain [24]. These factors deserve specific attention in the future.

In the next decade of Industrie 4.0, the continuing support of policymakers, trade unions, and civil society is needed, in addition to substantial funding for research and innovation. Only in this case can the economic, social, and ecological fruits of the significant investments in the first decade of the fourth industrial revolution emanating from Germany be harvested.

Author Contributions: H.K. and W.W. have developed all sections of the paper jointly. All authors have read and agreed to the published version of the manuscript.

Funding: This research was partially funded by the German Federal Ministry of Education and Research (BMBF).

Institutional Review Board Statement: Not applicable.

Informed Consent Statement: Not applicable.

Acknowledgments: We are grateful for the input and valuable discussions with members of Plattform Industrie 4.0 and Feldafinger Kreis.

Conflicts of Interest: The authors declare no conflict of interest.

References

1. Kagermann, H.; Lukas, W.-D.; Wahlster, W. Industrie 4.0: Mit dem Internet der Dinge auf dem Weg zur 4. Industriellen Revolution. VDI Nachrichten. 3 May 2011, p. 2. Available online: https://www.dfki.de/fileadmin/user_upload/DFKI/Medien/News_Media/Presse/Presse-Highlights/vdinach2011a13-ind4.0-Internet-Dinge.pdf (accessed on 26 April 2022).
2. Vierter Nationaler IT-Gipfel. Programm-Personen-Projekte. Bundesministerium für Wirtschaft und Technologie, Ed. Berlin, Germany. 2009. Available online: https://www.de.digital/DIGITAL/Redaktion/DE/IT-Gipfel/Publikation/2009/it-gipfel-2009-programm-personen-projekte.pdf?__blob=publicationFile&v=5 (accessed on 20 April 2020).
3. Kagermann, H.; Wahlster, W.; Helbig, J. *Recommendations for Implementing the Strategic Initiative Industrie 4.0: Final Report of the Industrie 4.0 Working Group*; Research Union of the German Government: Berlin, Germany, 2012.
4. Wahlster, W. (Ed.) *SemProM: Foundations of Semantic Product Memories for the Internet of Things*; Springer: Berlin/Heidelberg, Germany, 2013; ISBN 978-3-642-37376-3.
5. Wahlster, W.; Grallert, H.-J.; Wess, S.; Friedrich, H.; Widenka, T. (Eds.) *Towards the Internet of Services: The Theseus Research Program*; Springer: Berlin/Heidelberg, Germany, 2014. [CrossRef]
6. Schuh, G.; Anderl, R.; Dumitrescu, R.; Krüger, A.; ten Hompel, M. (Eds.) Using the Industrie 4.0 Maturity Index in Industry. Current Challenges, Case Studies and Trends. Acatech COOPERATION. 2020. Available online: https://www.acatech.de/publikation/der-industrie-4-0-maturity-index-in-der-betrieblichen-anwendung/download-pdf/?lang=en (accessed on 9 May 2022).
7. Forschungsbeirat der Plattform Industrie 4.0 (Ed.) Wertschöpfungsnetzwerke in Zeiten von Infektionskrisen. Expertise des Forschungsbeirats der Plattform Industrie 4.0. München, Germany. 2021. Available online: https://www.acatech.de/publikation/wertschoepfungsnetzwerke-in-zeiten-von-infektionskrisen-expertise/ (accessed on 26 April 2022).
8. Ullah, S. Modeling and simulation of complex manufacturing phenomena using sensor signals from the perspective of Industry 4.0. *Adv. Eng. Inform.* **2019**, *39*, 1–13. [CrossRef]
9. Sun, W.; Liu, J.; Yue, Y. AI-Enhanced Offloading in Edge Computing: When Machine Learning Meets Industrial IoT. *IEEE Netw.* **2019**, *33*, 68–74. [CrossRef]
10. Marcus, G. The Next Decade in AI: Four Steps Towards Robust Artificial Intelligence. *arXiv* **2020**, arXiv:2002.06177.
11. Bengio, Y. From System 1 Deep Learning to System 2 Deep Learning. In *NeurIPS*; 2019 Posner Lecture: Vancouver, BC, Canada, 11 December 2019. Available online: https://www.iro.umontreal.ca/~{}bengioy/NeurIPS-11dec2019.pdf (accessed on 26 April 2022).
12. Federal Ministry for Economic Affairs and Energy (BMWi) (Ed.) Shaping Digital Ecosystem Globally. 2030 Vision for Industrie 4.0; Berlin, Germany. 2019. Available online: https://www.plattform-i40.de/IP/Redaktion/EN/Downloads/Publikation/Vision-2030-for-Industrie-4.0.pdf?__blob=publicationFile&v=9 (accessed on 20 April 2022).
13. Parker, G.G.; Van Alstyne, M.W.; Choudary, S.P. *Platform Revolution: How Networked Markets Are Transforming the Economy and How to Make Them Work for You*; WW Norton & Company: New York City, NY, USA, 2016; ISBN 039-3-249-131.
14. Evans, P.; Gawer, A. *The Rise of the Platform Enterprise: A Global Survey*; The Center for Global Enterprise: New York, NY, USA, 2015.
15. Smart Service Welt. Recommendations for the Strategic Initiative Web-based Services for Businesses. Final Report. Kagermann, H., Riemensperger, F., Schuh, G., Scheer, A.-W., Spath, D., Leukert, B., Wahlster, W., Rohleder, B., Schweer, D., Eds.; 2015. Available online: https://www.acatech.de/publikation/abschlussbericht-smart-service-welt-umsetzungsempfehlungen-fuer-das-zukunftsprojekt-internetbasierte-dienste-fuer-die-wirtschaft/download-pdf/?lang=wildcard (accessed on 16 May 2022).
16. Dougherty, P.R.; Wilson, H.J. *Human + Machine: Reimagining Work in the Age of AI*; Harvard Business Press: Watertown, MA, USA, 2018; ISBN 163-3-693-864.
17. Acatech. Wegweiser Smart Service Welt. Smart Services im Digitalen Wertschöpfungsnetz. Available online: https://www.acatech.de/wp-content/uploads/2018/03/acatech2017_SSW_Wegweiser_de_bf.pdf (accessed on 20 April 2022).
18. Cusumano, M.A.; Gawer, A.; Yoffie, D.B. *The Business of Platforms: Strategy in the Age of Digital Competition, Innovation, and Power*; Harper Business: New York City, NY, USA, 2019; ISBN 006-2-896-326.
19. Goldfarb, A.; Tucker, C. Digital economics. *J. Econ. Lit.* **2019**, *57*, 3–43. [CrossRef]
20. Plattform Industrie 4.0 (Ed.) *Veranstaltungsbericht: Die Wertschöpfungsketten der Zukunft—Datenraum Industrie 4.0*; Plattform Industrie 4.0: Berlin, Germany, 2021. Available online: https://www.plattform-i40.de/IP/Redaktion/DE/Kurzmeldungen/2021/2021-11_Doku_Datenraum.html (accessed on 20 April 2022).
21. Stoica, I.; Shenker, S. From Cloud Computing to Sky Computing. In *Workshop on Hot Topics in Operating Systems (HotOS '21)*; Association for Computing Machinery: New York, NY, USA, 2021; pp. 26–32. [CrossRef]
22. Wahlster, W.; Winterhalter, C. (Eds.) *German Standardization Roadmap on Artificial Intelligence*; DIN/DKE: Berlin, Germany, 2020. Available online: https://www.din.de/resource/blob/772610/e96c34dd6b12900ea75b460538805349/normungsroadmap-en-data.pdf (accessed on 26 April 2022).

23. Plattform Industrie 4.0 (Ed.) Details of the Asset Administration Shell. From Idea to Implementation; Berlin, Germany. 2019. Available online: https://www.plattform-i40.de/IP/Redaktion/EN/Downloads/Publikation/vws-in-detail-presentation.pdf?__blob=publicationFile&v=12 (accessed on 2 May 2022).
24. Plattform Industrie 4.0 (Ed.) Factsheet: Joint Demonstrator on Interoperability. How the Exchange of CO_2 Data Along the Value Chain and across Countries Can Work on a Standardized Basis. Available online: https://www.plattform-i40.de/IP/Redaktion/EN/Downloads/Publikation/CESMII-Plattform-Demonstrator.html (accessed on 20 April 2022).

Concept Paper

Digital Twins in Manufacturing: A RAMI 4.0 Compliant Concept

Martin Lindner [1],*, **Lukas Bank [2]**, **Johannes Schilp [2]** and **Matthias Weigold [1]**

1   Institute of Production Management, Technology and Machine Tools (PTW), Technical University of Darmstadt, Otto-Berndt-Str. 2, 64287 Darmstadt, Germany; m.weigold@ptw.tu-darmstadt.de
2   Fraunhofer Institute for Casting, Composite and Processing Technology IGCV, Am Technologiezentrum 10, 86159 Augsburg, Germany; lukas.bank@igcv.fraunhofer.de (L.B.); johannes.schilp@igcv.fraunhofer.de (J.S.)
*   Correspondence: m.lindner@ptw.tu-darmstadt.de

Abstract: Digital twins are among the technologies that are considered to have high potential. At the same time, there is no uniform understanding of what this technology means. Definitions are used across disciplinary boundaries, resulting in a multitude of different interpretations. The concepts behind the terms should be clearly named to transfer knowledge and bundle developments in digitalization. In particular, the Reference Architectural Model for Industry (RAMI) 4.0, as the guiding concept of digitalization, should be in harmony with the terms to be able to establish a contradiction-free relationship. This paper therefore summarizes the most important definitions and descriptions from the scientific community. By evaluating the relevant literature, a concept is derived. The concept presented in this work concretizes the requirements and understanding of digital twins in the frame of RAMI 4.0 with a focus on manufacturing. It thus contributes to the understanding of the technology. In this way, the concept is intended to contribute to the implementation of digital twins in this context.

Keywords: digital twin; digital manufacturing; Industry 4.0

1. Introduction

Digital tools are becoming increasingly important in industrial production to improve decision-making processes and deal with increasing complexity [1]. The individualization of products and the resulting decrease in the number of units are a major complexity driver [2,3]. Currently, the high energy prices, at least in Western Europe, and thus the need to consider these in decisions are to be mentioned as an additional complexity factor. In the factory itself, heterogeneous production landscapes and many different systems are mentioned as a challenge in the management of complexity [4]. Digital twins (DTs), on the other hand, offer the opportunity to combine data from different sources to deal with high complexity and thus to support the decision-making process [4,5]. Although DT have been identified in many places as a technology with enormous potential, there is no uniform understanding of the term. This is partly due to the different application areas with their individual questions and requirements. Although DTs were originally developed as a safeguard for in-service objects in [6], most definitions refer to product development or are dedicated to a specific use, e.g., aviation [7]. However, the focus on the product has remained. Approaches to using existing models from development in further life cycle phases have existed for some time. Depending on the timeline in the life cycle of an object, the motivation and thus also the requirement for the DT changes. The classification is usually not considered in the definition, which means that definitions of DT are sometimes contradictory. Therefore, placing the definitions in their context is crucial. Furthermore, the definitions should be compatible with the concepts of digitalization.

This paper brings together the different definitions and provides an overview, and a concept of how DT can support factory operations. In the process of developing the concept, the different developments in connection with DT are addressed. For example, the Reference Architectural Model Industry 4.0 (RAMI 4.0) architecture is worth mentioning

Citation: Lindner, M.; Bank, L.; Schilp, J.; Weigold, M. Digital Twins in Manufacturing: A RAMI 4.0 Compliant Concept. *Sci* **2023**, *5*, 40. https://doi.org/10.3390/sci5040040

Academic Editor: Johannes Winter

Received: 12 July 2023
Revised: 18 September 2023
Accepted: 26 September 2023
Published: 10 October 2023

in German-speaking countries. This provides a framework for digitizing the factory, so a definition for the DT should be compatible with the RAMI 4.0 architecture. In this way, this work contributes to distinguishing the developments in the area of the digital twin from other digitization efforts, using a clear understanding and thus creating clarity. On the other hand, a superordinate concept is to be created that enables the development of digital twins and architectures based on RAMI 4.0 without contradictions.

2. State of the Art and Research

2.1. RAMI 4.0 Architecture

RAMI 4.0 is a cubic layer model and is defined in the DIN SPEC 91345 [8] (see Figure 1). The dimensions of the cube describe the architecture of assets, their life cycle, and their assignment to a hierarchy level.

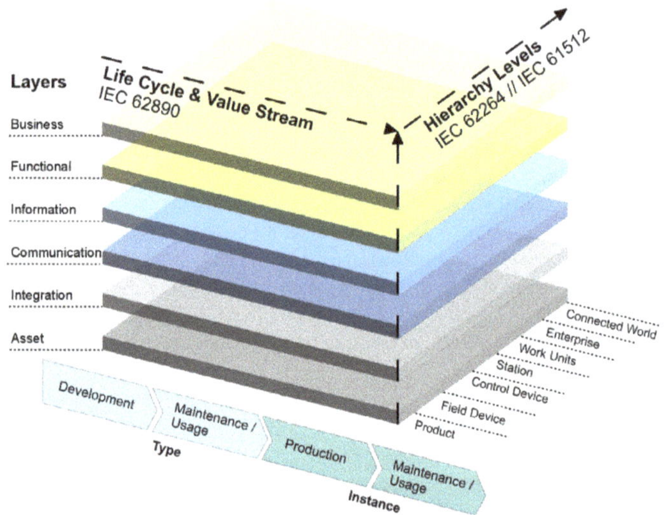

Figure 1. Figure of the Reference Architecture Model Industry 4.0 (RAMI 4.0) (source: [8]).

Layers describe the assets in their respective tasks or functions. The description categories are the classification in the business process, the function of the asset, the information, the communication, and the integration of the physical asset into the virtual world. Not all layers must be used at all times. For integration, the guideline provides the concept of the asset administration shell. In the AAS, the asset can be described digitally, with a communication interface to the physical system. In this way, the AAS can be understood not only as a digital representation but also as a gateway between the virtual and real world. The AAS manages all the essential data for an asset from creation to end of life [8,9].

The life cycle of assets is divided into two sections. The type section describes, as the name implies, the type of asset, i.e., in our example, a model series of a machine as shown in Figure 2. This section of the life cycle consists of a development phase and a utilization phase. In the development phase, when asset properties are defined, the AAS is created in parallel, which manages the general information for this type. As soon as an instance of this type is produced, an AAS is also derived for this instance, which contains information of that type and is additionally specific to this instance. Dynamic data are then added during the operation of the asset.

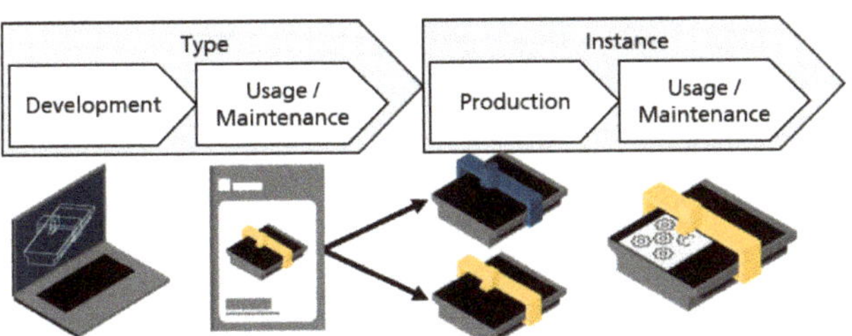

Figure 2. Life cycle of assets (source: own illustration adapted from ref. [8]).

The hierarchy levels place the asset in the factory structure. The axis starts with the product, so the output of an asset is integrated into the consideration as a component. On the other side, the highest level describes the connection to other assets or summarizing instances. In between are the organizational units of the manufacturing process.

2.2. Asset Administration Shell (AAS)

The AAS is part of RAMI 4.0 and defines the description of an asset in the digital space. The two main areas of an AAS are the header and body. The header contains the information required to identify the asset and the AAS. The unique identification is ensured by a uniform resource identifier (URI). The body contains the submodels that describe the functions and properties of the asset. The submodels can be added to the AAS according to the requirements [9]. In addition to the structure, it is advisable to use standardized data models as much as possible and to integrate these into the AAS and to only create submodels in fields in which no standards yet exist. In particular, data models that describe the operation are currently still rare. There is some work that uses RAMI or AAS as the basis for the implementation of digital twins. This includes work that concretizes the reference architecture and derives an architecture for digital twins. These include the work of Beregi et al. [10] and Steindl et al. [11]. Beregi et al. [10] take up the idea of AAS and define a production administration shell (PAS), which should allow plants to communicate with a manufacturing execution system. The idea is to build a modular and interoperable architecture in which resources can be integrated with little effort. Based on the architecture axes, Steindl et al. [11] develop a concrete implementation of an architecture for building a digital twin. Both works deal with specific aspects of RAMI to realize concrete implementations without focusing on the complexity of an entire factory or considering it over its entire life cycle. One work that takes a holistic view of the RAMI architecture is the work of Roscher [12], which applies the RAMI architecture to the energy information system application and develops its own reference architecture in the process. The developed reference architecture is called RAMEnIS6.0, where the life cycle axis is replaced by the energy production axis.

2.3. Digital Twin

Further publications and standards show a different understanding and descriptions of the term digital twin [13–16]. Furthermore, other papers show different stages of implementation or software by which a realization is possible. Concrete requirements for the realization in the context of RAMI 4.0 are not given extensively. This paper attempts to close this gap. On the one hand, Kritzinger et al. [17] show the division of definitions or descriptions of a DT into different categories. Thereby, his study focuses especially on integration levels and the areas within a production (e.g., product life cycle and production planning), and various tools and technologies are addressed (e.g., OPC UA and cloud computing), which are required for the use of DT.

Kritzinger et al. [17] also show in their study that most publications use the description of a **digital model** (DM) or a **digital shadow** (DS), rather than providing a clear definition of a DT (Table 1). This is based on their given understanding of the differences between the digital model, digital shadow, and digital twin, which is elaborated. This differentiation is referred to as the degree of integration. The differences are defined as follows by Kritzinger et al. [17] and is in the broadest sense also addressed by Stark and Damerau [7] and Grieves [18]:

> *A **DM** is a digital representation of an existing or planned physical object that does not use any form of automated data exchange between the physical object and the digital object. The digital representation might include a more or less comprehensive description of the physical object. These models might include, but are not limited to simulation models of planned factories, mathematical models, or any other models of a physical object, which do not use any form of automatic data integration. Digital data of existing physical systems might still be in use for the development of such models, but all data exchange is done in a manual way. A change in state of the physical object has no direct effect on the digital object and vice versa.*

Table 1. Different levels of integration found by Kritzinger et al. [17] in research on the topic of digital twin (source: [17]).

	Concept	Case-Study	Review	Definition
undefined	1.90%	4.76%	2.38%	0.00%
DM	14.29%	11.90%	0.00%	0.00%
DS	26.19%	7.14%	2.38%	0.00%
DT	2.38%	2.38%	9.52%	4.76%

Bearing this in mind, Kritzinger et al. [17] and Bauernhansl et al. [19] describe furthermore as follows:

> ***DS** based on the definition of a Digital Model, if there further exists an automated one-way data flow between the state of an existing physical object and a digital object, one might refer to such a combination as Digital Shadow. A change in state of the physical object leads to a change of state in the digital object, but not vice versa.*

Furthermore, Tao and Zhang [20] as well as Stark et al. [21] define the digital shadow as an essential part of a DT as follows:

> ***DS** is a data profile that couples with the corresponding entity throughout its life cycle, and carries all the data and knowledge to reflect the individual shape and historical, current, and expected future status.*

Based on this clarification, the concept of the three-dimensional DT is established. This concept describes the physical entity, the virtual models, and the data exchange between them as one dimension [18,20,22]. In extension, ref. [23] published the concept of a five-dimensional digital twin (see Figure 3), where services and data are also included in the DT. Colored in red are the parts of the three-dimensional concept, which contains the physical entity (PE), virtual entity (VE), and the connection between them (CN_{PV}). By adding software services (Ss) as well as considering digital twin data (DD) as further dimensions and describing the connections ($CN_{m,n}$) between these four parts as a further dimension, the five-dimensional concept of the DT is created. In addition, Zimmermann et al. [24] explain the term **digital master** as the functional combination of digital twin data (DD) and the virtual entity (VE). At this stage, it is clearly shown that there are many different understandings of the meaning, what a DT is, and which requirements it should fulfill. A literature review is therefore the basis for further discussions and to derive requirements for our approach.

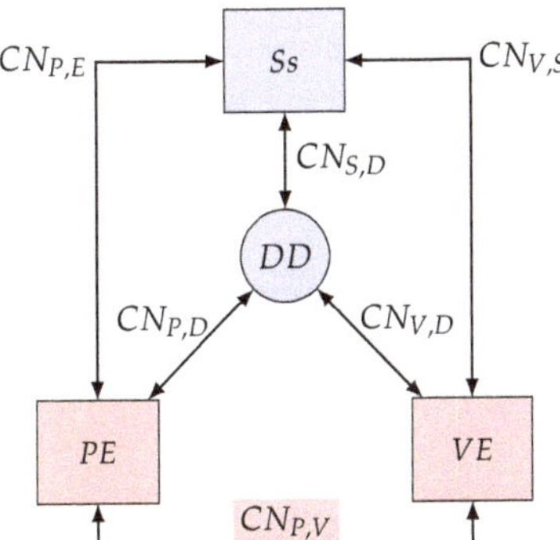

Figure 3. Concept of the five-dimension Digital Twin (source: own illustration adapted from ref. [7,16,23]).

2.4. Research Methodology

To show the need for clear requirements on a definition of a DT, which addresses the complexity of a production system, we conduct a literature review. This review contains a multi-step approach to find and classify relevant requirements and definitions. This approach is based on the procedure used in Glock and Hochrein [25], Hersi et al. [26] and Tawfik et al. [27]. The literature review includes the four superordinate steps:

1. Preparation
2. Planning
3. Screening
4. Classification

which are described in the following.

2.4.1. Step 1—Preparation

The need for a literature review is based on multiple facts. At first the large amount of descriptions of what a DT should be. This takes into account not only understanding but also naming. The terms Cyber Physical Twins [28] or Cyber Digital Twins also exist, which describe the same technological approach. Therefore, we want to collect the key requirements, that a DT must be fulfilled in the context of digital manufacturing. Second, we want to show the combination of the RAMI 4.0 model with the use of a DT, therefore we need a categorization, which is shown in step 4. This is needed to derivate necessary key requirements of a DT in the frame of RAMI 4.0.

2.4.2. Step 2—Planning

For our research, we choose as relevant databases ScienceDirect, IEEE Xplore, Springer-Link, and Web of Science to start a query. As boundaries of the querying only publications since 2000 were considered. For our classification, on the one hand, we cluster the relevant definitions into the topics, related to the RAMI 4.0 model, to product-related or process-related and into the topics general definitions or industry sector definitions. Therefore, our literature review focuses, but is not absolutely limited, on definitions from engineering fields, that means manufacturing, aerospace, electrical engineering, and Industry 4.0.

The screening itself contains multiple steps:

1. Check for the right research field of an article.
2. Review the title.
3. Verify the abstract.
4. Check the full text to find descriptions or definitions of the term "digital twin".
5. Check the references for additional sources.

An article was excluded from our study if the article did not fulfill one of these steps.

2.4.3. Step 3—Screening

The results of screening the databases are shown in Figure 4. At first, the relevance of the topic of "digital twin", exemplarily shown in Figure 4b for the query at Web of Science, is proved. It is clearly shown that, in the last six years, research on the topic of "digital twin" has risen significantly. Additionally, Figure 4a shows the distribution of the top 12 engineering fields, where publications with the topic of DT were made. It was found that most publications came from electrical and electronic engineering, followed by manufacturing and computer science. If one of these studies satisfied all steps, this study was considered for our classification.

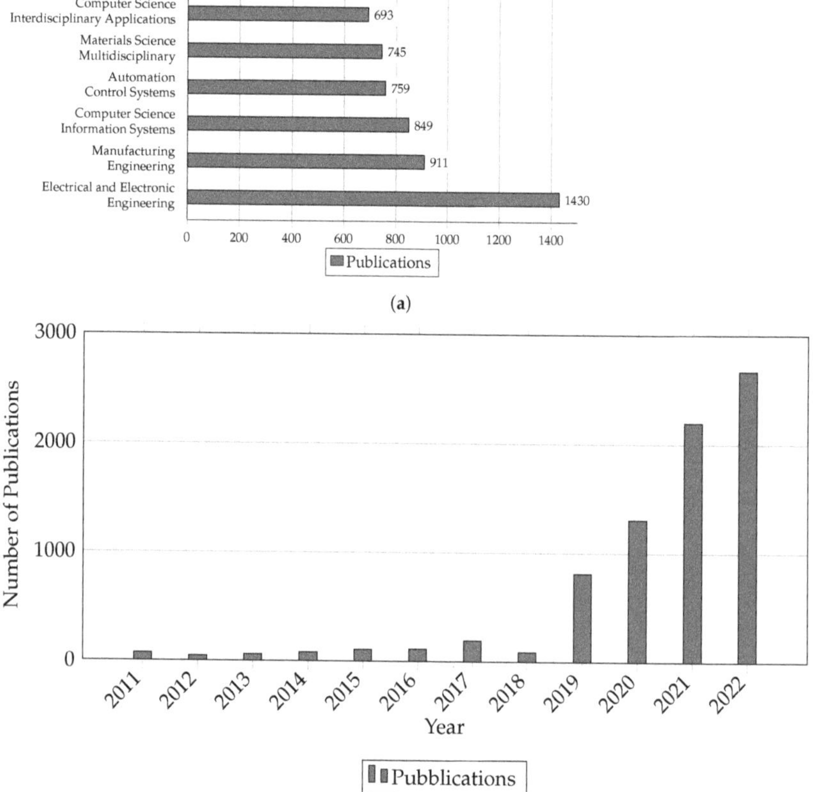

Figure 4. Statistical screening results of the literature review with (**a**) results from research fields and (**b**) the total number of publications per year (source: own illustration). (**a**) The amount of publications in different research fields is shown in this plot. Thereby, most of the publications were in the field of electrical and electronic followed by manufacturing engineering. (Web of Science 3 February 2023); (**b**) The figure shows the number of publications on the Web of Science database from 2010 to 2022, with the query term "digital twin". Clearly identifiable is the extreme rise of publications within the topic of "digital twin" since 2019. [Web of Science 3 February 2023].

2.4.4. Step 4—Classification

Within our classification, we sort the found descriptions oriented at the different lifecycle steps of the RAMI 4.0 model, production, or process-related descriptions of a DT. Here, seven product-related definitions and four process-related definitions were found. Additionally, 14 descriptions of the term DT in general were made, and 9 descriptions or definitions from industry were found. In the next step, we mark the content of each classified definition as a key element, in the sense that its high-level requirements or functionality is summarized.

2.5. Definitions of Digital Twin

Nearly all found definitions or descriptions of the term "digital twin" explain communication between a real physical asset or entity and a virtual representation, e.g., [14,15,22,29]. Some of them, e.g., [7,17,30], describe more in detail, that this representation is a model based on data which come from the physical asset. In addition, Garetti et al. [31], Kraft [32] and Schleich et al. [33] argue that these models can also handle information from software services. These services can handle different tasks, like prognoses, optimize or control the physical asset [6,22]. For this collaboration between the physical asset, model, and services, data are essentially those of [20,34],which can be measured from the physical asset or the services [32] and from other external sources [35]. This data communication between the data sources, physical asset, services, and models should be in real time [36,37] and use the standardization of all components [29]. An overview of the classified descriptions and definitions is given in Table 2.

Table 2. Table with the classified definitions of the term digital twin and the elements included in each. Marked definitions (*) are not retrievable as full text (not open source) by the authors and come from secondary sources (source: own illustration).

Topic	Reference	Definition	Year	Key Element
Product related definitions	Reifsnider and Majumdar [38] *	*Ultra-high fidelity physical models of the materials and structures that control the life of a vehicle.*	2013	virtual model
	Rios et al. [39]	*Product digital counterpart of a physical product.*	2015	virtual model, real asset
	Schroeder et al. [40]	*Virtual representation of a real product in the context of cyber-physical systems.*	2016	virtual model, real asset
	Manas Bajaj et al. [41] *	*A unified system model that can coordinate architecture, mechanical, electrical, software, verification, and other discipline-specific models across the system life cycle, federating models in multiple vendor tools and configuration-controlled repositories.*	2016	services, hierarchical, virtual models
	Abramovici et al. [30]	*A virtual twin is a model that integrates interdisciplinary (mechanics, electronics, software, and services) virtual product models and related real-time data of a product instance (physical twin). A virtual twin can be dynamically generated from a model and data space to fulfill a specific task (e.g., dynamic reconfiguration of a smart product during its use phase).*	2017	real asset, virtual models, data, services, hierarchical, real time
	Schleich et al. [33]	*In synthesis, the vision of the digital twin describes the vision of a bi-directional relation between a physical artifact and the set of its virtual models. In this context, the virtual "twinning", i.e., the establishment of such relations between physical parts and their virtual models, enables the efficient execution of product design, manufacturing, servicing, and various other activities throughout the product life cycle.*	2017	real asset, connection, virtual models, services

Table 2. *Cont.*

Topic	Reference	Definition	Year	Key Element
	Grieves [18]	*A digital twin is a distributed and decentralized approach to manage product information at product item level along its life cycle.*	2015	real asset, virtual model, connection, modularization
Process-related definition	Lee et al. [35]	*Coupled model of the real machine that operates in the cloud platform and simulates the health condition with an integrated knowledge from both data-driven analytical algorithms and other available physical knowledge.*	2013	hierarchical, real-time, real asset, virtual model
	Rosen et al. [42]	*Very realistic models of the process current state and its behavior in interaction with the environment in the real world.*	2015	connection, real asset, virtual models,
	Bauernhansl et al. [19]	*The digital shadow first transfers the real production process into the virtual world. Based on this, the Digital Twin can deliver an image of reality that is as identical as possible through a process model and simulation.*	2016	real asset, virtual model, services
	Garetti et al. [31]	*The DT consists of a virtual representation of a production system that is able to run on different simulation disciplines that is characterized by the synchronization between the virtual and real system, thanks to sensed data and connected smart devices, mathematical models and real time data elaboration. The topical role within Industry 4.0 manufacturing systems is to exploit these features to forecast and optimize the behavior of the production system at each lifecycle phase in real time.*	2012	virtual model, real asset, services, real time, connection, data, hierarchical, scalability
General definitions	Schluse and Rossmann [43]	*Virtual substitutes of real-world objects consisting of virtual representations and communication capabilities making up smart objects acting as intelligent nodes inside the Internet of Things and services.*	2016	virtual model, real asset, connection, services
	Canedo [44] *	*Digital representation of a real-world object with focus on the object itself.*	2016	virtual model, real asset
	Gabor et al. [45]	*The simulation of the physical object itself to predict future states of the system.*	2016	data, services, real asset, virtual model
	Gartner [34]	*A digital twin is a digital representation of a real-world entity or system. The implementation of a digital twin is an encapsulated software object or model that mirrors a unique physical object, process, organization, person or other abstraction. Data from multiple digital twins can be aggregated for a composite view across a number of real-world entities, such as a power plant or a city, and their related processes.*	2022	real asset, virtual model, modularization, hierarchical, services, data, scalability
	Kraft [32]	*An integrated multi-physics, multi-scale, probabilistic simulation of an as-built system, enabled by digital thread, which uses the best available models, sensor information, and input data to mirror and predict activities/performance over the life of its corresponding physical twin.*	2016	services, data, robustness, virtual model, hierarchical, real asset
	Söderberg et al. [46]	*Real-time optimization using digital copies of physical systems.*	2017	real-time, real asset, virtual model
	Bolton et al. [47]	*The dynamic virtual representation of a physical object or system throughout its life cycle, using real-time data to achieve understanding, learning, and reasoning.*	2018	virtual model, real asset, real time, data

Table 2. *Cont.*

Topic	Reference	Definition	Year	Key Element
	Tao et al. [23]	*Digital twin uses physical data, virtual data and interactive data between them to map all components in the product life cycle.*	2019	real asset, virtual model, data, connection
	Stark and Damerau [7]	*A digital twin is a digital representation of an active unique product (real device, object, machine, service, or intangible asset) or unique product-service system (a system consisting of a product and a related service) that comprises its selected characteristics, properties, conditions, and behaviors by means of models, information, and data within a single or even across multiple life cycle phases.*	2019	real asset, virtual model, services, data, hierarchical, modularization
	Rasheed et al. [22]	*A digital twin is defined as a virtual representation of a physical asset enabled through data and simulators for real-time prediction, optimization, monitoring, controlling, and improved decision making.*	2020	virtual model, real asset, services, real-time, hierarchical, data, connection
	Industrial Digital Association e. V. [48]	*Digital representation, sufficient to meet the requirements of a set of use cases.*	2022	virtual model, hierarchical,
	Claude Baudoin et al. [49]	*Digital model of one or more real-world entities, digital twin entities can be objects or processes, that is synchronized with those entities at a specified frequency and fidelity.*	2022	virtual model, modularization, connection, robustness, real asset
	Digital Twin Consortium [37]	*A digital twin is a virtual representation of real-world entities and processes, synchronized at a specified frequency and fidelity.*	2022	real asset, virtual model, connection, robustness
	Kritzinger et al. [17]	*If the data flows between an existing physical object and a digital object are further fully integrated in both directions, one might refer to it as a digital twin. In such a combination, the digital object might also act as controlling instance of the physical object. There might also be other objects, physical or digital, which induce changes of state in the digital object. A change in state of the physical object directly leads to a change in state of the digital object and vice versa.*	2018	data, real asset, virtual model, connection, hierarchical, modularization, real time, services
Industry-Sector definitions	Shafto et al. [50]	*An integrated multi-physics, multi-scale, probabilistic simulation of a vehicle or system that uses the best available physical models, sensor updates, fleet history, etc., to mirror the life of its flying twin. The digital twin is ultra-realistic and may consider one or more important and interdependent vehicle systems.*	2010	hierarchical, services, virtual model, modularization, scalability,
	Tuegel [51] *	*A cradle-to-grave model of an aircraft structure's ability to meet mission requirements, including submodels of the electronics, the flight controls, the propulsion system, and other subsystems.*	2012	modularization, hierarchical, virtual model
	Gockel et al. [52]	*Ultra-realistic, cradle-to-grave computer model of an aircraft structure that is used to assess the aircraft's ability to meet mission requirements.*	2012	virtual model, hierarchical, scalability
	Bielefeldt et al. [53]	*Ultra-realistic multi-physical computational models associated with each unique aircraft and combined with known flight histories.*	2016	virtual model, real asset, data
	Bazilevs et al. [54]	*High-fidelity structural model that incorporates fatigue damage and presents a fairly complete digital counterpart of the actual structural system of interest.*	2015	real time, virtual model, robustness

Table 2. *Cont.*

Topic	Reference	Definition	Year	Key Element
	El Saddik [36]	*Digital twin is digital copies of biological or non-biological physical entities. By bridging the physical and virtual worlds, data are seamlessly transferred, allowing virtual entities to exist simultaneously with physical entities.*	2018	virtual model, real asset, connection real time
	Negri et al. [29]	*Digital twins are digital representations based on semantic data models that allow running simulations in different disciplines, that support not only a prognostic assessment at the design stage (static perspective) but also a continuous update of the virtual representation of the object by a real-time synchronization with sensed data. This allows the representation to reflect the current status of the system and to perform real-time optimizations, decision making and predictive maintenance according to the sensed conditions.*	2017	virtual model, standardization, data, services, hierarchical, real time, services, connection, real asset
	ISO 23704-1:2022 [15]	*Digital replica of physical assets (physical twin), processes and systems that can be used for various purposes or a fit-for-purpose digital representation of something outside its own context with data connections that enable convergence between the physical and virtual states at an appropriate rate of synchronization.*	2022	virtual model, real asset, connection, data, standardization
	ISO 23247-1:2021 [14]	*Fit for purpose digital representation of an observable manufacturing element with synchronization between the element and its digital representation.*	2021	virtual model, real asset, connection, standardization

3. Research Gap

Consideration of the multitude of definitions has shown that there is no uniform understanding and requirements for the term "digital twin". The understanding varies across different areas, but even within the areas, there is often a divergence. Just in the area of manufacturing, a multitude of definitions can be found. Furthermore, there are already more concrete concepts, such as RAMI 4.0. However, RAMI 4.0 provides a reference architecture and with the AAS, a data format. The AAS is therefore not yet a digital twin, as is sometimes simplified. According to the five-dimensional model, for example, the simulation ability is not given in the AAS. The AAS can be more seen as a DS according to the definition by Tao and Zhang [20] and Stark et al. [21] to describe a certain asset. A concept therefore needs to derive how a factory with multiple assets can be modeled. Overall, there is a lack of a concept that links RAMI 4.0, AAS, and the understanding of a digital twin. To establish this connection, the first step is to define what a digital twin is before it can be placed in the context of the RAMI 4.0 standard. The goal should be to find an understanding that is as universally valid as possible and that is consistent with the existing and evaluated digitization concepts. The focus of the objective of the digital twin should not itself restrict the application of the definition. In concrete terms, this means that whatever the target value of optimization within manufacturing, the definition should be adaptable accordingly. In the consideration of the found definitions, some commonalities have become apparent, which are to be highlighted as a basis. Accordingly, based on the definitions given in Table 2, elements of a digital twin in the production area should be the following elements given in Table 3. In addition, a DT consisting of the above elements should have the following capabilities:

- Possibility to automatically control the real asset, also with results from the services.
- Possibility for real-time automatic data acquisition and control.

Furthermore, these eight general conditions, summarized by [20] and also described partly by other authors, should be fulfilled to achieve long-term and sustainable benefits through the use of a DT:

1. Data and knowledge-based (compare Table 3): the most up-to-date data and rules available should always be used in the modeling and in the use phase.
2. Modularization [6,7,17,18,20,34,49,51]: a DT should provide reusability, flexibility and interoperability for a system-of-systems approach.
3. Light weight [17,22,29–31,35,36,46,47,54]: models with maximally low complexity and thus with low computing time to enable the real-time capability of the DT.
4. Hierarchy [6,7,17,22,29–32,34,35,41,51,52]: a DT should use different hierarchical layers for the efficient use of different tasks.
5. Standardization [14,15,29]: to guarantee that all components of a DT communicate efficiently and securely with each other.
6. Servitization [6,7,17,19,22,29–34,41,43,45]: use of standard services for easy and convenient usage of the DT.
7. Openness and scalability [6,31,34,52]: open to interoperate with various resources and scalable, which enables functional extension.
8. Robustness [32,37,49,54]: a DT should be built with good robustness to deal with unpredictable changes.

Based on this list of requirements, an understanding of which elements and capabilities are needed is found. The elements and capabilities are to be concretized in a RAMI-compliant concept to be able to guarantee its implementability in industry. Furthermore, the existing definitions are to be taken into account, and thus a consistency is to be achieved.

Table 3. Elements of a DT which are derived from the different definitions in Table 2 (source: own illustration).

DT Element	Based on Reference
real physical asset/entity	[7,14,15,17–20,22,29–37,39,40,42–47,49,53]
virtual model/entity that describes the real asset (physical, math, 3D models, etc.) and is capable of predicting the behavior of the real asset	all references given in Table 2
data which describe the real physical asset (contains data harvested from the real asset and from services)	[7,15,17,22,23,29–32,34,45,47,53]
services (e.g., monitoring, simulations, prognosis, optimization, and control)	[6,7,17,19,22,29–34,41,43,45]
full connection between the elements (physical asset, virtual model, data and services) for data exchange	[14,15,17,18,20,22,29,31,33,36,37,42,43,49]

4. Conceptual Approach

The conceptual approach is based on the AAS and the result of the overview of the DT definitions. The AAS is used to include the factory assets and to realize bidirectional communication between assets and the factory control level, where the DT is located. The simulation ability on the factory level is realized by a factory model and corresponding services to fulfill these tasks. The factory model uses the information which is supplied by the AAS. The concept displayed in Figure 5 is based on the product-oriented approach to describe the life cycle of digital twins [7] and the RAMI reference architecture [8]. Unlike the DT of a product, a factory is a complex structure of many interrelated individual resources. Each resource, in turn, has its own life cycle. In this case, resources refer to assets that contribute directly or indirectly to the manufacture of the product and are thus modeled in a DT. These can be, for example, production facilities, such as a machine tool from the illustration, or a technical building equipment facility. Each resource has a representation that already includes information from the previous life cycles of this resource, which has been expanded accordingly for individual use in the factory. Furthermore, products and thus digital representations are created or extended anew in the factory. The products to be manufactured in the factory are also represented digitally. In this way, information can be documented in the creation phase, for example, process parameters related to the specific instance. The distinction between products and resources is made only because they are at different life cycle stages. Products are currently being produced, whereas the machines for them are in their usage phase. However, it follows the same idea that all relevant assets are represented digitally. This fact of the factory is taken into account in the following concept by choosing a modular approach in which each resource and product has an AAS. These individual representations provide the DT with asset-specific information. This information is then available in a distributed form and can be maintained by those responsible for the assets. This also allows individual factory components to be replaced at any time without the need for major changes to the factory model. Starting from the life cycle of the production system, which begins with engineering, the parallel development of the DT of the factory also begins. Once the resources have been selected, the digital representations, in this case, the AAS of the corresponding assets, can provide information about these resources. Ideally, the AAS is already set up by the machine supplier and is part of the machine delivery, and will be complemented if changes arise. In the factory, this AAS is integrated into the process and manages the machine in the factory system, including the documentation of the operation by storing process data. The product to be manufactured also receives a corresponding AAS in production, which can already contain relevant information in product creation and is now expanded by the process information in production and will then be delivered as part of the product. The AAS thus forms the interface to the physical world. This is performed, on the one hand, by the basic digitally stored description of the physical asset, which records the current state of an asset in each case with the help of sensor values. On the other hand, the AAS also offers the possibility to communicate with the physical elements of the digital world and to influence their state by control commands. Therefore, the concept can be seen as an enabler for the digital factory.

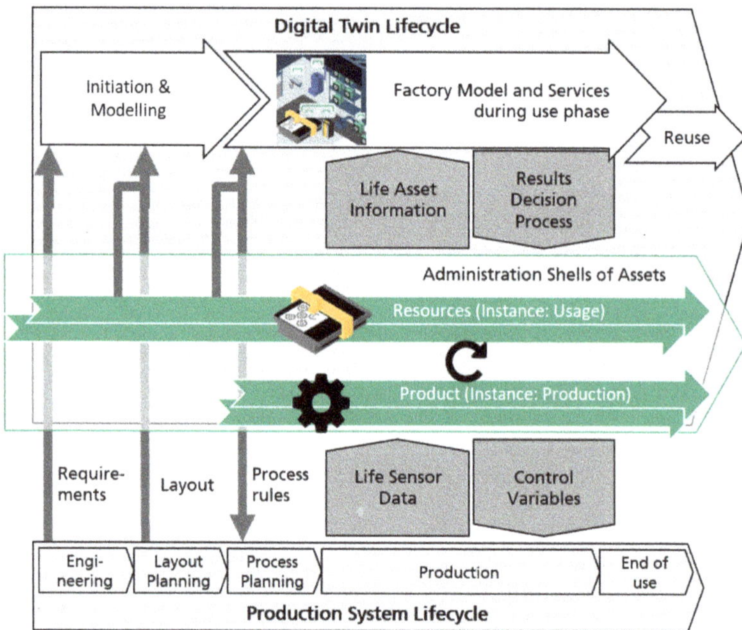

Figure 5. Lifecycle model for modular digital twins, where asset-specific information is modeled in asset administration shells (AASs), and asset interaction is modeled in a factory model (source: own illustration based on Ref. [7]).

The factory model, which turns the concept of a digital factory into a digital twin concept, is composed of individual models. Modeling effort is required here, but the information from the individual resources can already be accessed so that only the interrelationships need to be entered. Creating individual digital representations for each asset is much easier to handle in terms of complexity than for an entire factory system. The advantage is also that the manufacturers of the assets can already represent the basic data and behavior of the resource itself, and the operator can simply extend this with the data relevant to them.

As described, the concept is based on the derived considerations, elements, and conditions given in Section 3 and also by Stark and Damerau [7], who consider the DT in its life phase, as well as the considerations that are in the RAMI 4.0 reference architecture. Although the latter is intended for the concrete implementation of industrial applications, it is ultimately only a reference architecture. The concept is therefore to be concretized on the basis of the requirements given in Section 3 and checked for its ability to be implemented. To fulfill the derived requirements and conditions of a DT given in Section 3, our concept has the following characteristics:

1. Data and knowledge-based: The AAS provides current information about the resources in each case through an active connection to the physical resource.Due to the system's modular structure, the individual components can be kept up to date with little effort because the management of the resources lies with the respective experts and does not have to be carried out by a simulation expert.
2. Modularization: AAS provides modularization for individual resources. In the case of the replacement of resources, adaptation of the model is easily possible. Only the relationships of the resources to each other must be maintained in a factory model.
3. Light weight: The real-time capability depends on the technologies used but is not prevented by the concept.
4. Hierarchy: AAS offers the possibility to build a hierarchy. In the present concept, the product, the production facilities, and the factory model can already be called hierarchy levels. However, any units, e.g., production areas, can also be formed.
5. Standardization: The concept is based on DIN SPEC 91345 [8], which also includes the AAS. The AAS can be seen as a regulation that can work with standardized information models, etc. In addition, there are currently further standardization efforts in this area. It is crucial that modelers adhere to the existing standards.
6. Servitization: The use of standard services is highly dependent on the implementation of the factory model and the technologies used. The AASs on the resource level enable the standard protocols during communication and are therefore an enabler of a service-oriented architecture.
7. Openness and scalability: Modularization at the resource level enables easy extensibility of the model. Only the integration into the factory model depends on the concrete implementation and determines the effort required to integrate additional resources. The AAS is operated as an open-source project, so the work can be accessed here.
8. Robustness: This aspect must be considered, especially during implementation.

5. Conclusions

This paper first shows the need for a unified understanding of the requirements for the meaning of the term digital twin. This represents a central tool in the context of digital production. We show that there is no uniform definition or description of this term within the research community. This leads to the fact that the requirements which are derived from it are not clearly formulated. This gap is closed by the authors in this paper. Through a literature research, different definitions and descriptions of the term digital twin are classified and compared. From this, common properties are identified, which represent the essential requirements. Finally, an approach embedded in the RAMI 4.0 model is used to demonstrate how these derived requirements can be implemented. This enables the sustainable and scalable use of a DT in the context of digital production. The digitized production is an essential component and central element for future production to produce sustainably, flexibly, and efficiently. For example, further research must show how efficient modeling is possible for specific submodels, e.g., efficiency and flexibility models for production. In doing so, care must be taken to ensure that the requirements presented are met. The work dealt specifically with the life cycle and the higher-level architecture. RAMI was placed at the center of the consideration, and the agreement of an understanding for the digital twin and RAMI was considered. A comparison with other reference architectures, such as the industrial internet reference architecture (IIRA) or smart grid architecture model (SGAM), was left out. In the next step, however, it is appropriate to go deeper into the analysis and look more closely at the individual components of the architecture.

Author Contributions: M.L.: Conceptualization, Methodology, Project administration, Writing—Original Draft; L.B.: Conceptualization, Methodology, Writing—Original Draft; J.S.: Supervision, Funding acquisition, Writing—Review Editing; M.W.: Supervision, funding acquisition, writing—Review Editing. All authors have read and agreed to the published version of the manuscript.

Funding: The authors gratefully acknowledge the financial support of the Kopernikus-Projekt "SynErgie" (grant number 03SFK3A0-3) by the Federal Ministry of Education and Research of Germany (BMBF) and the project supervision by the project management organization Projektträger Jülich (PtJ).

Institutional Review Board Statement: Not applicable.

Informed Consent Statement: Not applicable.

Data Availability Statement: Not applicable.

Acknowledgments: The authors gratefully acknowledge the financial support of the Kopernikus-Projekt "SynErgie" (the grant number 03SFK3A0-3) by the Federal Ministry of Education and Research of Germany (BMBF) and the project supervision by the project management organization Projektträger Jülich (PtJ).

Conflicts of Interest: The authors declare no conflicts of interest.

References

1. Bracht, U.; Geckler, D.; Wenzel, S. *Digitale Fabrik: Methoden und Praxisbeispiele*, 2nd ed.; Aktualisierte und Erweiterte Auflage VDI-Buch; Springer Vieweg: Berlin/Heidelberg, Germany, 2018. [CrossRef]
2. Abele, E.; Reinhart, G. *Zukunft der Produktion: Herausforderungen, Forschungsfelder, Chancen*; Carl Hanser Fachbuchverlag: Munich, Germany, 2011. [CrossRef]
3. Westkämper, E.; Löffler, C. *Strategien der Produktion: Technologien, Konzepte und Wege in die Praxis*; Springer: Berlin/Heidelberg, Germany, 2016. [CrossRef]
4. Schuh, G.; Häfner, C.; Hopmann, C.; Rumpe, B.; Brockmann, M.; Wortmann, A.; Maibaum, J.; Dalibor, M.; Bibow, P.; Sapel, P.; et al. Effizientere Produktion mit Digitalen Schatten. *Zwf Z. Wirtsch. Fabr.* **2020**, *115*, 105–107. [CrossRef]
5. Schuh, G.; Gützlaff, A.; Sauermann, F.; Maibaum, J. Digital Shadows as an Enabler for the Internet of Production. In *Advances in Production Management Systems. The Path to Digital Transformation and Innovation of Production Management Systems*; Lalic, B., Majstorovic, V., Marjanovic, U., von Cieminski, G., Romero, D., Eds.; IFIP Advances in Information and Communication Technology; Springer International Publishing: Cham, Switzerland, 2020; Volume 591, pp. 179–186. [CrossRef]
6. Shafto, M.; Conroy, M.; Doyle, R.; Glaessgen, E.; Kemp, C.; LeMoigne, J.; Wang, L. NASA Technology Roadmap (Draft): Modeling, Simulation, Information Technology & Processing Roadmap Technology Area. *Natl. Aeronaut. Space Adm.* **2010**, *11*, 1–38.
7. Stark, R.; Damerau, T. Digital Twin. In *CIRP Encyclopedia of Production Engineering*; The International Academy for Production Engineering; Chatti, S., Tolio, T., Eds.; Springer: Berlin/Heidelberg, Germany, 2019; pp. 1–8. [CrossRef]
8. DIN Deutsches Institut für Normung e. V. DIN SPEC 91345: Referenzarchitekturmodell Industrie 4.0 (RAMI4.0). 2016. Available online: https://www.din.de/de/forschung-und-innovation/themen/industrie4-0/din-veroeffentlicht-din-spec-zu-rami4-0-1 58570 (accessed on 24 June 2023)
9. Adolphs, P.; Auer, S.; Bedenbender, H.; Billmann, M.; Hankel, M.; Heidel, R.; Hoffmeister, M.; Huhle, H.; Jochem, M.; Kiele-Dunsche, M.; et al. Ergebnispapier Struktur der Verwaltungsschale: Fortentwicklung des Referenzmodells für die Industrie 4.0-Komponente. 2016. Available online: https://www.zvei.org/fileadmin/user_upload/Presse_und_Medien/Publikationen/20 16/april/Struktur_der_Verwaltungsschale/Struktur-der-Verwaltungsschale.pdf (accessed on 12 January 2023).
10. Beregi, R.; Pedone, G.; Háy, B.; Váncza, J. Manufacturing Execution System Integration through the Standardization of a Common Service Model for Cyber-Physical Production Systems. *Appl. Sci.* **2021**, *11*, 7581. [CrossRef]
11. Steindl, G.; Stagl, M.; Kasper, L.; Kastner, W.; Hofmann, R. Generic Digital Twin Architecture for Industrial Energy Systems. *Appl. Sci.* **2020**, *10*, 8903. [CrossRef]
12. Roscher, M. Energieinformationssystemarchitektur für Produzierende Unternehmen. Ph.D. Thesis, Faculty of Mechanical Engineering, Aachen, Germany, 2018.
13. Cimino, C.; Negri, E.; Fumagalli, L. Review of Digital Twin Applications in Manufacturing. *Comput. Ind.* **2019**, *113*, 103130. [CrossRef]

14. *ISO 23247-1:2021*; Automation Systems and Integration—Digital Twin Framework for Manufacturing—Part 1: Overview and General Principles. International Organization for Standardization: Geneva, Switzerland, 2021.

15. *ISO 23704-1:2022*; General Requirements for Cyber-Physically Controlled Smart Machine Tool Systems (CPSMT)—Part 1: Overview and Fundamental Principles. International Organization for Standardization: Geneva, Switzerland, 2022.

16. He, B.; Bai, K.J. Digital Twin-Based Sustainable Intelligent Manufacturing: A Review. *Adv. Manuf.* **2021**, *9*, 1–21. [CrossRef]

17. Kritzinger, W.; Karner, M.; Traar, G.; Henjes, J.; Sihn, W. Digital Twin in Manufacturing: A Categorical Literature Review and Classification. *IFAC Pap.* **2018**, *51*, 1016–1022. [CrossRef]

18. Grieves, M. *Digital Twin: Manufacturing Excellence through Virtual Factory Replication*; Cocoa Beach, FL, USA, 2015. Available online: https://www.3ds.com/fileadmin/PRODUCTS-SERVICES/DELMIA/PDF/Whitepaper/DELMIA-APRISO-Digital-Twin-Whitepaper.pdf (accessed on 18 December 2022).

19. Bauernhansl, T.; Krüger, J.; Reinhart, G.; Schuh, G. *WGP-Standpunkt Industrie 4.0*; Wissenschaftliche Gesellschaft für Produktionstechnik: Darmstadt, Germany, 2016.

20. Tao, F.; Zhang, M. *Digital Twin Driven Smart Manufacturing*; Academic Press: Cambridge, MA, USA, 2019.

21. Stark, R.; Kind, S.; Neumeyer, S. Innovations in Digital Modelling for next Generation Manufacturing System Design. *CIRP Ann.* **2017**, *66*, 169–172. [CrossRef]

22. Rasheed, A.; San, O.; Kvamsdal, T. Digital Twin: Values, Challenges and Enablers From a Modeling Perspective. *IEEE Access* **2020**, *8*, 21980–22012. [CrossRef]

23. Tao, F.; Zhang, H.; Liu, A.; Nee, A.Y.C. Digital Twin in Industry: State-of-the-Art. *IEEE Trans. Ind. Inform.* **2019**, *15*, 2405–2415. [CrossRef]

24. Zimmermann, T.C.; Masuhr, C.; Stark, R. MBSE-Entwicklungsfähigkeit für Digitale Zwillinge. *Z. Wirtsch. Fabr.* **2020**, *115*, 51–54. [CrossRef]

25. Glock, C.H.; Hochrein, S. Purchasing Organization and Design: A Literature Review. *Bus. Res.* **2011**, *4*, 149–191. [CrossRef]

26. Hersi, M.; Traversy, G.; Thombs, B.D.; Beck, A.; Skidmore, B.; Groulx, S.; Lang, E.; Reynolds, D.L.; Wilson, B.; Bernstein, S.L.; et al. Effectiveness of Stop Smoking Interventions among Adults: Protocol for an Overview of Systematic Reviews and an Updated Systematic Review. *Syst. Rev.* **2019**, *8*, 28. [CrossRef] [PubMed]

27. Tawfik, G.M.; Dila, K.A.S.; Mohamed, M.Y.F.; Tam, D.N.H.; Kien, N.D.; Ahmed, A.M.; Huy, N.T. A Step by Step Guide for Conducting a Systematic Review and Meta-Analysis with Simulation Data. *Trop. Med. Health* **2019**, *47*, 46. [CrossRef] [PubMed]

28. Czwick, C.; Anderl, R. Cyber-Physical Twins - Definition, Conception and Benefit. *Procedia CIRP* **2020** *90*, 584–588. [CrossRef]

29. Negri, E.; Fumagalli, L.; Macchi, M. A Review of the Roles of Digital Twin in CPS-based Production Systems. *Procedia Manuf.* **2017**, *11*, 939–948. [CrossRef]

30. Abramovici, M.; Göbel, J.C.; Savarino, P. Reconfiguration of Smart Products during Their Use Phase Based on Virtual Product Twins. *CIRP Ann.* **2017**, *66*, 165–168. [CrossRef]

31. Garetti, M.; Rosa, P.; Terzi, S. Life Cycle Simulation for the Design of Product–Service Systems. *Comput. Ind.* **2012**, *63*, 361–369. [CrossRef]

32. Kraft, E.M. The Air Force Digital Thread/Digital Twin—Life Cycle Integration and Use of Computational and Experimental Knowledge. In Proceedings of the 54th AIAA Aerospace Sciences Meeting, San Diego, CA, USA, 4–8 January 2016; American Institute of Aeronautics and Astronautics: Reston, VA, USA, 2016. [CrossRef]

33. Schleich, B.; Anwer, N.; Mathieu, L.; Wartzack, S. Shaping the Digital Twin for Design and Production Engineering. *CIRP Ann.* **2017**, *66*, 141–144. [CrossRef]

34. Gartner. Definition of Digital Twin—Gartner Information Technology Glossary. Available online: https://www.gartner.com/en/information-technology/glossary/digital-twin (accessed on 18 December 2022).

35. Lee, J.; Lapira, E.; Bagheri, B.; Kao, H.a. Recent Advances and Trends in Predictive Manufacturing Systems in Big Data Environment. *Manuf. Lett.* **2013**, *1*, 38–41. [CrossRef]

36. El Saddik, A. Digital Twins: The Convergence of Multimedia Technologies. *IEEE Multimed.* **2018**, *25*, 87–92. [CrossRef]

37. Digital Twin Consortium. Definition Digital Twin. 2020. Available online: https://www.digitaltwinconsortium.org/glossary/glossary/ (accessed on 25 November 2022).

38. Reifsnider, K.; Majumdar, P. Multiphysics Stimulated Simulation Digital Twin Methods for Fleet Management. In Proceedings of the 54th AIAA/ASME/ASCE/AHS/ASC Structures, Structural Dynamics, and Materials Conference, Boston, MA, USA, 8–11 April 2013; American Institute of Aeronautics and Astronautics: Reston, VA, USA, 2013. [CrossRef]

39. Rios, J.; Hernández, J.C.; Oliva, M.; Mas, F. Product Avatar as Digital Counterpart of a Physical Individual Product: Literature Review and Implications in an Aircraft. In *Transdisciplinary Lifecycle Analysis of Systems*; IOS Press: Clifton, VA, USA, 2015; pp. 657–666. [CrossRef]

40. Schroeder, G.N.; Steinmetz, C.; Pereira, C.E.; Espindola, D.B. Digital Twin Data Modeling with AutomationML and a Communication Methodology for Data Exchange. *IFAC-PapersOnLine* **2016**, *49*, 12–17. [CrossRef]

41. Manas Bajaj.; Bjorn Cole.; Dirk Zwemer. Architecture To Geometry—Integrating System Models With Mechanical Design. 2016. Available online: https://arc.aiaa.org/doi/10.2514/6.2016-5470 (accessed on 20 November 2022).

42. Rosen, R.; von Wichert, G.; Lo, G.; Bettenhausen, K.D. About The Importance of Autonomy and Digital Twins for the Future of Manufacturing. *IFAC-PapersOnLine* **2015**, *48*, 567–572. [CrossRef]

43. Schluse, M.; Rossmann, J. From Simulation to Experimentable Digital Twins: Simulation-based Development and Operation of Complex Technical Systems. In Proceedings of the 2016 IEEE International Symposium on Systems Engineering (ISSE), Edinburgh, UK, 3–5 October 2016; pp. 1–6. [CrossRef]

44. Canedo, A. Industrial IoT Lifecycle via Digital Twins. In Proceedings of the Eleventh IEEE/ACM/IFIP International Conference on Hardware/Software Codesign and System Synthesis, New York, NY, USA, 2–7 October 2016; p. 1. [CrossRef]

45. Gabor, T.; Belzner, L.; Kiermeier, M.; Beck, M.T.; Neitz, A. A Simulation-Based Architecture for Smart Cyber-Physical Systems. In Proceedings of the 2016 IEEE International Conference on Autonomic Computing (ICAC), Wuerzburg, Germany, 17–22 July 2016; pp. 374–379. [CrossRef]

46. Söderberg, R.; Wärmefjord, K.; Carlson, J.S.; Lindkvist, L. Toward a Digital Twin for Real-Time Geometry Assurance in Individualized Production. *CIRP Ann.* **2017**, *66*, 137–140. [CrossRef]

47. Bolton, R.N.; McColl-Kennedy, J.R.; Cheung, L.; Gallan, A.; Orsinger, C.; Witell, L.; Zaki, M. Customer Experience Challenges: Bringing Together Digital, Physical and Social Realms. *J. Serv. Manag.* **2018**, *29*, 776–808. [CrossRef]

48. Industrial Digital Twin Association e. V. Digital Twin. 2022. Available online: https://industrialdigitaltwin.org/en/glossary/digital-twin (accessed on 11 November 2022).

49. Baudoin, C.; Bournival, E.; Buchheit, M.; Simmon, E.; Zarkout, B. The Industry IoT Vocabulary. 2022. Available online: https://www.iiconsortium.org/wp-content/uploads/sites/2/2022/04/Industry-IoT-Vocabulary.pdf (accessed on 11 November 2022).

50. Shafto, M.; Conroy, M.; Doyle, R.; Glaessgen, E.; Kemp, C.; LeMoigne, J.; Wang, L. *Modeling, Simulation, Information Technology & Processing Roadmap;* Technical report; National Aeronautics and Space Administration: Washington, DC, USA, 2010.

51. Tuegel, E. The Airframe Digital Twin: Some Challenges to Realization. In Proceedings of the 53rd AIAA/ASME/ASCE/AHS/ASC Structures, Structural Dynamics and Materials Conference, Honolulu, HI, USA, 23–26 April 2012. [CrossRef]

52. Gockel, B.; Tudor, A.; Brandyberry, M.; Penmetsa, R.; Tuegel, E. Challenges with Structural Life Forecasting Using Realistic Mission Profiles. In Proceedings of the 53rd AIAA/ASME/ASCE/AHS/ASC Structures, Structural Dynamics and Materials Conference, Honolulu, HI, USA, 23–26 April 2012; American Institute of Aeronautics and Astronautics: Reston, VA, USA, 2012. [CrossRef]

53. Bielefeldt, B.; Hochhalter, J.; Hartl, D. Computationally Efficient Analysis of SMA Sensory Particles Embedded in Complex Aerostructures Using a Substructure Approach. In Proceedings of the ASME 2015 Conference on Smart Materials, Adaptive Structures and Intelligent Systems, Colorado Springs, CO, USA, 21–23 September 2015; American Society of Mechanical Engineers Digital Collection: Little Falls, NJ, USA, 2016. [CrossRef]

54. Bazilevs, Y.; Deng, X.; Korobenko, A.; Lanza di Scalea, F.; Todd, M.D.; Taylor, S.G. Isogeometric Fatigue Damage Prediction in Large-Scale Composite Structures Driven by Dynamic Sensor Data. *J. Appl. Mech.* **2015**, *82*, 091008. [CrossRef]

Article

Towards a Democratization of Data in the Context of Industry 4.0

Tobias Harland [1], Christian Hocken [1,*], Tobias Schröer [2,*] and Volker Stich [2]

[1]   i4.0MC—Industrie 4.0 Maturity Center GmbH, 52074 Aachen, Germany; tobias.harland@i40mc.de

[2]   Institute for Industrial Management at RWTH Aachen University, 52074 Aachen, Germany;
    st@fir.rwth-aachen.de

*   Correspondence: christian.hocken@i40mc.de (C.H.); sr@fir.rwth-aachen.de (T.S.)

Abstract: Data-driven transparency in end-to-end operations in real-time is seen as a key benefit of the fourth industrial revolution. In the context of a factory, it enables fast and precise diagnoses and corrections of deviations and, thus, contributes to the idea of an agile enterprise. Since a factory is a complex socio-technical system, multiple technical, organizational and cultural capabilities need to be established and aligned. In recent studies, the underlying broad accessibility of data and corresponding analytics tools are called "data democratization". In this study, we examine the status quo of the relevant capabilities for data democratization in the manufacturing industry. (1) and outline the way forward. (2) The insights are based on 259 studies on the digital maturity of factories from multiple industries and regions of the world using the acatech Industrie 4.0 Maturity Index as a framework. For this work, a subset of the data was selected. (3) As a result, the examined factories show a lack of capabilities across all dimensions of the framework (IT systems, resources, organizational structure, culture). (4) Thus, we conclude that the outlined implementation approach needs to comprise the technical backbone for a data pipeline as well as capability building and an organizational transformation.

Keywords: data democratization; Industrie 4.0; fourth industrial revolution

Citation: Harland, T.; Hocken, C.; Schröer, T.; Stich, V. Towards a Democratization of Data in the Context of Industry 4.0. *Sci* **2022**, *4*, 29. https://doi.org/10.3390/sci4030029

Academic Editor: Johannes Winter

Received: 23 May 2022
Accepted: 4 July 2022
Published: 14 July 2022

Publisher's Note: MDPI stays neutral with regard to jurisdictional claims in published maps and institutional affiliations.

1. Introduction

The fourth industrial revolution is the only industrial revolution that was announced before it actually happened [1]. However, there were good reasons to believe that industry will go through an unprecedented development back in 2011. A set of new technologies matured and influenced each other in a synergetic way. Amongst others, technologies, such as cloud computing, the (industrial) internet of things, mobile computing and artificial intelligence, made the convergence of the digital and physical world possible and together, enabled the concept of "cyber-physical-systems" [2].

These technologies, indeed, have the potential to not only incrementally improve processes within the current organizational setup but to transform the entire way of working, as well as the business models of whole industries and, thus, should be considered an industrial revolution [3].

The three previous industrial revolutions all were, at least partly, driven by major advances in certain technological fields as well. Those were mechanization, electrification and computerization. Besides technological advances, the three industrial revolutions were accompanied by major changes to the way a company is organized [4]. Manual labor was substituted by machines (first industrial revolution), mass production and the scientific management divided processes in increments and optimized them (second industrial revolution) and automation of repetitive tasks led to the emergence of the knowledge worker (third industrial revolution) [5].

The implications of the abovementioned technologies in the fourth industrial revolution on labor are discussed controversially [6]. However, it is not arguable that there still will be humans working in industrial companies and that they will have the opportunity to work with much more data and will have access to much more sophisticated tools to gain insights from the data [7]. Currently, a promising concept is being discussed that is supposed to provide the organizational framework to utilize the newly available data and tools: data democratization [8].

At the core of the concept of "Data Democratization" is the idea to provide access to the company's data resources to all employees "given reasonable limitations on legal confidentiality and security" [9] (p. 1). This also implies that data should not only be available to technical experts but also to non-technical staff from the company [10]. Other authors highlight not only the access to data but the company culture of "willingness to share information" [11] (p. 5). The purpose of the concept is described as the "ability of users (. . .) to answer unexpected questions" [12] (pp. 1362–1368). This purpose is also well aligned with the overall purpose of the fourth industrial revolution, because many authors see agility and, thus, the quick and precise response to an unexpected event, as one of the key motivations to transform companies [13].

With this study, we intend to find out how advanced companies in the manufacturing sector operate in terms of a democratized use of their data and how the concept is related to what is commonly understood as Industry 4.0. Referring to a broadly used transformation framework (acatech Industrie 4.0 Maturity Index), the study focuses on a company's capabilities in the four structuring forces: "information systems", "resources", "organizational structure" and "culture".

2. State of the Art

Nowadays, the amount of information that is published is rapidly increasing [14]. Accordingly, the demand for experts who can process this data is high [15]. Data democratization allows not only data experts, but also non-specialists in companies to work with data [8]. Proper use of data can lead to increased revenue and brings the idea of data democratization back to the forefront [16], so that researchers have recently begun to look at data democratization [17]. Comprehensive research results are not available, yet. However, some focus areas must to be mentioned. Descriptive research with a focus on explaining the concepts and normative efforts has been conducted by numerous authors [8,18,19]. Based on those concepts, several case studies have been conducted on establishing the concepts in specific usage scenarios, including offshore drilling [20], banks and B2C businesses [9], real estate [21], medicine [22] and others. Surprisingly, no relevant studies on data democratization relating to manufacturing applications and Industry 4.0 are known to the authors. The processing and use of data play major roles in the context of Industry 4.0. By evaluating data, for example, errors can be detected at an early stage [23]. With 5G, larger volumes of data can be transmitted faster and more securely in the sense of Big Data [24]. The latest technologies, such as blockchain, can provide more data of higher quality and improve the operation of cyber–physical systems [25]. Through the use of smart contracts, processes can be established between IoT devices without intermediaries [26]. The IoT is related to data collection. It is associated with an object that is equipped with sensors to exchange data with other devices [27]. The generation, processing and use of data are, therefore, directly related to Industry 4.0 applications. In principle, the concept of data democratization is easily transferable on Industry 4.0 applications. However, it has not yet been examined from the perspective of manufacturing companies. Therefore, it seems reasonable to study the readiness of those companies for pursuing data democratization efforts.

3. Material and Methods

In order to answer the research question: "To what extent do industrial companies today have the necessary capabilities to apply the concept of data democratization?", an existing data set on the digital maturity of factories across the globe was analyzed. The data set is the result of 259 assessments of factories across the globe using the framework of the acatech Industrie 4.0 Maturity Index [13]. The framework was developed as a tool to measure the digital maturity of a factory. It focusses on the business process within the factories and describes the way towards a "learning, agile company". The framework is structured in six maturity levels and four structuring forces. The maturity levels can be summarized as follows:

- Computerization: information technologies are used in the core processes in an isolated way;
- Connectivity: information technology (IT) as well as operational technology (OT) systems offer connectivity and interoperability;
- Visibility: real-time data for end-to-end processes exist;
- Transparency: collected data are aggregated and contextualized, big data applications are deployed in parallel to the business applications;
- Predictability: the ability to simulate future scenarios for the core processes exists and it is possible to anticipate future developments;
- Adaptability: systems autonomously trigger actions and adapt to changing conditions.

The necessary capabilities per maturity level are organized along the structuring forces resources, information systems, organizational structure and culture. A central concept in the framework is that the development in the four structuring forces has to be synchronized to unfold the full potential [13].

The data were collected from 2017 to 2021. The factories originate from various industries with a certain focus on automotive, medical and food and beverage. A detailed distribution of industries is depicted in Figure 1.

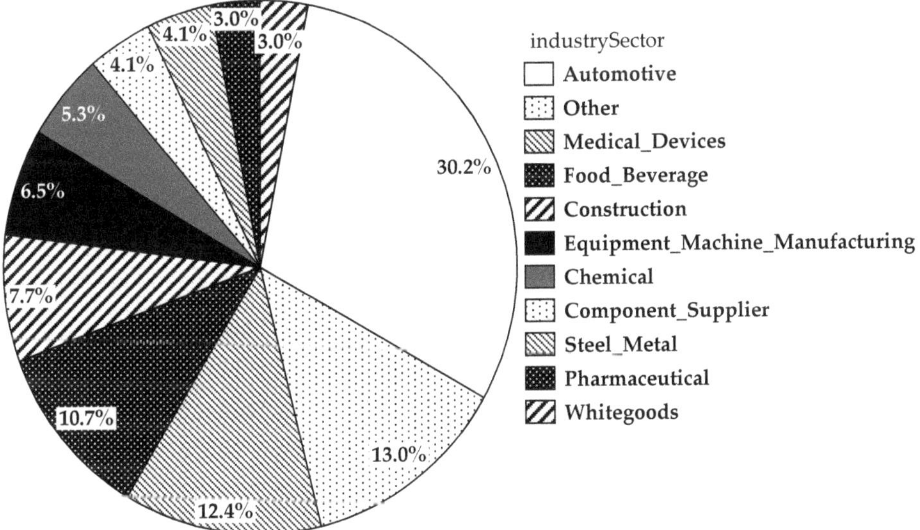

Figure 1. Distribution of industries.

In addition, the assessed factories are located across the globe with Turkey, Germany, the US and the Netherlands making up more than 50% (see Figure 2).

To be able to assess the dimensions of the framework in a standardized and assessor-independent way, a questionnaire was crafted. The questionnaire describes scenarios for each maturity level and each assessed dimension. The procedure for collecting the data was standardized and comprises interviews with experts from each core process in the assessed factory, Gemba walks following the value streams and group discussions to review the results.

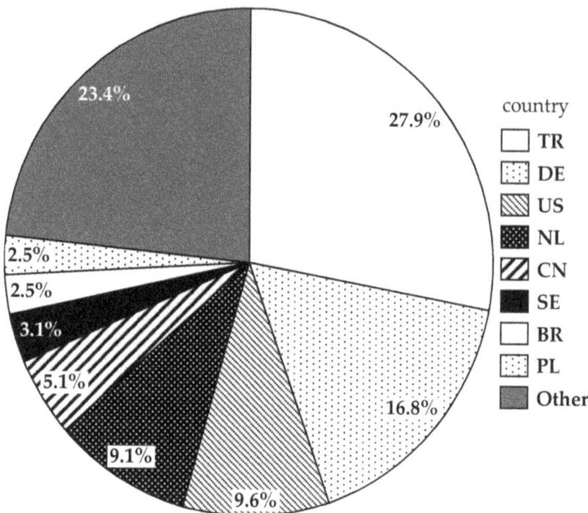

Figure 2. Country distribution.

Figure 3 shows that factories with more than 1500 employees, factories with 500–1500 employees and factories with 100–500 employees are evenly distributed. Only small factories with less than 100 employees are underrepresented.

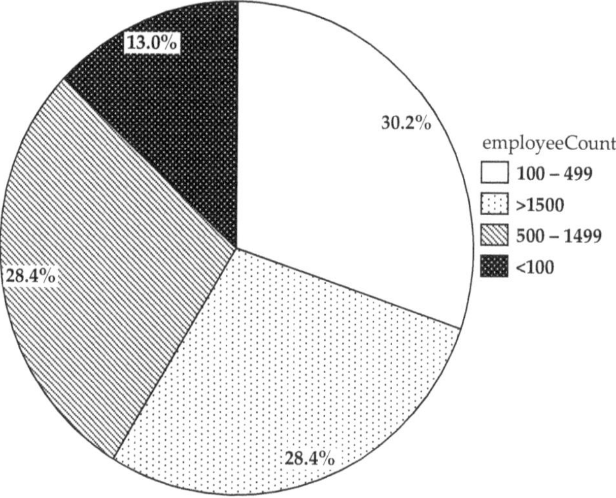

Figure 3. Number of employees.

The data set provides insights to many more capabilities than used for this study. In total, depending on the process to be analyzed, around 40 capabilities were assessed for each core process in the factories. However, we selected 10 of these capabilities for this study that provide insights to validate or falsify our hypotheses on the current state of the factories.

4. Results

The acatech Industrie 4.0 Maturity Index proposes that the digital maturity of a factory has to develop in its four structuring forces ("IT systems", "resources", "organizational structure" and "culture") in a synchronized way in order to unfold its full potential [13]

(p. 21). Following this idea, we formulated a hypothesis building on the abovementioned definitions of "data democratization" for each of the structuring forces accordingly. For the structuring force "IT systems" two hypotheses were examined.

4.1. IT Systems

Hypothesis 1. *The necessary data to provide a sufficiently detailed view on the current and past condition of processes and objects in the factory are available in a digital format and of an accurate quality.*

The existence of digital, accurate data for the area of the company under consideration is obviously an essential basis for the idea of data democratization. If there are no data to be shared to draw conclusions from, there is no sense in democratizing them.

To investigate this hypothesis, we examined the two capabilities "data quality" and "decision support (IT)". In this context, we understand data quality as the property of the data to be complete, accurate, current and consistent [28]. The maturity of this capability was characterized by the following statements (see Table 1). The percentage of answers per described scenario is depicted as well in the following table.

Table 1. Results for the capability "data quality".

Level	Scenario	Answers
1	"Data quality is not sufficient to further process the data."	22.92%
2	"The data quality is sufficient for further data processing, but there is a partly redundant collection of and storage of the data."	52.37%
3	"Redundant collection and storage of data with reference systems is avoided. The data quality allows a non-automatic further processing of the data."	19.35%
4	"The completeness, accuracy, currency and consistency of data sets are ensured. An automated data cleansing process is in place. High data quality ensures automatic further processing of the data ("fit for use")."	4.74%
5	"Regular profiling practices are used to identify errors in data sets and to ensure high data quality in the long term."	0.62%
6	"Systems for the self-healing of data sets as well as automated consistency checks and adjustments are in place."	0%

In order to comply with the concept of data democratization, the minimum target level is 4. However, most of the examined factories are currently at a lower level. More than 75% of the surveyed factories do not have the appropriate level of data quality to enable an automated use of data.

A second capability that is related to the first hypothesis is the decision support by IT systems. Decision support systems have been discussed in the literature for decades and are, for instance, defined as "interactive computer-based systems, which help decision makers utilize data and models to solve unstructured problems" [29] (pp. 1–26). Other authors confirm that these systems support decision-making in "semistructured tasks" [30] and emphasize that they "enhance the traditional information access and retrieval functions with support for model building and model-based reasoning" [31] (p. 6). The capability to create meaningful data-based insights to improve decision-making is at the center of the concept of data democratization and, thus, needs to be considered in this study.

Table 2 contains the possible scenarios for this capability. The minimum level for this capability that is necessary to implement the approach of data democratization is level 3. According to the described scenario, factories at level 3 are able to visualize operational data. At levels 1 and 2, data are not yet visually processed or decision-making is not supported with data. The data set shows that more than 68% of the assessed factories are not at that level. However, the fact that 47.82% of factories at least collect data on the current machine status indicates that some basics for databased decisions on their operations are present.

Table 2. Results for the capability "decision support (IT)".

Level	Scenario	Answers
1	"Decision-making is not supported by systems."	20.84%
2	"Machine status and condition data of the machine park are available in the leading system, but not visually processed and therefore not directly available for decisions."	47.82%
3	"Visualizations are created with dashboards (e.g., target / actual quantity, downtimes)."	24.59%
4	"Users are informed about standstills and limit exceedances by means of alerting and escalation processes."	6.23%
5	"In addition to level 4, effects are shown in advance."	0.51%
6	"Where possible, decisions are made automatically and presented to the user in a comprehensible manner."	0%

Comparing these results with other studies on data quality and databased decision support in manufacturing, the results even indicate a larger gap in these capabilities compared to the findings of other studies. For instance, Spath et al. found that 58.9% of surveyed companies are not able to detect relevant events in their production automatically and, thus, based on data, 43.7% of participants see the missing actuality of production data as the root cause of manual interventions to production planning [32]. Lanza et al. found, in their study on digital shopfloor management, that today, in more than 50% of the examined cases, operational KPIs were calculated manually; in more than 66%, the visualization of these KPIs done in an analog way [33].

In conclusion, the hypothesis cannot be considered validated. Thinking of this as a prerequisite for decisions based on democratized data, one central field of action for most companies is the systematic improvement in the quality of their operational data in combination with systems that make them available for decisions.

Hypothesis 2. *The data are organized in a way that it is accessible for ad-hoc analyses.*

"Data Democratization" and "Industrie 4.0" place the stakeholders at the center. They are expected to optimize their area of responsibility through the use of data, e.g., by making decisions themselves on the basis of their own data analyses. These analyses often arise spontaneously (ad hoc) due to unexpected events stakeholders need to cope with but can also be caused by the pursuit of general performance or quality improvements.

In many cases, analyses involve several data sources, e.g., order data, material flow data or quality data. Data mostly originate from IT systems that are used for controlling or monitoring processes. However, other data sources might also be used, e.g., feedback from operators on process or product quality. In hypothesis 1, we showed that these data must be of sufficient quality and that the type and scope of the data must be sufficient so that they can be used for data-driven decision-making. However, further requirements must be met in the area of data organization so that stakeholders can perform analyses on their own. Data from different sources must be associated with one another, and data must be accessible to the user. The organization of data must meet these principles so that data can be used for ad-hoc analysis.

To investigate this hypothesis based on the available data set, we focus on two capabilities. "Data model" describes the conceptual approach to structure data. It includes information on data objects, references between objects and it defines valid operations [34]. In this study, we focus primarily on the scope of the data model, which can encompass individual processes up to an entire production network. The maturity of the capability "data model" was characterized by the following statements. The percentage of answers per described scenario is depicted as well in the following table (see Table 3).

Table 3. Results for the capability "data model".

Level	Scenario	Answers
1	"There is no data model for the process and relevant data points are unknown."	18.27%
2	"There is only an isolated data model for the process."	59.27%
3	"There is an integrated cross-process data model. Example for production: orders, production parameters can be combined with quality and maintenance data."	20.78%
4	"An internal data model exists across all sites."	1.53%
5	"There is a cross-site data model in which suppliers and customers are integrated."	0.14%
6	No scenario defined	

The results show that in more than 18% of the performed assessments, no data model was available and data points to monitor or control the process were unknown. Companies that meet this maturity level are not yet able to offer their stakeholders the basis for their own data analyses. In almost 60% of cases, data models exist that refer to individual processes and, thus, enable analyses at process level. Practice shows that in production processes, Scada/Level 2 (ISA 95) data are often used (time-series data that can include machine states, sensor values or information on product quality). However, the analysis options are limited depending on the question. In cases where interfaces to other processes are to be examined or for the determination of performance indicators, data from higher layers (ISA 95), such as from Manufacturing Execution Systems and Enterprise Resource Planning Systems, are missing. This is made possible by cross-process data models, which are currently available to almost 21% of the plants examined. It can be observed that data models are limited to the plant itself in almost 98% of the cases surveyed. Only 1.5% state that they have integrated data models for all sites. The integration of suppliers and customers into company-wide data models has not yet played a role according to our data set.

The second capability we focus on is what we call "Collaboration in IT". This expresses whether social software or enterprise collaboration systems are being used for stakeholder communication and collaboration. A digital collaboration environment with a uniform user interface is important for collaboration between users, including working together on documents and other use cases [35]. The use of such solutions has two advantages. On the one hand, data are made available in a central location and can usually be accessed and consumed by users with the appropriate rights. On the other hand, they promote the formalization of tacit knowledge [36]. Table 4 contains the possible scenarios for this capability.

Table 4. Results for the capability "Collaboration in IT".

Level	Scenario	Answers
1	"There is no collaborative platform for employees."	7.98%
2	"Collaboration is supported through static IT systems, such as email accounts, file servers, and folder structures."	61.36%
3	"A collaborative platform is in place which facilitates the management of tasks and projects"	28.32%
4	"The collaborative platform is in addition used for the communication between employees. All internal communication takes place via the platform, so that email correspondence is reduced to a minimum."	2.26%
5	No scenario defined	0.08% *
6	No scenario defined	

* Due to false statement in one assessment.

The minimum level for this capability that is necessary to implement the approach of data democratization is level 4. The data set shows that, so far, only 2.6% of the companies surveyed have such a solution in place. However, just under a third (28.3%) are on their way there and have at least a solution in place that facilitates the management of tasks and projects. However, a large proportion of the companies surveyed still rely on classic file sharing via a network drive (61.36%). Just under 8% forgo the use of collaboration solutions altogether.

The research shows that we are still in the early stages of organizing data for use in ad-hoc analyses. Although data are available, they can usually only be correlated at the individual process level. Overarching models that encompass the entire value-creation process of a factory or production network are scarce. A similar picture can be drawn for the joint creation and use of data. Although data are shared in many cases, classic approaches, such as network drives, are used for this purpose. Our data set shows that comprehensive collaboration platforms have not been used much so far. However, due to the long duration of the study, it is possible that collaboration platforms are underrepresented in our results. Other studies seem to indicate that many companies adopted such platforms during the COVID-19 pandemic [37].

4.2. Organizational Structure

Hypothesis 3. *The organizational structure in the factory empowers employees to proactively improve their routines and initiate and implement improvements on their own.*

The effective use of data democratization requires that employees are involved in improving their working conditions, tasks and routines. It even goes one step further and relies on employees themselves identifying potential for improvement and implementing measures. The organization should not only tolerate this behavior but should actively promote it. It has long been recognized that improvements brought in by employees contribute to the company's success. Continuous improvement approaches are, for example, an important part of management and work organization methods, such as Lean [38] and Kaizen [39–41]. Our hypothesis is that the same mechanisms could be applied to the ideas of democratization of data. It is important here that stakeholders accompany both the analysis and the implementation of measures.

To examine our hypothesis, we look at two capabilities in the data set. First, we look at whether companies have implemented an "innovation process" and how far reaching it is. This capability provides a good understanding of whether and how employees are involved in improvement processes. The maturity level of the "innovation process" capability was characterized by the following state characteristics. The percentage of responses per described scenario is also shown in the following table (see Table 5).

In order to comply with the concept of data democratization, the minimum target level is 4. About 23% of the examined factories currently are, at least, at this level. We observe that almost all of the factories studied have implemented an innovation process, with only around 8% not doing so at all. In just under 5% of the plants surveyed, idea generation and idea development are based on the open innovation principle. Employees at all levels contribute to idea generation and evaluation. Measures and their effectiveness are systematically monitored, documented and, if necessary, adjusted in accordance with the PDCA cycle. As such, 33% have at least one process for collecting suggestions for improvement (level 2), and another 35% circulate information about ideas submitted, evaluation and implementation (level 3). Thus, in more than half of the companies, there is an active innovation process that leads to implemented improvements. Only the participation of employees in the implementation of measures is not yet sufficiently well developed. Overall, however, these are good conditions for data democratization.

"Decision power and responsibility" is the second capability we examine. This is about how and by whom decisions are made in the organization and how decision-making is supported. We are particularly interested in the hierarchical level at which decisions are made for operational activities. The defined maturity levels and the associated scenarios are listed in Table 6.

Table 5. Results for the capability "innovation process".

Level	Scenario	Answers
1	"Improvements and innovations are not considered responsibilities of operative employees."	8.16%
2	"A process exists through which employee ideas can be collected and evaluated (e.g., suggestion scheme, CIP)."	33.07%
3	"In addition to level 2, there is visibility about the ideas submitted, their evaluation and their implementation."	35.56%
4	"There is a process in place to evaluate and subsequently implement ideas. Employees are involved in the implementation of their ideas."	17.60%
5	"Idea generation and idea development take place according to the open innovation principle. Employees at all levels contribute to the generation and assessment of ideas. Measures and their effectiveness are systematically monitored, documented, and, possibly, modified in accordance with the PDCA cycle."	4.61%
6	"Employees are given sufficient leeway to develop new ideas and concepts and test them."	0.99%

Table 6. Results for the capability "Decision power and responsibility".

Level	Scenario	Answers
1	"Operative employees do not have the power to make decisions. All decisions are made at the management level. Clear hierarchies with inflexible reporting lines have been established."	15.19%
2	"Decisions on operations activities are made by operative employees (within the group). There is no support in the decisions process by IT systems. Uncertainties are clarified and conflicts are resolved by (first-line) managers or supervisors."	35.09%
3	"Decisions on operative activities are made by operative employees (within the group). Information to support the decision-making process are provided by IT systems. Uncertainties are clarified and conflicts are resolved by (first-line) managers or supervisors."	45.79%
4	"Decisions, including those on operational activities, are made by the employee with the most expertise. This is not necessarily a supervisor or manager. A holocratic approach is taken.	3.92%
5	No scenario defined	
6	No scenario defined	

At least level 3 is required to use data democratization. Decisions concerning operational activities must be able to be made by operators. In addition, however, there must also be support from IT systems that hold data relevant to decision-making. We see that just under half of the factories studied meet these requirements (49.71%). Among them, 3.92% even choose holocratic approaches for decision-making. In around 35% of the factories surveyed, decisions are made as a team (level 2). However, this group lacks the necessary IT support. Only slightly more than 15% still rely on distinct hierarchical decision-making structures.

Both capabilities studied show that we are at a turning point. It is apparent that organizations want to involve employees more and are increasingly granting them freedom to do so. Based on the available data, it can be seen that both capabilities can still be expanded in the majority of the factories surveyed. The most important factors include employee participation in the implementation of measures and the creation of organizational conditions for the use of data in decision-making processes. The potential of employee participation in terms of data democracy still relies on the maturity of the data dimension, though.

4.3. Culture

Hypothesis 4. *The culture among the associates in factories encourages data-driven decision-making and an open sharing of knowledge and insights.*

Besides formal processes, organizational structure and assigned responsibilities, the culture among the associates has a paramount influence on how the concept of data democratization is embraced in a factory. In general, we see culture as the commonly accepted set of values within the organization that guides the actions of employees [17,42]. Specifically focusing on the concept of data democratization, these values need to comprise the "willingness to share information" [11] (p. 5). For that reason, the capability "knowledge and knowledge responsibility" is selected from the data set. In addition, the capability "data-based decision processes" was analyzed to understand to what extent the companies base their decisions on data [43].

In Table 7, the possible scenarios for the capability "Knowledge and knowledge responsibility" are listed. With every level, the importance to formalize and to share knowledge increases. Thus, this capability describes to what extent "tacit knowledge" or "explicit knowledge" guide the course of actions of the employees. "Tacit knowledge" solely exists in the minds of individuals and, thus, is not shareable and available to the rest of the organization [44]. For the idea of data democratization, the willingness to share knowledge is crucial. Due to that, the minimum level of this capability is level 3. At this level, a company collects and shares knowledge in a structured way. In nearly 30% of the assessed companies, such a culture can be observed. However, about two-thirds of the companies have a culture that does not actively encourage the formalization and sharing of knowledge.

Table 7. Results for the capability "Knowledge and knowledge responsibility".

Level	Scenario	Answers
1	"Knowledge is not shared, as it helps the individual to become indispensable. Thus, employees typically rely on their own experience."	4.75%
2	"Employees network within their respective areas and from expert groups. Employees are not overly willing to formalize their knowledge and make it available to the entire company."	63.87%
3	"Experience and practical knowledge are systematically collected and transformed into explicit knowledge. Employees are willing to share their knowledge in a formalized way with the company."	29.80%
4	"Knowledge guides the actions of all employees. Employees are willing to participate in an intensive exchange of information and voluntarily seek to understand relationships and generate new knowledge, even outside of work. The provision of knowledge is supported by appropriate IT solutions, such as knowledge management systems and collaborative platforms."	1.58%
5	No scenario defined	
6	No scenario defined	

The results in Table 8 for the capability "data-based decision processes" confirm the impression of the previous capability: About 40% of the companies already use data as the basis for their decision-making. Still, around 60% rely on intuition and individual knowledge or only partly consider data in their decision-making processes.

Table 8. Results for the capability "Data-based decision processes".

Level	Scenario	Answers
1	"Decisions are made on the basis of the knowledge of the intuition of individual employees."	14.17%
2	"Data from IT systems is taken into account in the decision-making process, but decisions are adapted case by case on the basis of personal knowledge or the intuition of individual employees."	45.08%
3	"Current and historical data are used to support the decision-making process."	27.05%
4	"Data analyses are made to support the decision-making process."	11.86%
5	"The decision-making process is supported by simulations and scenarios."	1.83%
6	"Decision-making processes are largely automated; employees supervise the decisions and intervene in exceptional cases only."	0%

Based on the analysis of the two capabilities, the hypothesis cannot be confirmed. Even if around two-thirds of the companies have already adopted a culture of data-based decision-making and of an initiative-taking sharing of insights, still, most of the evaluated companies are lacking such a culture. Other studies partly confirm these findings. For instance, Lanza et al. found that in five out of nine analyzed case studies on digital shopfloor management, there is an exchange of knowledge, but it is an analog exchange without the support of digital tools [33]. A recent study of YouGov, with more than 3500 surveyed decision-makers, found that 56% of the companies consider themselves as data driven, which is a significantly higher share than in the analyzed data set [45]. Another study with more than 1000 participants, conducted by Capgemini in 2020, found that 50% consider their decision-making as data driven [16]. However, the results in the present analysis are based on external, independent observations, whereas the cited studies rely on a self-evaluation of the surveyed participants.

As a conclusion, it is a central field of action for these companies to create such a culture on their way towards a democratization of data, to not only have the technology ready, but also create a "pull" for data in their workforce.

4.4. Resources

Hypothesis 5. *The employees have the appropriate capabilities to work with data and interpret them in the context of their domain.*

As Belli et al. state in their definition of data democratization, it is the "ability of users to access all data using well-defined and easily used analytic patterns to answer unexpected questions" [12] (pp. 1362–1368). The definition implies, besides organizational aspects, such as the access to data, and technological aspects, such as the easily used analytic patterns, that the users themselves need to be capable. To examine the necessary skills of the employees, the two capabilities "IT competencies" and "Interdisciplinary skills" were analyzed.

As depicted in Table 9, the surveyed scenarios range from a rudimentary utilization of existing system functionalities with manual bypasses to the system to a full utilization of the features and an independent covering of information needs. In only 3.69% of the companies, the employees can cover their information needs themselves, which would be the required level for the concept of data democratization. The vast majority are only capable of handling basic system features. Still, in around one-quarter of companies, it can be observed that the associates are able to understand the logics and dependencies of the systems they are using. This leads to the conclusion that, at least in this group of companies, a certain awareness for the multiple data in their IT systems and its value exists.

Table 9. Results for the capability "IT competencies of employees".

Level	Scenario	Answers
1	"Employees can only partly handle the functions that are necessary for their operational activities. This leads to an avoidance of systems or using Excel-solutions although application systems exist."	9.02%
2	"Employees can only handle functions that are necessary for their operational activities."	59.36%
3	"Employees are aware of the logics and dependencies of the used system functions."	27.94%
4	"Employees are able to independently operate systems beyond their standard functionalities in order to satisfy their information needs, e.g., analytical applications required for their daily work."	3.69%
5	No scenario defined	
6	No scenario defined	

Table 10 contains the scenarios for the capability "interdisciplinary skills of employees". The scenarios range from employees having a very limited scope for their own considerations to employees who understand the dependencies with and their impact on other processes and activities in the company. This capability is important for a successful implementation of a data democratization approach, as the whole idea is based on accessing the relevant data from the whole organization and not only from their own limited domain. In order to make sense of these data, it is inevitable that the context the data is sourced from is understood.

Table 10. Results for the capability "Interdisciplinary skills of employees".

Level	Scenario	Answers
1	"Employees focus on subject- or domain-specific questions and problems."	24.59%
2	No scenario defined	3.04% *
3	"Employees know about neighboring process steps/activities, include them in their considerations and exchange information."	61.67%
4	"Employees are aware of the impact of their activities on neighboring process steps/activities, include them in their considerations and proactively address interdisciplinary problems and questions on their own."	10.70%
5	No scenario defined	
6	No scenario defined	

* Due to revision questionnaire.

The results show that the associates of only around 25% of the companies strictly focus on their own domain. A large group either is aware and considers neighboring processes or even proactively involves them in their activities.

Looking at these two capabilities, a slightly higher readiness can be observed compared to other examined capabilities in this study. In addition, it is not visible in the data whether the capability "IT competencies of employees" would be on an even higher level if the companies were to provide their employees with more capable systems. After all, employees were not able to demonstrate the capabilities required for the higher levels, even if they had had them. It is obvious that the competencies required to work with data effectively rely on the data itself being available and structured, which cannot be considered true, as already mentioned earlier.

The comparatively high maturity level in the capability "Interdisciplinary skills of employees" might be related to an increasingly high maturity level in lean methodologies [46]. The concepts of interdisciplinary collaboration and process-oriented thinking are core ideas of lean management [47]. This, once again, supports the importance of considering Industry 4.0 and Lean Management or comparable approaches in an integrated way [48].

5. Discussion

The presented study was conducted to answer the question "To what extent do industrial companies today have the necessary capabilities to apply the concept of data democratization?" We translated the overall concept of data democratization in the first step to a set of hypotheses on the existence of the relevant capabilities in manufacturing companies. Theses hypotheses were evaluated by utilizing an existing data set containing detailed data on the current digital maturity of 259 factories from multiple industries and regions of the world. The results of this study were related and compared with existing studies (see description of results per examined capability). In most of the cases, other studies could confirm the findings. However, in general, the findings of other studies were slightly more optimistic regarding the existence of certain capabilities in the industry. We relate this difference to the fact that many other studies rely on a self-evaluation of the surveyed person. We believe that the used data set for this study provides a much more realistic view on the status of manufacturers, since all data were collected from an external team of assessors.

Still, this study has its limitations. The used data set was collected over the period of four years and, thus, does not, for all evaluated companies, reflect the latest development stage. In addition, the evaluation process requires a certain investment of time and money from each company. Due to that, companies who feel the need themselves to improve in terms of digital maturity would typically go through the process in order to identify opportunities to improve. Companies that already perform at a high level tend to not go through the evaluation. Due to that, the data set might be biased. Furthermore, there might be capabilities that are relevant for the concept of data democratization but that are not part of the used data set.

Thus, further research can focus on finding and filling these potential existing gaps and, with this, complement the results of this study. In addition, the results of this study can be used as a starting point for the development of a more detailed concept of "Data Democratization for Manufacturing" that is closely aligned with the current state of relevant capabilities in the industry and, thus, is tailored to the needs of practitioners.

Lastly, the results can be used directly in manufacturing companies to benchmark their status quo and to start shaping their own roadmap towards a democratization of data in their own factories.

6. Conclusions

The conducted analysis clearly observed a strong interdependency between the maturity of a company in terms of Industry 4.0 and the ability to utilize data in a democratized manner. Considering this fact, it becomes an obvious ambiguity to think of data democracy as an enabler for monetizing data in terms of business cases, on the one hand, and having many aspects of Industry 4.0 as a prerequisite, on the other hand. In short terms: structured, consistent and available data are necessary to start a data democracy—an initiative of which the business value can hardly be estimated ex ante. Therefore, we suggest a set of essential capabilities based on our studies (Figure 4).

We suggest starting the process from the dimension "culture", in order to first create a "pull" from the organization instead of first developing the technology with the risk of missing essential user requirements or overengineering. Based on the findings, most companies still base many decisions on individual experience and intuition and not on data. Thus, procedures to obligatorily base decisions on data need to be established. To initiate this cultural change, an option might be to let the executives serve as role models for this kind of decision-making and let a data-driven culture grow from this nucleus [49].

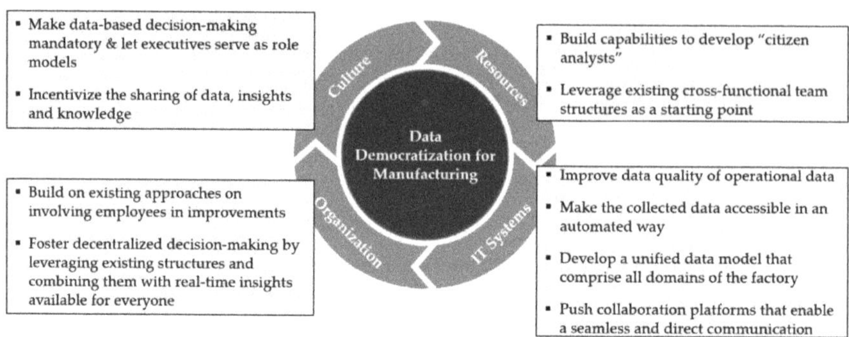

Figure 4. Essential fields of Data Democratization for Manufacturing.

Another finding of the study is that the motivation to formalize and share knowledge can become a roadblock for the democratization of data. Therefore, a field of action is to set specific incentives to foster a mindset of knowledge sharing. For instance, existing practices to incentivize the participation in suggestion programs or in lessons-learned programs can be used as a reference [50].

As a second step, we recommend considering the dimension "resources". Here, the study shows that in most companies, the employees are able to use the existing IT systems to perform their day-to-day work but lack the capabilities to cover their individual information needs. Thus, capability building to so-called "citizen analysts" might be a field of action to enable the associates. These are employees who are qualified in self-service analytics tools to cover their data needs own their own [10]. In addition, the results indicate that the existing awareness of the importance of cross-functional problem solving can be capitalized on. It is, for instance, part of many problem-solving techniques to involve experts from multiple domains to contribute to the process. Such groups may provide fertile ground for promoting the concept of data democratization.

After having a first set of associates selected and qualified, the capabilities of the IT systems have to be expanded. The examination of the related capabilities points out that the relevant data on the factory's operations are not of sufficiently good quality, nor can they be easily accessed, nor are data from different domains logically linked. That leads, on the one hand, to the conclusion that companies need to launch initiatives to improve the data quality. These initiatives might include organizational measures in the field of data governance [51] but also measures to collect data in an automated way, for instance, directly from the machine PLCs [52]. Furthermore, the accessibility of the data in the systems seems to be another roadblock. Typical measures in this field comprise the implementation of a middleware or an enterprise service bus [53]. In order to logically link data from multiple sources in an automated way, it is necessary to have a unified data model across all domains that defines the database schema for the data platform and builds the basis to map the data from various sources.

Besides the mentioned fields of actions related to data quality and data access, the analyses of the dimension IT systems show that the roll-out of social software and collaboration platforms is another field of action. They are supposed to catalyze the distribution of and communication about insights in the organization.

In order to utilize the insights from the "citizen analysts" the examined capabilities in the dimension "organization" can serve as a basis. The results indicate that many companies already have procedures in place to involve employees in the continuous improvement process. The insights gained through the analysis of data are another source to identify opportunities to improve the mentioned existing procedures to channel and route these opportunities and build a solid foundation. The field of action here would be to enable the integration of both initiatives, the continuous improvement programs and the data democratization program.

Another examined capability in the dimension organization is the decentralization of decision-making. The rationale here is that if associates are able to gain insights on their own, they also need to be able to act on these insights on their own. The results indicate that many companies are organized in a way that enables these decentralized decisions. Thus, this is another capability that acts synergistically on a successful data democratization program.

In total, as Figure 4 shows, an organization will iterate multiple times through these fields of action. An iterative approach that starts in a limited area of the organization and from there, is scaled to other domains, helps to avoid risks and generates the first tangible results earlier.

Author Contributions: Conceptualization, C.H., T.S. and T.H.; methodology, T.H., C.H. and T.S.; formal analysis, T.H.; writing—original draft preparation, T.H.; writing—review and editing, C.H., T.S. and V.S.; supervision, V.S. All authors have read and agreed to the published version of the manuscript.

Funding: This research received no external funding.

Institutional Review Board Statement: Not applicable.

Informed Consent Statement: Informed consent was obtained from all subjects involved in the study.

Data Availability Statement: The data set is the result of 259 assessments of factories across the globe using the framework of the acatech Industrie 4.0 Maturity Index. The data were collected in from 2017 to 2021.

Conflicts of Interest: The authors declare no conflict of interest.

References

1. Bauernhansl, T.; ten Hompel, M.; Vogel-Heuser, B. *Industrie 4.0 in Produktion, Automatisierung und Logistik*; Springer Fachmedien Wiesbaden: Wiesbaden, Germany, 2014; ISBN 978-3-658-04681-1.
2. Pishdad-Bozorgi, P.; Gao, X.; Shelden, D.R. Introduction to cyber-physical systems in the built environment. In *Construction 4.0*; Sawhney, A., Riley, M., Irizarry, J., Riley, M., Eds.; Routledge: London, UK, 2020; pp. 23–41. ISBN 9780429398100.
3. Schwab, K. The Fourth Industrial Revolution. *J. Int. Aff.* **2016**, *72*, 17–22.
4. Veza, I.; Mladineo, M.; Peko, I. *Analysis of the Current State of Croatian Manufacturing Industry with Regard to Industr 4.0*; Croatian Association of Production Engineering: Vodice, Croatia, 2015.
5. Kagermann, W.H.; National Academy of Science; Deutsche Post AG; German Research Center for Artificial Engineering. *Recommendations for implementing the strategic initiative INDUSTRIE 4.0: Final report of the Industrie 4.0*; Working Group, Acatech: Frankfurt, Germany, 2013.
6. Sima, V.; Gheorghe, I.G.; Subić, J.; Nancu, D. Influences of the Industry 4.0 Revolution on the Human Capital Development and Consumer Behavior: A Systematic Review. *Sustainability* **2020**, *12*, 4035. [CrossRef]
7. Anderson, J.; Rainie, L. Artificial Intelligence and the Future of Humans. *Pew Res. Cent.* **2018**, *10*, 1–9.
8. Lefebvre, H.; Legner, C.; Fadler, M. Data democratization: Toward a deeper understanding. In Proceedings of the International Conference on Information Systems (ICIS), Austin, TX, USA, 12 December 2021.
9. Awasthi, P.; George, J.J. A case for Data Democratization. In Proceedings of the AMCIS, Salt Lake City, UT, USA, 10–14 August 2020.
10. Loshin, D. Data Intelligence: Empowering the Citizen Analyst with Democratized Data. *Erwin* **2019**, *1*, 2–3.
11. Kim, H.; Kim, H.R.; Kim, S.; Kim, E.; Kim, S.Y.; Park, H.-Y. Public Attitudes Toward Precision Medicine: A Nationwide Survey on Developing a National Cohort Program for Citizen Participation in the Republic of Korea. *Front. Genet.* **2020**, *11*, 283. [CrossRef]
12. Bellin, E.; Fletcher, D.D.; Geberer, N.; Islam, S.; Srivastava, N. Democratizing information creation from health care data for quality improvement, research, and education-the Montefiore Medical Center Experience. *Acad. Med.* **2010**, *85*, 1362–1368. [CrossRef]
13. Schuh, G.; Anderl, R.; Gausemeier, J.; ten Hompel, M.; Wahlster, W. *Industrie 4.0 Maturity Index: Die Digitale Transformation von Unternehmen Gestalten (Acatech Studie)*; Herbert Utz Verlag: Munich, Germany, 2017.
14. Statista Research Department. Volume of data/information created, captured, copied and consumed worldwide from 2010 to 2025. *Statista* **2022**, *1*. Available online: https://www.statista.com/statistics/871513/worldwide-data-created/ (accessed on 19 May 2022).

15. Chea, J.; Shen Lai, P. Accelerated Digital Transformation: Research, perspective and guidance from data scientists and thought leaders. *SAS* **2022**, *1*, 4–15.
16. Thieullent, A.-L.; Jiang, Z.; Perhirin, V.; Tolido, R.; Jones, S.; Svahn, T.; Clarke, N.; Buvat, J.; Manchanda, N.; Puttur, R. The data-powered enterprise: Why organizations must strengthen their data mastery. *Capgemini* **2020**, *1*, 3–29.
17. Legner, C.; Lefebvre, H. How Communities of Practice Enable Data Democratization Inside the Enterprise. *ResearchGate* **2022**, *2*, 1796.
18. Fahey, S. The democratization of big data. *J. Natl. Sec. L. Poly.* **2014**, *7*, 325–332.
19. Crosby, L.A.; Johnson, S.L. The democratization of data. *Mark. Manag.* **2001**, *10*, 8–15.
20. Yoder, R.T. Digitalization and Data Democratization in Offshore Drilling. In Proceedings of the Offshore Technology Conference, Houston, TX, USA, 6–9 May 2019.
21. McLaughlin, R.; Young, C. *Data Democratization and Spatial Heterogeneity in the Housing Market*; Harvard Joint Center for Housing Studies: Cambridge, MA, USA, 2018.
22. Alexander, J.C.; Joshi, G.P. Smartphone Application-based Medical Devices: Twenty-first Century Data Democratization or Anarchy? *Anesth. Analg.* **2016**, *123*, 1046–1050. [CrossRef]
23. Dong, W.; Kwok-Leung, T.; Qiang, M. Prognostics and Health Management: A Review of Vibration Based Bearing and Gear Health Indicators. In Proceedings of the 2017 IEEE International Conference on Prognostics and Health Management (ICPHM), Dallas, TX, USA, 19–21 June 2017; IEEE: Piscataway, NJ, USA, 2017.
24. Duan, L.; Da Xu, L. Data Analytics in Industry 4.0: A Survey. *Inf. Syst. Front.* **2021**, *1*, 1–17. [CrossRef] [PubMed]
25. De, D.; Bhattacharyya, S.; Rodrigues, J.J.P.C. *Blockchain Based Internet of Things*; Springer Singapore: Singapore, 2022; ISBN 978-981-16-9259-8.
26. Da Xu, L.; Viriyasitavat, W. Application of Blockchain in Collaborative Internet-of-Things Services. *IEEE Trans. Comput. Soc. Syst.* **2019**, *6*, 1295–1305. [CrossRef]
27. Ramasamy, L.K.; Kadry, S. (Eds.) *Blockchain in the Industrial Internet of Things*; IOP Publishing: Bristol, UK, 2021; ISBN 978-0-7503-3663-5.
28. Rohweder, J.P.; Kasten, G.; Malzahn, D.; Piro, A.; Schmid, J. Informationsqualität—Definitionen, Dimensionen und Begriffe. In *Daten-und Informationsqualität*; Hildebrand, K., Gebauer, M., Hinrichs, H., Mielke, M., Eds.; Springer Fachmedien Wiesbaden: Wiesbaden, Germany, 2015; pp. 25–46. ISBN 978-3-658-09213-9.
29. Sprague, R.H. A Framework for the Development of Decision Support Systems. *MIS Q.* **1980**, *4*, 1–26. [CrossRef]
30. Keen, P.G.W.; Morton, M.S.S. *Decision Support Systems: An Organizational Perspective*; Addison-Wesley: Reading, MA, USA, 1978; ISBN 0201036673.
31. Druzdzel, F.; Flynn, R.R. *Decision Support Systems*; Encyclopedia of Library and Information Science: New York, NY, USA, 2002; Volume 2.
32. Spath, D.; Ganschar, O.; Gerlach, S.; Hämmerle, M.; Krause, T.; Schhlund, S. *Produktionsarbeit der Zukunft—Industrie 4.0*; Fraunhofer Verlag: Stuttgart, Germany, 2013.
33. Lanza, G.; Hofmann, C.; Stricker, N.; Biehl, E.; Braun, Y. Auf dem Weg zum digitalen Shopfloor Management: Eine Studie zum Stand der Echtzeitentscheidungsfähigkeit und des Industrie 4.0-Reifegrads. *Karlsr. Inst. Für Technol. (KIT) Karlsr.* **2018**, *1*, 34–36.
34. Yarlagadda, R.T.; Syed, H.H. Data Models In Information Technology. *ResearchGate* **2016**, *1*, 1–8.
35. Schwade, F.; Schubert, P. Social Collaboration Analytics for Enterprise Collaboration Systems: Providing Business Intelligence on Collaboration Activities. In Proceedings of the Conference on System Sciences, Hawaii International Conference on System Sciences, Hilton Waikoloa Village, HI, USA, 4 January 2017.
36. Kucharska, W.; Kowalczyk, R. Trust, Collaborative Culture and Tacit Knowledge Sharing in Project Management—A Relationship Model. In Proceedings of the 13th International Conference on Intellectual Capital, Knowledge Management & Organisational Learning, Ithaca, NY, USA, 2016.
37. Gartner. Gartner Forecasts Worldwide Social Software and Collaboration Market to Grow 17% in 2021. Available online: https://www.gartner.com/en/newsroom/press-releases/2021-03-23-gartner-forecasts-worldwide-social-software-and-collaboration-market-to-grow-17-percent-in-2021 (accessed on 19 May 2022).
38. Melović, B.; Mitrović, S.; Zhuravlev, A.; Braila, N. The role of the concept of LEAN management in modern business. *MATEC Web Conf.* **2016**, *86*, 5029. [CrossRef]
39. Imai, M. *Kaizen: (Ky'zen): The Key to Japan's Competitive Success*, 1st ed.; McGraw-Hill: New York, NY, USA, 1986; ISBN 007554332X.
40. Malik, S.; YeZhuang, T. Execution of Continuous Improvement Practices in Spanish and Pakistani Industry: A Comparative Analysis. In Proceedings of the 2006 IEEE International Conference on Management of Innovation & Technology, Singapore, 21 June 2006; Wilde, O., Ed.; John Wiley: Hoboken, NJ, USA, 2007; pp. 761–765, ISBN 1-4244-0147-X.
41. Otsuka, K.; Jin, K.; Sonobe, T. *Applying the Kaizen in Africa*; Springer International Publishing: Cham, Switzerland, 2018; ISBN 978-3-319-91399-5.
42. Kucharska, W.; Wildowicz-Giegiel, A. Company Culture, Knowledge Sharing and Organizational Performance: The Employee's Perspective. In Proceedings of the 18th European Conference on Knowledge Management, Barcelona, Spain, 7 September 2017.
43. Michael, M. Organisational Culture: Definitions and Trends. *Res. Gate* **2018**, *1*, 1–7.
44. Gamble, J.R. Tacit vs explicit knowledge as antecedents for organizational change. *JOCM* **2020**, *33*, 1123–1141. [CrossRef]

45. YouGov. Data-Driven Companies Are Resilient Companies. Available online: https://public.tableau.com/app/profile/tableau.research/viz/YouGovSurvey/Data-DrivenCompaniesareResilient (accessed on 14 May 2022).

46. Lerch, C.; Jäger, A.; Heimberger, H. Lean 4.0: Smart und schlank produzieren—Potenziale und Effekte einer neuen Managementdisziplin: Potenziale und Effekte einer neuen Managementdisziplin. *Fraunhofer ISI* **2021**, *80*, 2–8.

47. Bertagnolli, F. *Lean Management: Einführung und Vertiefung in die japanische Management-Philosophie*; Springer Gabler: Wiesbaden, Germany, 2020; Volume 1, pp. 256–264. ISBN 978-3-658-31239-8.

48. Dillinger, F.; Formann, F.; Reinhart, G. Lean Production und Industrie 4.0 in der Produktion. *ZWF* **2020**, *115*, 738–741. [CrossRef]

49. Harvard Business Review. How CEOs Can Lead a Data-Driven Culture. Available online: https://hbr.org/2020/03/how-ceos-can-lead-a-data-driven-culture (accessed on 17 May 2022).

50. Soulejman Janus, S. *Becoming A Knowledge Sharing Organization: A Handbook for Scaling Up Solutions through Knowledge Capturing and Sharing*; World Bank Publications: Washington, DC, USA, 2016.

51. Brüning, A.; Gluchowski, P.; Kaiser, A. Data Governance—Einordnung, Konzepte und aktuelle Herausforderungen. *Chemnitz Econ. Pap.* **2017**, *15*, 1–8.

52. Ehrlinger, L.; Wöß, W. Automated Data Quality Monitoring. In Proceedings of the ICIQ 2017, Little Rock, AR, USA, 6–7 October 2017.

53. Rti. Data-Centric Middleware: Modular Software Infrastructure to Monitor, Control and Collect Real-Time Equipment Analytics. Rti Whitepaper. Available online: https://www.google.com/url?sa=t&rct=j&q=&esrc=s&source=web&cd=&ved=2ahUKEwiFh87Ph_b4AhWdX_EDHcRcDPwQFnoECAMQAQ&url=https%3A%2F%2Fwww.rti.com%2Fhubfs%2Fdocs%2F2RTI_Data_Centric_Middleware.pdf&usg=AOvVaw16O9ISxjihOowvXW_snImY (accessed on 17 May 2022).

Article

Scarce Data in Intelligent Technical Systems: Causes, Characteristics, and Implications

Christoph-Alexander Holst * and Volker Lohweg

inIT—Institute Industrial IT, Technische Hochschule Ostwestfalen-Lippe, Campusallee 6, 32657 Lemgo, Germany
* Correspondence: christoph-alexander.holst@th-owl.de

Abstract: Technical systems generate an increasing amount of data as integrated sensors become more available. Even so, data are still often scarce because of technical limitations of sensors, an expensive labelling process, or rare concepts, such as machine faults, which are hard to capture. Data scarcity leads to incomplete information about a concept of interest. This contribution details causes and effects of scarce data in technical systems. To this end, a typology is introduced which defines different types of incompleteness. Based on this, machine learning and information fusion methods are presented and discussed that are specifically designed to deal with scarce data. The paper closes with a motivation and a call for further research efforts into a combination of machine learning and information fusion.

Keywords: scarce data; machine learning; information fusion

check for updates

Citation: Holst, C.-A.; Lohweg, V. Scarce Data in Intelligent Technical Systems: Causes, Characteristics, and Implications. *Sci* **2022**, *4*, 49. https://doi.org/10.3390/sci4040049

Academic Editor: Johannes Winter

Received: 31 October 2022
Accepted: 30 November 2022
Published: 12 December 2022

Publisher's Note: MDPI stays neutral with regard to jurisdictional claims in published maps and institutional affiliations.

1. Introduction

In modern industrial applications, data are generated in increasing amounts due to better availability, accessibility, and cost-effectiveness of technical sensors. In fact, modern methods for data analysis often assume the availability of big data. Many machine learning methods not only assume big data but also require it. This is also the case in many industrial use-cases [1], such as predictive maintenance [2] or machine fault diagnosis [3].

However, the reality—also in industrial applications—is that data is not always available in sufficient quantities. It may also be that data is recorded in large quantities, but the data are repetitive containing the same information repeatedly. The presence of only a few data sources or data points is summarised by the term scarce data or data scarcity [4]. The goal in dealing with scarce data must nevertheless be to obtain as much information and as much knowledge as possible from the little data that is available. Causes of scarce data are, for example, measured variables that are difficult to collect, costly measurement methods, or a low number of measurement objects that need to be collected. However, an explicit definition and detailed specification of different types of data scarcity is rare in the current literature. For example, Wang et al. [5] define two types: scarce data due to a limited number of samples and sparse data (e.g., sparse time series or matrices).

The problem of scarce data is recognised in the state of the art of machine learning [6,7]. Approaches to addressing data scarcity include inherently data-efficient algorithms and methods for enabling data-hungry algorithms to be used on scarce data—as identified recently by Adadi [8] in their survey on data-efficient algorithms. Regarding the former, it is generally considered that low-complexity models, such as decision trees or linear regression, require less data than high-complexity models, such as deep neural networks. Regarding the second, various methods have been devised and proposed for highly complex models that are intended to be applicable to scarce data, such as data augmentation [9] or transfer learning [10].

In current machine learning approaches, data scarcity is often only implicitly taken into account by extending and adapting existing algorithms [11–13]. Another research area which focuses on data scarcity is information fusion. Information fusion has developed independently from machine learning. Fusion methods specifically expect data to be uncertain due to scarcity (as well as other data imperfections) [14]. In case of multiple uncertain information sources, e.g., sensors, experts, or machine learning models, fusion aims to create a single output with increased certainty. To achieve this, uncertainties based on data scarcity are explicitly modelled, quantified, and considered.

This article addresses scarce data due to its frequency of occurrence in industrial applications and the implications for data processing methods. The aim of this article is (i) to more specifically detail scarce data in its causes and subtypes, and (ii) to provide an overview of both machine learning and information fusion methods that address scarce data. Towards this end, the following contributions are presented in this article:

- A closer look into the causes and implications of scarce data is provided. A typology is presented which categorises the subtypes of scarce data.
- An overview of data augmentation, transfer learning, and information fusion methods is given.
- A combination of machine learning and fusion techniques is discussed and further research efforts in this area are motivated.

The further structure of this paper follows these contributions.

2. A Typology of Scarce Data

Scarce or incomplete data is a form of data imperfection that affects the ability of algorithms, machine-learned models, or human engineers to extract information and induce knowledge. Incomplete data represents uncertainty in the data, but also leads to uncertainty in the process of induction. In this sense, it is closely related to uncertainties—especially to the notion of epistemic uncertainty. It follows an introduction of epistemic uncertainty together with its counterpart aleatoric uncertainty.

Definition 1 (Aleatoric Uncertainty). *Aleatoric uncertainty refers to the inherent variation of an object, concept, process, or phenomenon. It is random and non-deterministic in nature [15]. Even if data is complete and the underlying process is completely understood, the outcome of this process cannot be predicted with absolute certainty [16,17]. Consequently, gathering more data—or adding new data or information sources—does not reduce aleatoric uncertainty. Take, for example, a classification problem. In such a problem, aleatoric uncertainty is the intra-class distance or variance.*

Definition 2 (Epistemic Uncertainty). *In contrast, epistemic uncertainty results from a lack of knowledge about a phenomenon. This lack is caused by incomplete—not available—or inconsistent information. Epistemic uncertainty is, in principle, reducible by gathering additional information. In practice, reducing epistemic uncertainty is often not possible, feasible, or valuable [15–17]. In technical or industrial systems, this is due to one or more of the following reasons.*

- *Sensors are not available or limited in their functionality. They are technically infeasible, too costly, or not obtainable. The engineering effort to design and plan sensor systems is too complex or too expensive. The sensors' properties are limited, for example, their sampling rate or operating range.*
- *The observation period or sampling size is insufficient. Observations do not cover certain concepts or phenomena (Data does not capture the Black Swan [18]). The operation of a sensor is too costly, takes too much time, or is destructive.*
- *Blind ignorance of human engineers prevents all potential data from being obtained. Missing knowledge about real-world phenomena or the availability of sensors limits the amount of data gathered.*

Scarce Data is, therefore, itself a form of epistemic uncertainty. Handling epistemic uncertainty is one of the major challenges in data analysis. This is also recognised in machine learning research very recently [19–21]. To overcome this challenge, it is crucial to understand the various types of scarcity, their causes, and their interactions.

Several taxonomies and typologies have been proposed in the literature to categorise and relate data or information imperfections, uncertainties, and quality [15,22–26]. An overview of taxonomies and typologies is given by Jousselme et al. [27], which includes some of the works just mentioned. An overview of data quality in databases provided by de Almeida et al. [28] is also of interest. The authors identify data completeness as a major data quality issue. However, work limited to databases will not be discussed further here. Instead, we summarise taxonomies and typologies which focus on or at least address incompleteness, missing data, or missing information in Table 1. Most of the works referenced in Table 1 rely on the term incompleteness which is used interchangeably with scarcity in the table.

Table 1. Taxonomies of uncertainty, imperfection, ignorance, and quality which address the topic of data or information incompleteness (in the sense of missing data or information, i.e., scarcity). Incompleteness is recognised as the main concept of imperfection throughout the referenced works. However, a categorisation of the various kinds of missing data or information is not carried out.

Authors	Focus	Builds upon	Relies on Incompleteness	Details Subcategories of Incompleteness
Smithson [22]	Ignorance	-	yes	Partially. *Incompleteness* is subcategorised into *Uncertainty* (including *Vagueness, Probability, Ambiguity*) and *Absence*. *Absence* of information is not further detailed.
Smets [23]	Imperfection	-	yes	No
Krause and Clark [29]	Uncertainty	-	yes	No
Ayyub and Klir [15]	Ignorance	[22]	yes	Partially. Similar to Smithson.
Bosu and MacDonell [24]	Data Quality	-	yes	No
Rogova [25]	Information Quality	[23]	yes	No
Raglin et al. [26]	Uncertainty	-	yes	No

This survey shows that incompleteness is recognised broadly as a type of data imperfection, a kind of uncertainty, and a source of ignorance. In nearly all referenced taxonomies, incompleteness is not further subcategorised. A detailed look into various forms of incompleteness and missing data is not provided.

In the following, we present a more detailed typology of incompleteness as a form of imperfection (see Figure 1) based on Smets' [23] taxonomy. This typology perceives incompleteness as a form of data imperfection along with imprecision and inconsistency.

The proposed typology subdivides incompleteness into six categories.

Undersampled: Data points always represent only a sample of a distribution or the characteristics of a phenomenon. Sensors only provide a window into the real world. Their observations are a fragmented representation. A phenomenon is undersampled if there is insufficient data available to make sound and significant findings about its characteristics. Due to undersampled data, information remains partially hidden. The aleatoric uncertainty of a phenomenon can only be described inadequately. Figure 2 illustrates two cases of undersampling using a scatter plot in a one-dimensional and a two-dimensional feature space.

As a consequence, training with machine learning methods does not lead to satisfactory results. The generalisation ability of the trained models is questionable at best. Probabilistic methods rely on the availability of statistically sound data or knowledge about prior distributions [30]. Kalman filters, for example, assume zero-mean Gaussian distributed data [31]. In the case of undersampled data, this knowledge cannot be derived from the data itself. Few data points also increase the risk of finding spurious correlations in the data [32]—especially when many features or data sources are involved. Another threat of undersampled data is that machine-learned models tend to easily overfit [33].

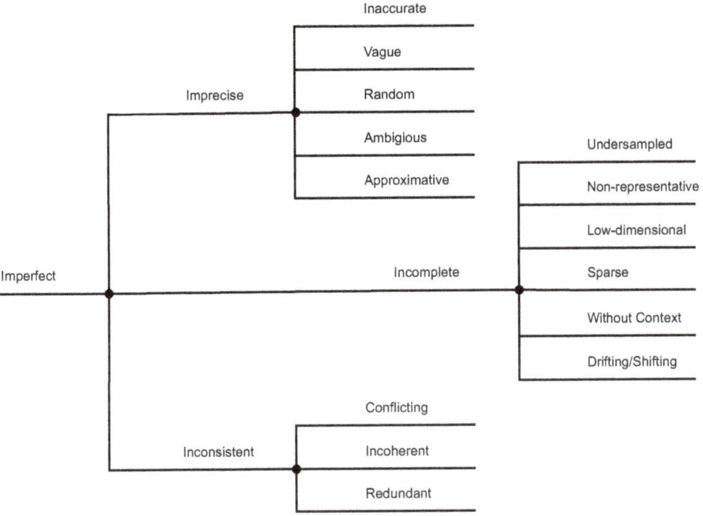

Figure 1. A typology of data and information imperfection with a detailed subcategorisation of incompleteness. The typology is based on the work of Smets' [23]. It recognises incompleteness as one of three major sources of imperfection – besides inconsistency and imprecision. Imprecision captures deficiencies that prevent unambiguous statements from being made based on individual data points. Inconsistency refers to situations in which a piece of information is contradictory to existing knowledge or with other information sources. Incompleteness is lacking, absent, or non-complete data and information.

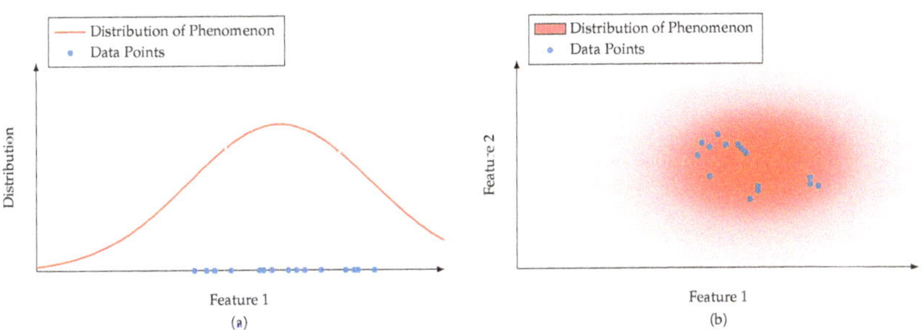

Figure 2. Two examples showcasing undersampled data: (**a**) an ill-represented one-dimensional distribution and (**b**) an ill-represented two-dimensional distribution. A two-dimensional scatter plot showcasing undersampled data. The plots show the distributions of phenomena in feature space (red). The distributions are unknown and represent the aleatoric uncertainty of the phenomena. In both examples, the sampled data points (blue) are insufficient to draw conclusions about the distributions. The missing data points are a form of epistemic uncertainty.

Non-representative: Data or information is non-representative when only certain parts or subconcepts of a phenomenon are observable or represented in the data. Other subconcepts may be very well represented. Take, for example, a bi-modal distribution of a phenomenon's characteristics. One of the modes may be very well sampled, whereas the other is absent in the data. In extreme cases, complete concepts are missing. In less extreme cases, subconcepts may merely be undersampled. Data in which subconcepts are undersampled are often also referred to as biased. The observation of industrial machines (condition monitoring) often produces non-representative data. Machines are specifically built to run as smoothly and faultlessly as possible. Consequently, data obtained during normal operation is often available in abundance. In contrast, data on fault states or unusual operating conditions are often rare. Reducing this kind of epistemic uncertainty is difficult in practice since running a machine in fault states is either costly or infeasible. Figure 3 shows the multi-modal and condition monitoring examples as a form of non-representative data.

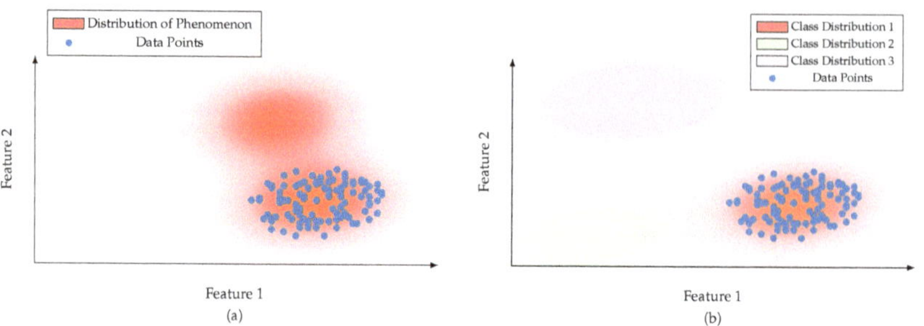

Figure 3. Two cases of non-representative data. In (**a**) a bi-modal distribution is shown (red, unknown). One mode is very-well sampled; the second is missing in the data. Plot (**b**) shows a multi-class classification problem, in which certain classes are missing in the data. Such missing data can, for example, be due to unseen fault states of a machine.

Low-dimensional: Real-world processes can only be observed by a finite number of sensors. Data may be incomplete due to missing data sources – in this case, the data space is too low-dimensional. A low-dimensional space may be insufficient to handle the aleatoric uncertainty of the phenomenon at hand. Figure 4 illustrates a case where data is scarce with respect to the number of available sources.

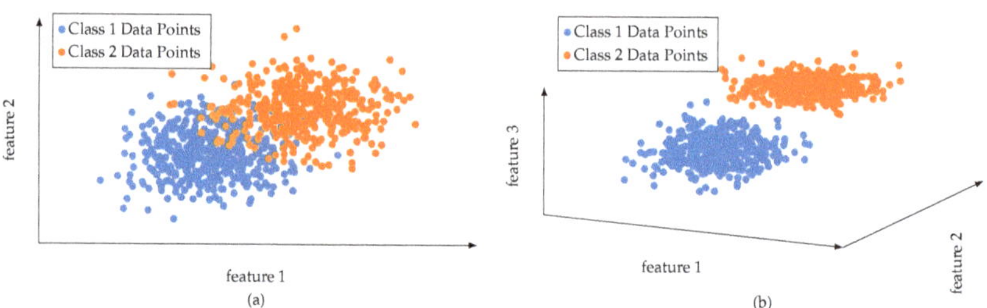

Figure 4. A classification example in which the addition of a new data source allows us to distinguish two classes perfectly (**b**). In the two-dimensional space shown in (**a**), the aleatoric uncertainty prevents a clear separation of classes. Low-dimensional data is still a form of epistemic uncertainty as it is unknown how the class distributions evolve with new sources.

This epistemic uncertainty is reducible by adding new sources although it is crucial to carefully select new sources that are meaningful.

Sparse: Sparse data is caused by sensors or data sources which do not provide data continuously. For example, data is missing over certain time periods or data from different sources cannot be synchronised with each other. Missing data can be caused by defective sensors. This leads to data gaps. Take, for instance, data which is organised in a two-dimensional table. Its rows represent data instances and its columns are data sources. Sparse data is then characterised by missing entries throughout this table (think of a sparse matrix).

Without Context: Context is needed to extract information and knowledge from data. Roughly speaking, context is itself information that surrounds the phenomenon of interest and its data-generating process [34]. Context aids in understanding the phenomenon. It can be provided by domain knowledge. Examples of context are labels in classification applications or maps in applications of autonomous driving. Context, and specifically labels, are often costly to produce or provide. If in large datasets only a fraction of data instances are labelled, then the problem relates to undersampled data.

Drifting/Shifting: The effectiveness of machine learning algorithms relies heavily on the assumption that training and test data are taken from the same or at least similar distributions [35]. In reality, concepts and phenomena often drift in their distribution over time, e.g., data clusters move through feature space. As a consequence, models which have learned from training data are outdated as soon as significant drift occurs. Adaptation or retraining is usually necessary. Because the drifting data distribution over time is not known, drift is categorised as a form of incomplete information.

These six types of incomplete data have different causes, characteristics, and effects on machine learners or other data-processing algorithms. To overcome the associated challenges, algorithms have to specifically consider each type. This has to be kept in mind in designing data analyses.

3. An Overview of Methods for Working with Scarce Data

The challenges associated with scarce data have been known and intensively discussed in the research community for some time. Various methods and approaches exist that can deal with scarce data. In the following, we discuss methods of transfer learning, data augmentation, and information fusion that act in very different ways on scarce data. This survey is closely related to the work of Adadi [8], who studied machine learning methods for scarce data. We extend this survey with an insight into information fusion methods. We mainly focus on the problem of undersampled data and non-representative data. In the ensuing discussion, we motivate further research efforts on the combination of machine learning and information fusion methods.

3.1. Transfer Learning

Transfer learning is a machine learning method in which a model that has been trained in one domain is reused in a related domain. The model is not completely retrained but only adapted by post-training [36,37]. The purpose of transfer learning is to be able to use machine learners even with scarce data. Transfer learning requires a model which has learned as many basic concepts of a domain as possible. For example, these may be geometric shapes in image data, basic patterns such as a Mexican hat in time series, or basic pronunciations or sounds in human speech. Once basic concepts are known to a model, few training examples are required to adapt to a new domain—even zero-shot learning is possible under specific circumstances and depending on the application [38]. Most commonly neural networks and convolutional neural networks are used to transfer learning, but other machine learning methods have been adapted for transfer learning, such as Markov logic networks [39] and Bayesian networks [40]. Transfer learning has been applied to many domains. A survey on machine diagnostics in industrial applications is provided by Yao et al. [41].

Transfer learning comes with several drawbacks and pitfalls. Because a source model is required to know as many concepts as possible, large datasets and resources are necessary to train the source model in the first place. Such a model needs to be trained on a general dataset, which is at best not domain-specific. Secondly, the target domain is still charac-terised by scarce data. Therefore, some risks remain even if the transfer is learned. Models are still at risk to overfit or detect spurious correlations [37,42]. Finally, performance is affected negatively if the source and target domain do not cover the same concepts or focus on different concepts. This is referred to as negative transfer [43,44]. For example, recent studies have shown that models trained on the ImageNet (https://www.image-net.org/, accessed on 9 November 2022) dataset favour texture over shape [45]. Transferring these models into domains in which textural information is less important and objects are mostly defined by shape—such as object recognition of machinery parts, screws, or nuts [46–48]—will not result in optimal performance.

3.2. Data Augmentation

Data augmentation refers to methods that artificially increase the amount of available data. The aim is to facilitate machine learners to train on even small amounts of training data. Augmentation creates slightly modified copies of existing data or completely new synthetic data [49]. Data augmentation techniques have been successfully applied to image [49,50], text and natural language [51–53], and time series data [54]. Augmentation has a regularising effect on machine learning models, helps to reduce overfitting, and can improve the generalisability of models [50]. Industrial applications of data augmentation are, for example, given by Dekhtiar et al. [46], Židek et al. [47,48], Parente et al. [55], or Shi et al. [56].

Additional data instances are usually created by applying various transformations to data. In image datasets, these are, e.g., rotations, scaling, cropping, colour transformations, distortions, or erasing random parts of an image [50]. In natural language, parts of a text are randomly swapped, inserted, deleted, or replaced synonymously [52]. Time series transformations take place either in the time or frequency domain. These include cropping, slicing, jittering, or warping among others [54]. These transformations aim to teach a machine learner which information is important for defining a concept. For example, additional rotated images teach that rotation is not important to a concept or class. It is still the same class. By replacing the background in images, models learn to focus on objects in the foreground. Thus, augmenting data by selected transformations allows us to integrate expert knowledge into the machine learning process. However, it is crucial to apply the right transformation for a particular application in order for the data augmentation to be useful. Often data augmentation seems to be carried out in an "ad-hoc manner with little understanding of the underlying theoretical principles"—as stated by Dao et al. [57].

Another approach to data augmentation is to create additional data automatically by generative models such as generative adversarial networks [58]. The expectation is that expert's knowledge will no longer be necessary or will be at least less crucial. A major drawback of generative augmentation is that it is susceptible to perpetrate bias in data [59].

With all these methods, there is a risk of losing important information in the augmentation process. Information may be discarded, e.g., by cropping an image, or may be overwritten by erasing parts of a text randomly [50,52]. It follows that patterns or classes are not correctly preserved. The data instance and its label may then no longer match (The label is not preserved). This problem is aggravated if small details in a data instance are crucial for a concept. Slight changes to the original data may then already be enough to distort or destroy concepts.

3.3. Information Fusion

Scarce data and epistemic uncertainty are intensively addressed in the research field of information fusion. Information fusion has been researched since the midst of the 20th century as a distinct field in parallel to machine learning [60,61]. While information fusion has similar goals and applications as machine learning—such as classification, regression, detection, or recognition—its focus differs. The aim of information fusion methods is to extract and condense high-quality information from a set of low-quality data sources [62]. Information fusion explicitly assumes that sources provide incomplete or imprecise information. The task of information fusion is to make the best of what imperfect data is available [14]. Fusion methods include a strong focus on modelling uncertain, error-prone, imprecise, and vague information [63]. For instance, fuzzy information is modelled via fuzzy set theory. Missing information or ignorance are modelled via evidence theories, such as the Dempster-Shafer theory. Fusion methods address scarce data with possibility theory. In direct comparison to probability theory, possibility theory is characterised by the fact that incomplete information is represented qualitatively [64]. The possibility theory requires a smaller amount of data but is less expressive in the final analysis [63,64]. Established methods of machine learning, on the other hand, rarely model missing information or epistemic uncertainty explicitly. Instead, they rely on a quantitative evaluation of data. In the following, we provide an overview of the mathematical tools fusion relies on, that is, the Dempster-Shafer theory, the fuzzy set theory, and the possibility theory.

3.3.1. Dempster-Shafer Theory

The Dempster-Shafer theory of evidence (DST) has been proposed by Shafer [65] on the foundation of Dempster's works on a framework for expressing upper and lower probabilities [66]. In the DST, available evidence forms the basis to express a degree of belief in a proposition that quantifies incomplete knowledge [67]. In this basic sense, it is comparable to Bayesian probability theory. It is motivated by the fact that probability theory is not able to distinguish between ignorance (epistemic uncertainty) and well-informed uncertainty (aleatoric uncertainty) natively [65].

Probability theory (ProbT) operates on a frame of discernment Ω which includes all given propositions or hypotheses X as singletons, i.e., $\Omega = \{X_1, X_2, \ldots, x_n\}$. Each proposition is given a probability $0 \le p(X) \le 1$ to be true with the restriction of $\sum_{X \in \Omega} p(X) = 1$. In the case of total ignorance, one tends to distribute probabilities uniformly over Ω but this is arbitrary. A uniform distribution is not distinguishable from a situation in which it is known that propositions are actually equally likely. DST allows us to assign evidence to sets of combined propositions. It operates on the power set of the frame of discernment, i.e., $\mathcal{P}(\Omega) = \{\varnothing, X_1, X_2, \ldots, \{X_1, X_2\} \ldots, \Omega\}$. By assigning evidence m to combined propositions (e.g., $\{X_1, X_2\}$), a state of incomplete knowledge is expressed. In case of $\{X_1, X_2\}$, it is unclear whether evidence favours X_1 or X_2. Belief in a proposition is then obtained by $\mathrm{Bel}(X) = \sum_{A \subseteq X} m(A)$. The usage of the power set allows DST to handle incomplete knowledge due to scarce data better and more properly than probability theory. An example of the difference between ProbT and DST is given in Figure 5.

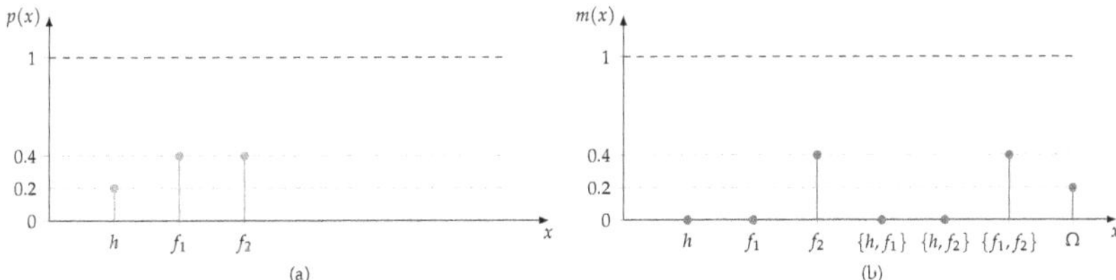

Figure 5. Probability theory versus Dempster-Shafer's theory in a condition monitoring example. The basic propositions are h: the monitored object is healthy and f_1, f_2: the object is in one of two fault states. The distribution modelled with ProbT (**a**) is ambiguous since it cannot distinguish between ignorance (epistemic uncertainty) and well-informed uncertainty (aleatoric uncertainty). Using DST (**b**), it turns out that the expert or model is indeed partly ignorant. This is expressed by $m(\{f_1, f_2\}) = 0.4$ (a fault occurred but it is unknown which one) and by $m(\Omega) = 0.2$ (nothing is known).

DST is designed with a fusion of independent multiple sources in mind. Having multiple partially ignorant and uncertain sources, the aim is to get to a single estimation with reduced ignorance and increased certainty. To achieve this, most fusion rules involve a reinforcement effect. If, for example, $m_1(X) = m_2(X)$, then the fused mass $m_{12}(X) > m_1(X)$. Several fusion rules have been proposed over the years, for example, Dempster's rule of combination [66,68], Yager's rule [69], Campos' rule [70], or the Balanced Two-Layer Conflict Solving rule [61], to name just a few.

DST fusion achieves that—if a group of sensors, experts, or machine learning models is uncertain in their assessments because of scarce data—to increase certainty. A popular approach in machine learning is to apply ensemble learners [71]. In ensemble learning, multiple weak learners are trained simultaneously. Their outputs are fused into a single one. An example of an ensemble is random forests. Although this seems to be an exemplary area of application for DST fusion, most ensemble learners rely on majority votings or averaging functions [72–74]. This motivates further research efforts in combining DST and machine learning methods as a way to handle the effects of scarce data.

3.3.2. Fuzzy Set Theory

Fuzzy set theory (FST) was proposed by Zadeh [75] motivated by the intrinsic vague nature of language. The fuzzy set theory facilitates the modelling of imprecise and vague information (cf. Figure 1). Although FST is not focused on incomplete information, it brings benefits when it comes to scarce data. Zadeh introduces sets with vague boundaries in contrast to crisp sets known from probability theory or Dempster-Shafer theory. In a crisp set, an element either belongs to this set or not. Its membership function μ is a mapping of all elements belonging to the frame of discernment Ω to a boolean membership $\mu : \Omega \to \{0, 1\}$. Fuzzy sets allow degrees of memberships, that is, $\mu : \Omega \to [0, 1]$.

The inherent vagueness of fuzzy membership functions can be exploited to learn class distributions from only a few data instances [76]. If class borders are only needed to be modelled imprecisely and vaguely, then less effort has to be put into a training process than learning precise class borders. The fuzzy membership of a data instance is then interpreted as the uncertainty of the classification model. This blurring of class borders results in weaker models with the upside of less data demand.

An approach for this kind of classification is fuzzy pattern classifiers (FPC). Fuzzy pattern classifiers have been introduced and advanced by Bocklisch [77,78]. An FPC learns a unimodal potential function for each data source. This function serves as a membership function. Each membership function is a weak classifier in itself. Seen as a group, the membership functions are similar to an ensemble. They output each a gradual estimate for the predicted membership. This allows to apply fuzzy aggregation rules to fuse the outputs into a singular class membership (see for example previous works by Holst and Lohweg [79–82]).

Unimodal potential functions were proposed by Aizerman et al. [83] as a pattern recognition tool. It was only later that they were applied as membership functions for fuzzy sets. Unimodal potential functions are used to model the distribution of compact and convex classes. Lohweg et al. [84] described a resource-efficient variant optimised for limited hardware:

$$\mu(x) = \begin{cases} 2^{-d(x,\mathbf{p}_l)} & \text{if } x \leq \overline{x}, \\ 2^{-d(x,\mathbf{p}_r)} & \text{if } x > \overline{x}, \end{cases}$$

$$\text{with } d(x, \mathbf{p}_l) = \left(\frac{|x - \overline{x}|}{C_l} \right)^{D_l},$$

$$d(x, \mathbf{p}_r) = \left(\frac{|x - \overline{x}|}{C_r} \right)^{D_r}, \text{ and}$$

x a data instance (measurement value).

The unimodal potential function has several advantages for the use of scarce data. The function is parameterizable with few parameters. The number of parameters scales with data sources linearly. The parameters are relatively easy to train in data. Training methods can be found in [76,81,84]. The parameters are intuitive to interpret. Therefore, expert knowledge can be integrated easily. On the other hand, FPCs require unimodal and convex data distributions. In this regard, Hempel [85] proposed a multi-modal FPC, although his approach requires more training data in general.

3.3.3. Possibility Theory

The possibility theory (PosT) was introduced by Zadeh in 1978 as an extension of fuzzy set theory [86]. It is designed as a counterpart to probability theory because of its limited ability to represent epistemic uncertainty.

Possibility theory is based on possibility distributions π—similar to probability distributions p. The possibility $0 \leq \pi(x) \leq 1$ conveys how plausible the event x is. A value $\pi(x) = 1$ means completely plausible; $\pi(x) = 0$ completely implausible. At least one x is required to be fully plausible (normality requirement). But more than one x can be fully plausible. This leads to $\sum_{x \in \Omega} \pi(x) \geq 1$ or $\int_{x \in \Omega} \pi(x) \geq 1$.

Possibility distributions are similarly defined as fuzzy membership functions, that is, $\pi(x) = \mu(x)$ [16]. This has the advantage that mathematical operations defined on fuzzy sets can be directly applied to possibility distributions [87]. Though it has to be verified first if this is sensible. Fuzzy membership functions and possibility distributions differ in interpretation. Let x be an alternative for an unknown value v and A be a fuzzy set. The $\pi(x)$ expresses the possibility of $x = v$ knowing that $x \in A$. In contrast, $\mu(x)$ expresses the degree of membership of x to A knowing that $x = v$.

Possibility distributions are also a less expressive and weaker model than probability distributions. Roughly speaking, it is easier to conclude that a proposition is possible rather than probable. Moreover, for a proposition to be probable it must preliminarily be possible. This leads to the probability/possibility consistency principle stating that $\pi(x) \geq p(x)$. In return, possibility distributions] require less effort – meaning training data or expert's knowledge – to construct [88]. They do not require statistically sound data because they model incomplete information qualitatively; whereas probability distributions model random phenomena quantitatively. This distinction is highlighted in Figure 6.

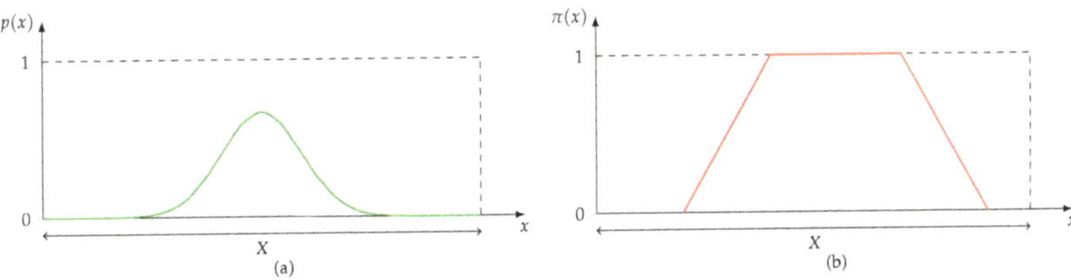

Figure 6. A continuous probability (**a**) and a continuous possibility distribution (**b**). The probability distribution models a random phenomenon quantitatively; the possibility of distribution of incomplete information qualitatively. The following applies: $\int_{x \in \Omega} p(x) = 1$, $\int_{x \in \Omega} \pi(x) \geq 1$, and $\pi(x) \geq p(x)$.

This leads to the conclusion that possibility theory is well-suited to be used in the case of epistemic uncertainty and scarce data.

3.4. Discussion

Scarce data and epistemic uncertainty remain major challenges to machine learning and data analysis approaches. Missing information in data obstructs inherent aleatoric uncertainty.

In the area of machine learning, several techniques for coping with few training data have been thoroughly studied. Some of the most important are data augmentation, transfer learning, and interpretable models. While data augmentation and transfer learning focus on undersampled data mainly, interpretable models address also non-representative data. But only recently has epistemic uncertainty come into focus. Researchers have begun to explicitly define and quantify epistemic uncertainty of machine learning models [17,20,21,89].

In contrast, the research field of information fusion focuses on scarce data and epistemic uncertainty since its emergence in the mid-twentieth century. Fusion methods apply evidence theories such as DST, fuzzy set theory, and possibility theory to either quantify epistemic uncertainty or reduce its impact on performance.

However, combining fusion and machine learning methods is rare in the state of the art, although research need has been recognised recently [90–92]. Several works have been published that attempt to fill this open research topic. Among these are approaches which apply fusion techniques as a preprocessing step before machine learning [93,94]. These works focus on providing a machine learner with a more robust and condensed data basis through prior fusion. They do not focus on incomplete information though. Further works devise classifiers based on the Dempster-Shafer theory [95–97]. Finally, machine learning in a possibilistic setting exists but is very rare. A small survey is conducted by Dubois et al. [98]. This leads to the conclusion that further research is needed to more successfully and formally address scarce data in machine data analysis.

4. Conclusions

Despite the increasing number of sensors and measuring devices, data is often scarce in industrial applications. The scarcity of data stems from limited sensor availability and functionality, limited observation periods, hidden concepts, and the inevitable blind ignorance of engineers. This leads to challenges in data analysis. In this paper, we have typologized missing data and information in more detail based on the works of Smets [23]. According to this new typology, incomplete data is categorised into (1) undersampled, (2) non-representative, (3) low-dimensional, (4) sparse, (5) without context, and (6) drifting data. Existing typologies did not or only insufficiently detail the category of incompleteness [15,22–26,29]. In this respect, we have filled an open gap in existing works.

This paper also explored machine learning and information fusion methods that deal with scarce data and incomplete information. As such, this paper complements Adadi's survey [8], which is limited to machine learning methods. Regarding machine learning, we focused on methods enabling data-hungry algorithms to be used on scarce data. Such methods are data augmentation [9] and transfer learning [10], among other methods. The idea behind transfer learning is to reuse and adapt models which have been trained on large, preferably general, datasets. However, efforts for training a source model are substantial and the risk of negative transfer has to be considered. Data augmentation creates new data points artificially by modifying existing ones. Data augmentation can reduce overfitting at the risk of destroying information.

Information fusion, on the other hand, relies on evidence theories, fuzzy set theory, and possibility theory to model, quantify, and cope with epistemic uncertainty [14]. This paper motivates and calls for further research efforts in combining fusion and machine learning approaches.

Author Contributions: C.-A.H. conceptualised the paper, conducted the research, and wrote the article. V.L. supervised the research activity and revised the article. All authors have read and agreed to the published version of the manuscript.

Funding: This work was partly funded by the Ministry of Economic Affairs, Innovation, Digitalisation and Energy of the State of North Rhine-Westphalia (MWIDE) within the project AI4ScaDa, grant number 005-2111-0016.

Institutional Review Board Statement: Not applicable.

Informed Consent Statement: Not applicable.

Data Availability Statement: Not applicable.

Conflicts of Interest: The authors declare no conflict of interest. The funders had no role in the design of the study; in the collection, analyses, or interpretation of data; in the writing of the manuscript, or in the decision to publish the results.

Abbreviations

The following abbreviations are used in this manuscript:

DST	Dempster-Shafer theory of evidence
FST	Fuzzy set theory
PosT	Possibility theory
ProbT	Probability theory

References

1. Sharp, M.; Ak, R.; Hedberg, T. A survey of the advancing use and development of machine learning in smart manufacturing. *J. Manuf. Syst.* **2018**, *48*, 170–179. [CrossRef] [PubMed]
2. Carvalho, T.P.; Fabrízzio A. A. M. N. Soares.; Vita, R.; Da Francisco, R.P.; Basto, J.P.; Alcalá, S.G.S. A systematic literature review of machine learning methods applied to predictive maintenance. *Comput. Ind. Eng.* **2019**, *137*, 106024. [CrossRef]
3. Lei, Y.; Yang, B.; Jiang, X.; Jia, F.; Li, N.; Nandi, A.K. Applications of machine learning to machine fault diagnosis: A review and roadmap. *Mech. Syst. Signal Process.* **2020**, *138*, 106587. [CrossRef]
4. Babbar, R.; Schölkopf, B. Data scarcity, robustness and extreme multi-label classification. *Mach. Learn.* **2019**, *108*, 1329–1351. [CrossRef]
5. Wang, Q.; Farahat, A.; Gupta, C.; Zheng, S. Deep time series models for scarce data. *Neurocomputing* **2021**, *456*, 504–518. [CrossRef]
6. Shu, J.; Xu, Z.; Meng, D. Small Sample Learning in Big Data Era. *arXiv* **2018**, arXiv:1808.04572.
7. Qi, G.J.; Luo, J. Small Data Challenges in Big Data Era: A Survey of Recent Progress on Unsupervised and Semi-Supervised Methods. *IEEE Trans. Pattern Anal. Mach. Intell.* **2022**, *44*, 2168–2187. [CrossRef]
8. Adadi, A. A survey on data–efficient algorithms in big data era. *J. Big Data* **2021**, *8*, 24. [CrossRef]
9. Andriyanov, N.A.; Andriyanov, D.A. The using of data augmentation in machine learning in image processing tasks in the face of data scarcity. *J. Phys. Conf. Ser.* **2020**, *1661*, 012018. [CrossRef]
10. Hutchinson, M.L.; Antono, E.; Gibbons, B.M.; Paradiso, S.; Ling, J.; Meredig, B. Overcoming data scarcity with transfer learning. *arXiv* **2017**, arXiv:1711.05099.
11. Chen, Z.; Liu, Y.; Sun, H. Physics-informed learning of governing equations from scarce data. *Nat. Commun.* **2021**, *12*, 6136. [CrossRef] [PubMed]
12. Vecchi, E.; Pospíšil, L.; Albrecht, S.; O'Kane, T.J.; Horenko, I. eSPA+: Scalable Entropy-Optimal Machine Learning Classification for Small Data Problems. *Neural Comput.* **2022**, *34*, 1220–1255. [CrossRef] [PubMed]
13. Bhouri, M.A.; Perdikaris, P. Gaussian processes meet NeuralODEs: a Bayesian framework for learning the dynamics of partially observed systems from scarce and noisy data. *Philos. Trans. R. Soc. A Math. Phys. Eng. Sci.* **2022**, *380*, 20210201. [CrossRef]
14. Dubois, D.; Liu, W.; Ma, J.; Prade, H. The basic principles of uncertain information fusion. An organised review of merging rules in different representation frameworks. *Inf. Fusion* **2016**, *32*, 12–39. [CrossRef]
15. Ayyub, B.M.; Klir, G.J. *Uncertainty Modeling and Analysis in Engineering and the Sciences*; Chapman & Hall/CRC: Boca Raton, FL, USA, 2006.
16. Lohweg, V.; Voth, K.; Glock, S. A possibilistic framework for sensor fusion with monitoring of sensor reliability. In *Sensor Fusion*; Thomas, C., Ed.; IntechOpen: London, UK, 2011.
17. Hüllermeier, E.; Waegeman, W. Aleatoric and epistemic uncertainty in machine learning: an introduction to concepts and methods. *Mach. Learn.* **2021**, *110*, 457–506. [CrossRef]
18. Taleb, N.N. *The Black Swan: The Impact of the Highly Improbable*; Incerto, Random House Publishing Group: New York, NY, USA, 2007.
19. Abdar, M.; Pourpanah, F.; Hussain, S.; Rezazadegan, D.; Liu, L.; Ghavamzadeh, M.; Fieguth, P.; Cao, X.; Khosravi, A.; Acharya, U.R.; et al. A review of uncertainty quantification in deep learning: Techniques, applications and challenges. *Inf. Fusion* **2021**, *76*, 243–297. [CrossRef]
20. Huang, Z.; Lam, H.; Zhang, H. Quantifying Epistemic Uncertainty in Deep Learning. *arXiv* **2021**, arXiv:2110.12122.
21. Bengs, V.; Hüllermeier, E.; Waegeman, W. Pitfalls of Epistemic Uncertainty Quantification through Loss Minimisation. *arXiv* **2022**, arXiv:2203.06102.
22. Smithson, M. *Ignorance and Uncertainty: Emerging Paradigms*; Cognitive science; Springer: New York, NY, USA; Heidelberg, Germany, 1989.
23. Smets, P. Imperfect Information: Imprecision and Uncertainty. In *Uncertainty Management in Information Systems: From Needs to Solutions*; Motro, A.; Smets, P., Eds.; Springer: New York, NY, USA, 1997; pp. 225–254.
24. Bosu, M.F.; MacDonell, S.G. A Taxonomy of Data Quality Challenges in Empirical Software Engineering. In Proceedings of the 2013 22nd Australian Software Engineering Conference, Melbourne, VIC, Australia, 4–7 June 2013, pp. 97–106.
25. Rogova, G.L. Information quality in fusion-driven human-machine environments. In *Information Quality in Information Fusion and Decision Making*; Bossé, É.; Rogova, G.L., Eds.; Springer: Cham, Switzerland, 2019; pp. 3–29.
26. Raglin, A.; Emlet, A.; Caylor, J.; Richardson, J.; Mittrick, M.; Metu, S. *Uncertainty of Information (UoI) Taxonomy Assessment Based on Experimental User Study Results*; Human-Computer Interaction. Theoretical Approaches and Design Methods; Kurosu, M., Ed.; Springer: Cham, Switzerland, 2022; pp. 290–301.
27. Jousselme, A.L.; Maupin, P.; Bosse, E. Uncertainty in a situation analysis perspective. In Proceedings of the Sixth International Conference of Information Fusion, Cairns, QSL, Australia, 8–11 July 2003; Volume 2, pp. 1207–1214.
28. de Almeida, W.G.; de Sousa, R.T.; de Deus, F.E.; Daniel Amvame Nze, G.; de Mendonça, F.L.L. Taxonomy of data quality problems in multidimensional Data Warehouse models. In Proceedings of the 2013 8th Iberian Conference on Information Systems and Technologies (CISTI), Lisbon, Portugal, 19–22 June 2013; pp. 1–7.
29. Krause, P.; Clark, D. *Representing Uncertain Knowledge: An Artificial Intelligence Approach*; Springer: Dordrecht, The Netherlands, 2012.
30. Huber, W.A. Ignorance Is Not Probability. *Risk Anal.* **2010**, *30*, 371–376. [CrossRef]

31. Kim, Y.; Bang, H. Introduction to Kalman Filter and Its Applications: 2. In *Introduction and Implementations of the Kalman Filter*; Govaers, F., Ed.; IntechOpen: London, UK, 2018.

32. Calude, C.; Longo, G. The deluge of spurious correlations in big data. *Found. Sci.* **2017**, *22*, 595–612. [CrossRef]

33. Horenko, I. On a Scalable Entropic Breaching of the Overfitting Barrier for Small Data Problems in Machine Learning. *Neural Comput.* **2020**, *32*, 1563–1579. [CrossRef] [PubMed]

34. Snidaro, L.; Herrero, J.G.; Llinas, J.; Blasch, E. Recent Trends in Context Exploitation for Information Fusion and AI. *AI Mag.* **2019**, *40*, 14–27. [CrossRef]

35. Gama, J.; Žliobaitė, I.; Bifet, A.; Pechenizkiy, M.; Bouchachia, A. A Survey on Concept Drift Adaptation. *ACM Comput. Surv.* **2014**, *46*, 44:1–44:37. [CrossRef]

36. Pan, S.J.; Yang, Q. A Survey on Transfer Learning. *IEEE Trans. Knowl. Data Eng.* **2010**, *22*, 1345–1359. [CrossRef]

37. Weiss, K.; Khoshgoftaar, T.M.; Wang, D. A survey of transfer learning. *J. Big Data* **2016**, *3*, 9. [CrossRef]

38. Oreshkin, B.N.; Carpov, D.; Chapados, N.; Bengio, Y. Meta-learning framework with applications to zero-shot time-series forecasting. *arXiv* **2020**, arXiv:2002.02887.

39. Mihalkova, L.; Huynh, T.N.; Mooney, R.J. *Mapping and Revising Markov Logic Networks for Transfer Learning*; AAAI: Menlo Park, CA, USA, 2007.

40. Niculescu-Mizil, A.; Caruana, R. Inductive Transfer for Bayesian Network Structure Learning. In Proceedings of the Eleventh International Conference on Artificial Intelligence and Statistics, San Juan, Puerto Rico, 21–24 March 2007 ; Meila, M.; Shen, X., Eds.; PMLR Proceedings of Machine Learning Research: New York City, NY, USA, 2007; Volume 2, pp. 339–346.

41. Yao, S.; Kang, Q.; Zhou, M.; Rawa, M.J.; Abusorrah, A. A survey of transfer learning for machinery diagnostics and prognostics. *Artif. Intell. Rev.* **2022**. [CrossRef]

42. Sun, Q.; Liu, Y.; Chua, T.S.; Schiele, B. Meta-Transfer Learning for Few-Shot Learning. In Proceedings of the IEEE/CVF Conference on Computer Vision and Pattern Recognition (CVPR), Long Beach, CA, USA, 15–20 June 2019.

43. Wang, Z.; Dai, Z.; Poczos, B.; Carbonell, J. Characterizing and Avoiding Negative Transfer. In Proceedings of the IEEE/CVF Conference on Computer Vision and Pattern Recognition (CVPR), Long Beach, CA, USA, 15–20 June 2019.

44. Zhang, W.; Deng, L.; Zhang, L.; Wu, D. A Survey on Negative Transfer. *IEEE/CAA J. Autom. Sin.* **2022**, *9*, 1. [CrossRef]

45. Geirhos, R.; Rubisch, P.; Michaelis, C.; Bethge, M.; Wichmann, F.A.; Brendel, W. ImageNet-trained CNNs are biased towards texture; increasing shape bias improves accuracy and robustness. In Proceedings of the International Conference on Learning Representations, New Orleans, LA, USA, 6–9 May 2019.

46. Dekhtiar, J.; Durupt, A.; Bricogne, M.; Eynard, B.; Rowson, H.; Kiritsis, D. Deep learning for big data applications in CAD and PLM – Research review, opportunities and case study. *Emerg. Ict Concepts Smart Safe Sustain. Ind. Syst.* **2018**, *100*, 227–243. [CrossRef]

47. Židek, K.; Lazorík, P.; Piteľ, J.; Hošovský, A. An Automated Training of Deep Learning Networks by 3D Virtual Models for Object Recognition. *Symmetry* **2019**, *11*, 496. [CrossRef]

48. Židek, K.; Lazorík, P.; Piteľ, J.; Pavlenko, I.; Hošovský, A. Automated Training of Convolutional Networks by Virtual 3D Models for Parts Recognition in Assembly Process. In *ADVANCES IN MANUFACTURING*; Trojanowska, J.; Ciszak, O.; Machado, J.M.; Pavlenko, I., Eds.; Lecture Notes in Mechanical Engineering; Springer: Cham, Switzerland, 2019; Volume 13, , pp. 287–297.

49. Perez, L.; Wang, J. The effectiveness of data augmentation in image classification using deep learning. *arXiv* **2017**, arXiv:1712.04621.

50. Shorten, C.; Khoshgoftaar, T.M. A survey on Image Data Augmentation for Deep Learning. *J. Big Data* **2019**, *6*, 60. [CrossRef]

51. Feng, S.Y.; Gangal, V.; Wei, J.; Chandar, S.; Vosoughi, S.; Mitamura, T.; Hovy, E. A Survey of Data Augmentation Approaches for NLP. *arXiv* **2021**, arXiv:2105.03075.

52. Shorten, C.; Khoshgoftaar, T.M.; Furht, B. Text Data Augmentation for Deep Learning. *J. Big Data* **2021**, *8*, 101. [CrossRef] [PubMed]

53. Bayer, M.; Kaufhold, M.A.; Reuter, C. A Survey on Data Augmentation for Text Classification. *ACM Computing Surveys* **2022**, *accept.* [CrossRef]

54. Wen, Q.; Sun, L.; Yang, F.; Song, X.; Gao, J.; Wang, X.; Xu, H. Time Series Data Augmentation for Deep Learning: A Survey. In Proceedings of the Thirtieth International Joint Conference on Artificial Intelligence, Virtual, 19–27 August 2021; International Joint Conferences on Artificial Intelligence Organization: Menlo Park, CA, USA, 2021.

55. Parente, A.P.; de Souza Jr, M.B.; Valdman, A.; Mattos Folly, R.O. Data Augmentation Applied to Machine Learning-Based Monitoring of a Pulp and Paper Process. *Processes* **2019**, *7*, 958. [CrossRef]

56. Shi, D.; Ye, Y.; Gillwald, M.; Hecht, M. Robustness enhancement of machine fault diagnostic models for railway applications through data augmentation. *Mech. Syst. Signal Process.* **2022**, *164*, 108217. [CrossRef]

57. Dao, T.; Gu, A.; Ratner, A.J.; Smith, V.; de Sa, C.; Ré, C. A Kernel Theory of Modern Data Augmentation. *arXiv* **2018**, arXiv:1803.06084.

58. Antoniou, A.; Storkey, A.; Edwards, H. Data Augmentation Generative Adversarial Networks. *arXiv* **2017**, arXiv:1711.04340.

59. Jain, N.; Manikonda, L.; Hernandez, A.O.; Sengupta, S.; Kambhampati, S. Imagining an Engineer: On GAN-Based Data Augmentation Perpetuating Biases. *arXiv* **2018**, arXiv:1811.03751.

60. Hall, D.; Llinas, J. Multisensor Data Fusion. In *Handbook of Multisensor Data Fusion*; Electrical Engineering & Applied Signal Processing Series; Hall, D.; Llinas, J., Eds.; CRC Press: Boca Raton, FL, USA, 2001; Volume 3.

61. Mönks, U. *Information Fusion Under Consideration of Conflicting Input Signals*; Technologies for Intelligent Automation; Springer: Berlin/Heidelberg, Germany, 2017.
62. Bloch, I.; Hunter, A.; Appriou, A.; Ayoun, A.; Benferhat, S.; Besnard, P.; Cholvy, L.; Cooke, R.; Cuppens, F.; Dubois, D.; et al. Fusion: General concepts and characteristics. *Int. J. Intell. Syst.* **2001**, *16*, 1107–1134. [CrossRef]
63. Dubois, D.; Everaere, P.; Konieczny, S.; Papini, O. Main issues in belief revision, belief merging and information fusion. In *A Guided Tour of Artificial Intelligence Research: Volume I: Knowledge Representation, Reasoning and Learning*; Marquis, P.; Papini, O.; Prade, H., Eds.; Springer: Cham, Switzerland, 2020; pp. 441–485.
64. Denœux, T.; Dubois, D.; Prade, H. Representations of uncertainty in artificial intelligence: Probability and possibility. In *A Guided Tour of Artificial Intelligence Research: Volume I: Knowledge Representation, Reasoning and Learning*; Marquis, P.; Papini, O.; Prade, H., Eds.; Springer: Cham, Switzerland, 2020; pp. 69–117.
65. Shafer, G. *A Mathematical Theory of Evidence*; Princeton University Press: Princeton, NJ, USA, 1976.
66. Dempster, A.P. Upper and lower probabilities induced by a multivalued mapping. *Ann. Math. Stat.* **1967**, *38*, 325–339. [CrossRef]
67. Salicone, S.; Prioli, M. *Measuring Uncertainty within the Theory of Evidence*; Springer Series in Measurement Science and Technology; Springer: Cham, Switzerland, 2018.
68. Shafer, G. Dempster's rule of combination. *Int. J. Approx. Reason.* **2016**, *79*, 26–40. [CrossRef]
69. Yager, R.R. On the dempster-shafer framework and new combination rules. *Inf. Sci.* **1987**, *41*, 93–137. [CrossRef]
70. Campos, F. Decision Making in Uncertain Situations: An Extension to the Mathematical Theory of Evidence. (Ph.D. Dissertation), Dissertation.Com., Boca Raton, FL, USA, 2006.
71. Polikar, R. Ensemble Learning. In *Ensemble Machine Learning: Methods and Applications*; Zhang, C.; Ma, Y., Eds.; Springer: New York, NY, USA, 2012; pp. 1–34.
72. Sagi, O.; Rokach, L. Ensemble learning: A survey. *WIREs Data Min. Knowl. Discov.* **2018**, *8*, e1249. [CrossRef]
73. Dong, X.; Yu, Z.; Cao, W.; Shi, Y.; Ma, Q. A survey on ensemble learning. *Front. Comput. Sci.* **2020**, *14*, 241–258. [CrossRef]
74. Zhou, Z.H. Ensemble Learning. In *Machine Learning*; Springer: Singapore, 2021; pp. 181–210.
75. Zadeh, L.A. Fuzzy sets. *Inf. Control* **1965**, *8*, 338–353. [CrossRef]
76. Mönks, U.; Petker, D.; Lohweg, V. Fuzzy-Pattern-Classifier training with small data sets. *Information Processing and Management of Uncertainty in Knowledge-Based Systems. Theory and Methods*; Hüllermeier, E.; Kruse, R.; Hoffmann, F., Eds.; Springer: Berlin/Heidelberg, Germany, 2010; pp. 426–435.
77. Bocklisch, S.F. *Prozeßanalyse mit unscharfen Verfahren*, 1st ed.; Verlag Technik: Berlin, Germany, 1987.
78. Bocklisch, S.F.; Bitterlich, N. Fuzzy Pattern Classification—Methodology and Application—. In *Fuzzy-Systems in Computer Science*; Kruse, R.; Gebhardt, J.; Palm, R., Eds.; Vieweg+Teubner Verlag: Wiesbaden, Germany, 1994; pp. 295–301.
79. Holst, C.A.; Lohweg, V. A conflict-based drift detection and adaptation approach for multisensor information fusion. In Proceedings of the 2018 IEEE 23rd International Conference on Emerging Technologies and Factory Automation (ETFA), Torino, Italy, 1–4 September 2018, pp. 967–974.
80. Holst, C.A.; Lohweg, V. Improving majority-guided fuzzy information fusion for Industry 4.0 condition monitoring. In Proceedings of the 2019 22nd International Conference on Information Fusion (FUSION), IEEE, Ottawa, ON, Canada, 2–5 July 2019.
81. Holst, C.A.; Lohweg, V. A redundancy metric set within possibility theory for multi-sensor systems. *Sensors* **2021**, *21*, 2508. [CrossRef]
82. Holst, C.A.; Lohweg, V. Designing Possibilistic Information Fusion—The Importance of Associativity, Consistency, and Redundancy. *Metrology* **2022**, *2*, 180–215. [CrossRef]
83. Aizerman, M.A.; Braverman, E.M.; Rozonoer, L.I. Theoretical foundations of the potential function method in pattern recognition learning. *Autom. Remote Control* **1964**, *25*, 821–837.
84. Lohweg, V.; Diederichs, C.; Müller, D. Algorithms for hardware-based pattern recognition. *EURASIP J. Appl. Signal Process.* **2004**, *2004*, 1912–1920. [CrossRef]
85. Hempel, A.J. Netzorientierte Fuzzy-Pattern-Klassifikation nichtkonvexer Objektmengenmorphologien. Ph.D. Thesis, Technische Universität Chemnitz, Chemnitz, Germany, 2011.
86. Zadeh, L.A. Fuzzy sets as a basis for a theory of possibility. *Fuzzy Sets Syst.* **1978**, *1*, 3–28. [CrossRef]
87. Solaiman, B.; Bossé, É. *Possibility Theory for the Design of Information Fusion Systems*; Information Fusion and Data Science; Springer: Cham, Switzerland, 2019.
88. Dubois, D.; Prade, H. Practical methods for constructing possibility distributions. *Int. J. Intell. Syst.* **2016**, *31*, 215–239. [CrossRef]
89. Wang, G.; Li, W.; Aertsen, M.; Deprest, J.; Ourselin, S.; Vercauteren, T. Aleatoric uncertainty estimation with test-time augmentation for medical image segmentation with convolutional neural networks. *Neurocomputing* **2019**, *338*, 34–45. [CrossRef]
90. Diez-Olivan, A.; Del Ser, J.; Galar, D.; Sierra, B. Data fusion and machine learning for industrial prognosis: Trends and perspectives towards Industry 4.0. *Inf. Fusion* **2019**, *50*, 92–111. [CrossRef]
91. Blasch, E.; Sullivan, N.; Chen, G.; Chen, Y.; Shen, D.; Yu, W.; Chen, H.M. Data fusion information group (DFIG) model meets AI+ML. In *Signal Processing, Sensor/Information Fusion, and Target Recognition XXXI*; Kadar, I.; Blasch, E.P.; Grewe, L.L., Eds.; SPIE: Bellingham, WA, USA, 2022; Volume 12122, p. 121220N.
92. Holzinger, A.; Dehmer, M.; Emmert-Streib, F.; Cucchiara, R.; Augenstein, I.; Del Ser, J.; Samek, W.; Jurisica, I.; Díaz-Rodríguez, N. Information fusion as an integrative cross-cutting enabler to achieve robust, explainable, and trustworthy medical artificial intelligence. *Inf. Fusion* **2022**, *79*, 263–278. [CrossRef]

93. Holst, C.A.; Lohweg, V. Feature fusion to increase the robustness of machine learners in industrial environments. *at-Automatisierungstechnik* **2019**, *67*, 853–865. [CrossRef]
94. Kondo, R.E.; de Lima, E.D.; de Freitas Rocha Loures, E.; dos Santos, Eduardo Alves Portela.; Deschamps, F. Data Fusion for Industry 4.0: General Concepts and Applications. In Proceedings of the 25th International Joint Conference on Industrial Engineering and Operations Management—IJCIEOM, Novi Sad, Serbia, 15–17 July 2019; Anisic, Z.; Lalic, B.; Gracanin, D., Eds.; Springer International Publishing: Cham, Switzerland, 2020; pp. 362–373.
95. Denœux, T.; Masson, M.H. Dempster-Shafer Reasoning in Large Partially Ordered Sets: Applications in Machine Learning. In *Integrated Uncertainty Management and Applications*; Huynh, V.N.; Nakamori, Y.; Lawry, J.; Inuiguchi, M., Eds.; Springer: Berlin/Heidelberg, Germany, 2010; pp. 39–54.
96. Hui, K.H.; Ooi, C.S.; Lim, M.H.; Leong, M.S. A hybrid artificial neural network with Dempster-Shafer theory for automated bearing fault diagnosis. *J. Vibroengineering* **2016**, *18*, 4409–4418. [CrossRef]
97. Peñafiel, S.; Baloian, N.; Sanson, H.; Pino, J.A. Applying Dempster–Shafer theory for developing a flexible, accurate and interpretable classifier. *Expert Syst. Appl.* **2020**, *148*, 113262. [CrossRef]
98. Dubois, D.; Prade, H. From possibilistic rule-based systems to machine learning—A discussion paper. In *Scalable Uncertainty Management*; Davis, J.; Tabia, K., Eds.; Springer International Publishing: Cham, Switzerland, 2020; pp. 35–51.

Article

The Digital Calibration Certificate (DCC) for an End-to-End Digital Quality Infrastructure for Industry 4.0

Siegfried Hackel [1,*], **Shanna Schönhals** [1], **Lutz Doering** [1], **Thomas Engel** [2] and **Reinhard Baumfalk** [3]

1   Physikalisch-Technische Bundesanstalt (PTB), 38116 Braunschweig, Germany
2   Siemens Aktiengesellschaft, 81739 München, Germany
3   Sartorius Lab Instruments GmbH & Co. KG, 37079 Göttingen, Germany
*   Correspondence: siegfried.hackel@ptb.de; Tel.: +49-531-592-1017

Abstract: This article depicts the role of the Digital Calibration Certificate (DCC) for an end-to-end digital quality infrastructure and as the basis for developments that are designated by the keyword "Industry 4.0". Furthermore, it describes the impact the DCC has on increasing productivity in the manufacturing of products and in global trade. The DCC project is international in its scope. Calibration certificates document the measurement capability of a measurement system. They do this independently and by providing traceability to measurement standards. Therefore, they do not only play an important role in the world of metrology, but they also make it possible for manufacturing and commercial enterprises to exchange measurement values reliably and correctly at the national and at the international level. Thus, a DCC concept is urgently needed for the end-to-end digitalization of industry for the era of Industry 4.0 and for Medicine 4.0. A DCC brings about important advantages for issuers and for users. The DCC leads to the stringent, end-to-end, traceable and process-oriented organization of manufacturing and trading. Digitalization is thus a key factor in the field of calibration as it enables significant improvements in product and process quality. The reason for this is that the transmission of errors will be prevented, and consequently, costs will be saved as the time needed for distributing and disseminating the DCCs and the respective calibration objects will be reduced. Furthermore, it will no longer be necessary for the test equipment administration staff to update the data manually, which is a time-consuming, tedious and error-prone process.

Keywords: D-SI; DCC; digital signature; calibration; Industry 4.0

Citation: Hackel, S.; Schönhals, S.; Doering, L.; Engel, T.; Baumfalk, R. The Digital Calibration Certificate (DCC) for an End-to-End Digital Quality Infrastructure for Industry 4.0. *Sci* **2023**, *5*, 11. https://doi.org/10.3390/sci5010011

Academic Editors: Johannes Winter and Claus Jacob

Received: 28 November 2022
Revised: 19 January 2023
Accepted: 9 February 2023
Published: 6 March 2023

1. Introduction

The analogue calibration certificate is currently still used and issued in paper form or as a closed PDF document. These closed documents hardly contribute to improving the production process of a company. The reason for this is that using analogue calibration data from a calibration in subsequent processes is very time-consuming and prone to errors as these data have to be converted once again into the digital formats of the specific manufacturing plant. The DCC [1] overcomes this disadvantage of its analogue counterpart. The objective was to create an internationally recognized DCC format which acts as an interface (exchange format) in the whole field of metrology, and especially in the field of machine-to-machine communication. On the basis of the DCC, we will be able to develop further exchange formats in the future. These might be in the field of legal metrology, for digital type examination certificates, for the **D**igital **T**win (DT) or for developments in many other fields—for example, the **D**igital **C**alibration **R**equest (DCR) format or, in the field of automated accident notification, in the format of the "**I**nternational **S**tandard **A**ccident **N**umber (ISAN): Linking data in accidents and emergencies." The DCC serves for the electronic storage, the authenticable and—if necessary—the encrypted and signed dissemination and the uniform interpretation of the calibration results. Due to the DCC schema to be applied, the DCC is thus both machine-readable and machine-interpretable when using Good Practice (GP) DCCs. Further information on machine interpretability can be found in [2]. The DCC has been developed and agreed upon in a broad community and is constantly being developed further in order to achieve worldwide acceptance. The target groups are all those bodies, authorities and companies worldwide which require proof of the metrological traceability of their measurement results and use these results in modern manufacturing processes in the field of IIoT/Industry 4.0. These bodies, authorities and companies include the metrology institutes, designated institutes, national calibration centres, calibration laboratories and the many companies in industry that require traceable measurement results for their quality management systems.

According to the International Vocabulary of Metrology (VIM), the term "calibration" is defined as follows [3]:

"Operation that, under specified conditions, in a first step, establishes a relation between the quantity values with measurement uncertainties provided by measurement standards and corresponding indications with associated measurement uncertainties and, in a second step, uses this information to establish a relation for obtaining a measurement result from an indication" [4], p. 3.

The ISO IEC 17025 standard [5] describes the general requirements that are placed on qualified calibrations. Section 7.8 of this standard deals with "reporting on results." Although digitalization is increasing worldwide, calibration certificates are still mainly issued in analogue form, meaning that they are generated in paper form. This does not only lead to numerous problems, but it also prevents, in many fields, the complete digitalization of the value chain, not least due to the fact that analogue calibration certificates are often designed differently, even if the calibrations are the same. A uniform design for certificates for calibrations which are basically equal in nature is shown in a Good Practice (GP) DCC. Work on the GP-DCC is currently in process in the most diverse fields (temperature, humidity, ambient pressure, mass, weighing instruments, force, torque...). By means of the GP-DCC, the results are becoming machine-interpretable (as explained above) as, thanks to the GP-DCC, it is exactly known what has been stored, where it has been stored and in which way it has been stored.

Calibrations—and thus also calibration certificates—play an important role in many branches of industry (e.g., in the automotive industry and the pharmaceutical industry). A pharmaceutical company may require several hundreds of thousands of calibrations certificates every year. It is easy to imagine which efforts currently have to be made to archive analogue calibration certificates (either exclusively in paper form or also in other forms). In this publication, the DCC concept is presented. It will solve the abovementioned problems and will bring about many additional advantages for companies and other users. Furthermore, it will allow machines to communicate with each other (M2M). All of these aspects are described in more detail below.

2. The Role of Calibration and Calibration Certificates

2.1. Quality Infrastructure

The quality infrastructure of a country is of fundamental importance for the services that are provided to the public as well as for consumer protection. This was published for the first time in 2007 in a report of the World Bank [6]. Meanwhile, various contributions have been issued which were developed on the basis of this report (see also [7–10]). Figure 1 shows the general set-up of an arbitrary national infrastructure, and it also shows digitalization by means of DCC. From this figure, it becomes obvious what an important role calibration certificates play.

Calibration certificates appear several times in Figure 1. It should be stated that besides the "normal" digitalization effect which normally occurs, for example, in the field of eGovernment, an exorbitant increase in quality can be observed, which is due to the fact that the calibration information is more precise and can be used in an automated way. The data are transferred in M2M communication without any transmission errors. The step in which data are transferred manually according to the four-eyes principle and which so far has been necessary, for example, in the pharmaceutical industry, can then be omitted. Thereby, the time expenditure can be substantially reduced, and errors during data transmission will thus be ruled out.

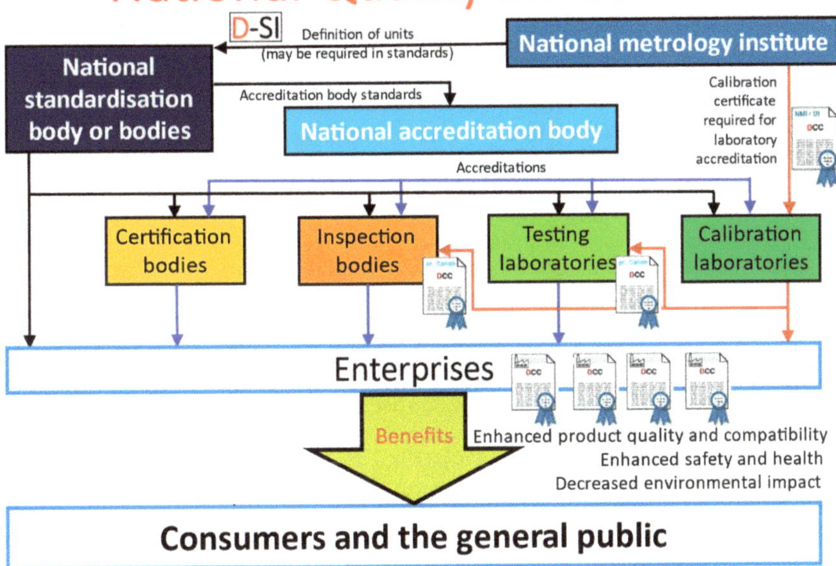

Figure 1. Presentation of the national quality infrastructure (based on [6]).

After the data have been transferred, the reject rate in manufacturing can immediately be reduced and products can be manufactured in a more resource-saving and more sustainable way. Manufacturing is increasingly helping to save many raw materials, as well as time and energy, and thus often reduces CO_2 emissions to a high extent. This, in turn, will contribute essentially to environmental protection and to achieving the global UN sustainable development goals not only in the industrial countries, but also in developing countries and in countries in transition. The latter can often even skip an innovation cycle (which would normally still be based on a paper-based calibration system) and thus participate more quickly in an international economic system. In addition, they could contribute effectively to the conservation of nature especially in their own countries.

The topic of "data as economic goods" has become particularly important in the value chain of industrial companies and will become more and more important in the future due to the advancing digital transformation. In this regard, the quality of the data which have been gained, among others, via sensors and actuators and the verifiability of the statements and conclusions derived from this data are of central importance especially for industry. It will only be possible to use data successfully as economic goods if the quality of the data is reliably verified and safeguarded.

For more than 125 years, maintaining the high level of quality of German products has been one of the fundamental pillars of the quality label "Made in Germany" which is recognized worldwide today. One of the fundamental challenges of the national quality infrastructure (QI) is to transfer this quality label to the digitalized world and to establish it there. This includes, in particular, the development of safe and robust calibrated measurement systems to be able to ensure data quality and the trust people have in the data. Metrology plays a decisive role in gaining the trust people have in measurements and in ensuring the quality of the measurement data and measurement results. Figure 1 shows how the different elements of QI interact with each other.

The quality infrastructure of a country consists of seven elements. The **National Metrology Institute** (NMI) disseminates the SI units to the national standardization institute. The standardization institute ensures that the level of the enterprises (see the level "enterprises" in the diagram above) is provided with norms and standards. At the same time, it determines the norms and standards for the national accreditation body. The accreditation body has the task of accrediting calibration and testing laboratories, inspection bodies and certification bodies and of supervising the accreditation. In this way, the accreditation bodies are able to supply the level of the enterprises with verified expertise.

The second task of the national metrology institutes is to provide the accredited calibration laboratories with traceable systems. In this way, it is ensured that the measurement standards of the accredited calibration laboratories are linked up with the national measurement standards. The accredited calibration laboratories will then ensure the traceability of the measurement standards of the test laboratories, of the inspection bodies and of the other calibration laboratories. Calibration thus has a special position in the national quality infrastructure (see the red arrows in Figure 1).

So far, the certification bodies have not been particularly active in the field of digitalization. However, it has turned out that especially in the field of software development, considerable damage has occurred. This is due to the fact that different systems of units are used. The software certification bodies will therefore play a particularly important role in this field.

DCCs have not yet been the subject of considerations. However, using DCCs, and thanks to the fact that the error-free transmission of data becomes possible using DCCs, a high increase in quality will be achieved as the transmission problems will be eliminated.

In addition, there is the advantage that in a DCC, larger amounts of data can be transmitted, and that the calibration information can be integrated in a **Digital Twin** (DT) in an automated and secure way.

2.2. Task of the Calibration

The calibration of measurement instruments is the main pillar of measurements which are comparable with each other and whose contents are correct. It is also one of the main pillars of the distributed manufacturing processes that are common in industry today. Without calibration, it is impossible to make a sound statement on the quality of a manufactured product. The calibration is based on the measurement standards provided by the NMIs and the **Designated Institutes** (DIs). These measurement standards are embedded in the SI system of units [11]. Figure 2 illustrates this relationship. In many calibration fields, the factors indicated in the figure are clearly higher. It can be assumed that in Germany alone, several millions of calibrations are carried out every year. Via large-scale comparison measurements (called "interlaboratory comparisons"), as well as via audits and other procedures, the NMIs and the DIs ensure the provided measurement standards agree within the scope of the conventional and/or stated measurement uncertainties. In addition to this, the NMIs and the DIs of different countries stay in close contact with each other in order to ensure there are harmonized measurement standards worldwide. The calibration pyramid shows that the measurement uncertainty increases from the top to the bottom.

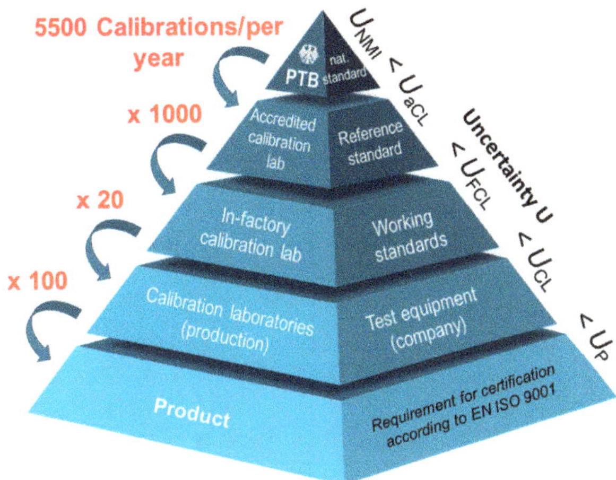

Figure 2. Representation of the calibration pyramid of the individual national metrology institutes/designated institutes and their integration in international interlaboratory comparisons with other NMIs/DIs.

3. Research for Practical Applications

Both industry and the economy are waiting for—and urgently require—a successful and lawful digital transformation so that data, information and certificates can securely and robustly be transmitted throughout the QI processes. Such a digital transformation is the precondition for both industry and the economy to be able to keep their promise of quality also efficiently and effectively in a digitalized world. Since August 2020, a project consortium has been established in Germany which is known as GEMIMEG-II [12] and is funded by the BMWK (Federal Ministry for Economic Affairs and Climate Action). It consists of 12 partners as well as of further institutions (the latter in a consultative capacity). The partners are contributing research achievements in the field of DCC [13,14] and are setting up a digital calibration infrastructure. The overriding objective and the central challenge of the GEMIMEG-II project is to bring the framework requirements of QI and the practical demands of industry together in a joint, holistic approach which is applicable in practice.

The focus of this project is on metrology (calibration, referencing, measurement, i.e., on trustable information on the quality of measuring instruments, sensor networks, digital twin and data analysis methods). The overall objective is to make information for the realization of reliable, connected measuring systems available in a secure, lawful and legally compliant end-to-end way. The results of the research fields are implemented in four so-called "RealBeds" (i.e., demonstrators which are geared to different fields of application), and these demonstrators thus take on the status of a prototype. The four RealBeds are connected calibration facilities, Industry 4.0 applications, the pharmaceutical/process industry and autonomous driving. The legal questions of the research fields and of the RealBeds will be investigated in addition in a legal simulation study.

In these investigations, two aspects are important: (1) In how far is reliable and trustworthy information on measuring instruments and measurement data digitally available, and (2) how can a secure and robust orchestration of the measuring systems be achieved? The investigations also include the development of a digital metrology system in the sense of an end-to-end, digitalized, traceable and legally secure measuring and calibration chain for complex sensor networks. Thereby, the sensors—which, so far, have been considered individually—will increasingly be connected with each other in a network by the IIoT (Industrial Internet of Things) in Industry 4.0. A basic set-up of a generic, massive, connected sensor system is shown in Figure 3. The processing of the sensor data follows the fog/edge computing approach. This allows computing functions and memory assignments to take place on suitable nodes of a network, but completely decentralized solutions for use in mobile systems are also possible by means of identical stack and software technologies.

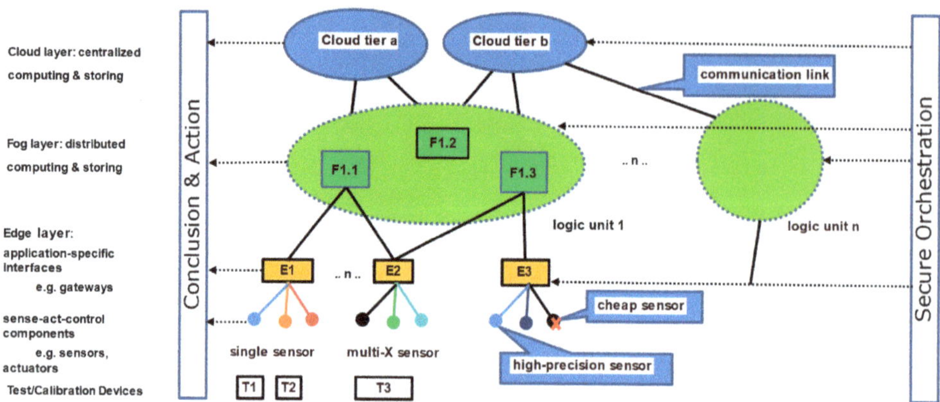

Figure 3. Architecture for a massive sensor network based on the OpenFog reference architecture [15,16].

This technical and automated interconnection in a network significantly increases the requirement to determine the quality of data and their availability if, on the one hand, the integrity of the citizens and of the environment and, on the other hand, the economic success of the companies is to be ensured. It can thus be expected that the need for developing and setting up sensor networks as holistically calibrated, digital measuring systems will continuously increase.

At this point, special requirements will not only arise for the quality of data, but also for the communication infrastructure and for the management of data sources and data sinks. Figure 4 shows a schematic overview of the GEMIMEG-II concept.

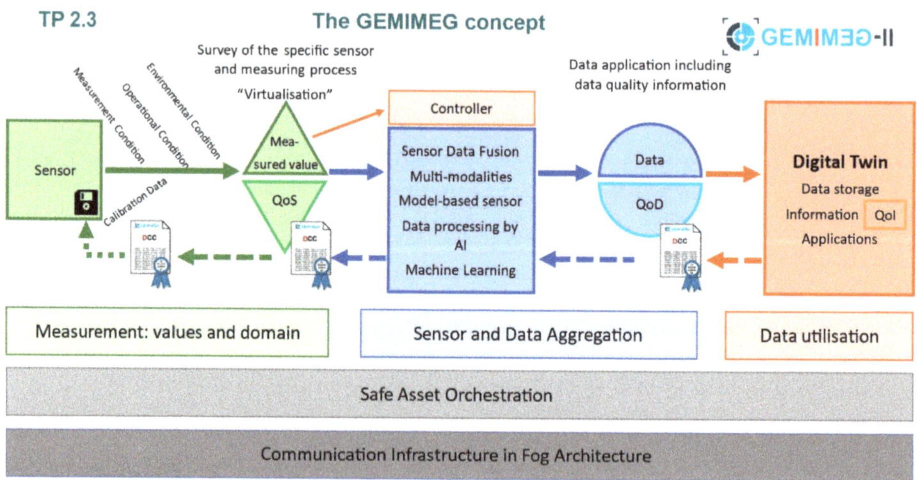

Figure 4. GEMIMEG concept in a schematic overview.

This concept extends from the sensor measurement values, including calibration and the characteristic numbers QoS (QoS: **Q**uality **of S**ensing), via sensor fusion and data aggregation with the characteristic numbers QoD (**Q**uality **of D**ata) to data application (e.g., in the digital twin) with the characteristic numbers QoI (**Quality of Information**). The quality characteristic numbers can also be described as QoX, where the X stands for **Sensing**, **Data**, or **Information**.

That step of the measurement process which contains the sensor measurement values and the calibration of the sensors is supposed to abstract the measurement values (by exploiting the domain knowledge during the measurement process) and to decouple them from the specific characteristics of the actually used sensor. The qualified measurement values and the quality characteristic numbers which have been obtained in this way in the subsequent step (sensor and data aggregation) can be processed without the concrete measurement system being known any further. The measurement values are thus decoupled from the corresponding measurement modalities in the best possible way and associated QoS trust measures have been determined. After these input quantities have been processed during the sensor and data aggregation step, new data are generated as output quantities—after data fusion, or via software-based sensors, or via data analysis using neural networks, or via artificial intelligence—which, in turn, can then have the quality characteristic numbers QoD. This step can be available several times. It can be arranged in parallel (for different measurement values) and/or in a cascaded way (multi-step processing). The data with QoD can be used in different digital twins which, in turn, can access the data with QoD from the multi-step sensor and data aggregation.

If pieces of information are then derived from these data (in an application such as a digital twin), this information can also be supplemented by means of quality characteristic numbers for the QoI information—where the QoI describes, for example, a level of trust or the reliability of this information.

The DCCs, which have already been addressed several times in this article, bring about several conceptual advantages for the project. Avoiding changes between different media—and thus ruling out transmission errors—is an obvious advantage. The calibration results are immediately available for further processing. It is thus, for example, possible to extract different measuring uncertainties automatically in the areas that are important for the manufacturing process.

4. The DCC Concept

In its digital form, the DCC is defined via an XSD schema [13]. For the information obtained in a calibration, the schema contains clearly defined fields for entering mandatory specifications, as well as open text fields in which the calibration laboratory can enter further information. In this way, the entire information content obtained during a calibration will be included in the DCC and transmitted.

The Digital System of Units (D-SI [17,18]) is used to express the units for the measurement values. It supports both the seven base units and the measurement units derived from these—including decimal multiples or dividers, among others—as well as the disseminated imperial units. Even unusual systems of units are possible as long as they have been clearly defined. Actually, this is necessary today in order to ensure, on the one hand, international acceptance, and to be able, on the other hand, to cope without difficulties with all the applications that exist today.

Figure 5 shows an initial rough generic view of the way in which the DCC is embedded in the existing environment of norms, standards, terms and technical regulations.

In addition to the DCC, Figure 5 also shows the **Digital Calibration Request** (DCR). This is a standardized digital document by means of which a desired calibration can be requested and specified. Furthermore, Figure 5 shows a document named the **Digital Calibration Answer** (DCA). In the DCA, the calibration laboratory can enter further information on the calibration which—for formal reasons (e.g., according to the ISO IEC 17025 standard [5])—is not supposed to appear in an official calibration document. The aim is that all documents should be based on a joint structured document schema DX which, in turn, must take the norms, standards, nomenclatures and technical regulations into account which are shown in Figure 5 below the level "Digital Document Schema DX." The purpose of this is to ensure that the calibration is documented digitally in a way that is legally valid and in compliance with the norms and standards.

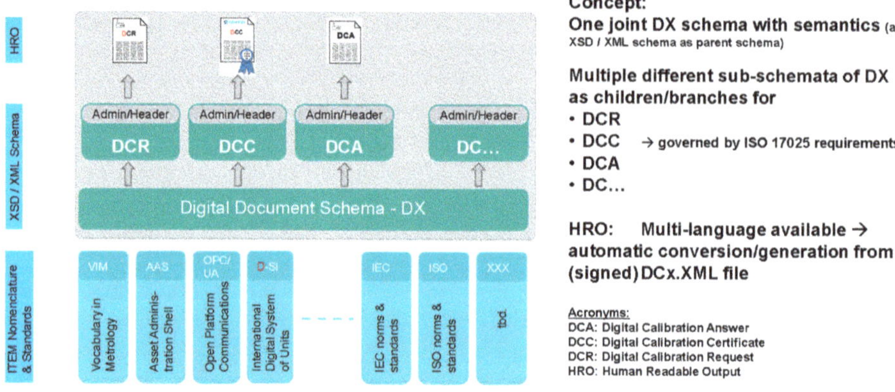

Figure 5. Generic view of the digital calibration document "ecosystem".

4.1. Structure of the DCC Files

Technically, the DCC is a text document which has an XML structure. XML documents are stored as plain text files, based on the Unicode character set (UTF-8 format [19]), and are thus suitable for long-term storage. The files can be opened and processed with an arbitrary text editor. Version 1.0, which is used for DCC, is widespread and has been standardized for more than 20 years now, which is why XML is ideally suited for the use of such important documents as calibration reports. The calibration reports can be ported from XML to other data formats (e.g., JSON).

More information on the structure of a DCC can be found on the DCC homepage [20].

4.2. Prologue including a Processing Instruction

Every XML document must start with a single-spaced prologue. The prologue is an instruction which has the form (<?xml...?>) and is named XML **P**rocessing **I**nstruction (PI). The prologue gives the reading program instructions on what must be observed during reading. Apart from the version (which can be indicated via the "version" attribute), it is also the agreed character set which can be stored in the document file. The characters in documents are encoded according to various ISO standards. The ISO 8859 standard [21] specifies the characters of different languages in ten subdivisions. Further characters are defined in ISO standard IEC 10646 [22]. DCC uses the character set UTF-8. This character set can map all official languages in the world.

One of the main advantages of XML is that it can be read and interpreted by machines. In addition, it is also readable by humans. A decisive factor is furthermore that XML is a data format that is suitable for long-term storage. This is of special importance as it must still be possible to read the files in several decades. Further information on data formats that are suitable for long-term storage can be found, for example, in [23,24].

4.3. Cryptographic Signatures in the DCC

In order to ensure the authenticity and the originality of each calibration certificate, the XML document can be provided by the issuing body (the calibration body) with a corresponding hash value and can be signed digitally. According to ISO standard 17025, adding a signature is **not mandatory**. As a measure to create trust between the calibration institute und the customer, however, adding a signature is recommended. As the signature can be verified at any time, the recipients of a DCC are in a position to verify the authenticity of the content of a DCC as well as the issuer of a DCC themselves. For this purpose, public software tools are available free of cost [25]. The cryptographic procedures can be applied to XML data structures in a robust and probative way [23,26]. Further information on XML is to be found in the literature, e.g., in [27]. The VDI/VDE has already developed a very broad approach for data exchange [28] and—although this is a national standard for Germany—it is already used in various other countries. In this approach, the workflow in industry, in which the DCC can be integrated without any problems, is taken into account. On the one hand, the integrity and authenticity of a DCC must be ensured. On the other hand, however, electronically stored data can easily be changed and/or copied as often as desired. The use of cryptographic security procedures for the DCC is therefore highly recommended. Good summaries on the topic of cryptography can be found in [29,30].

Unfortunately, this report cannot describe this concept in more detail.

5. The DCC within the Scope of the Manufacturing Process

In this section, we, first of all, show the processes which take place within the scope of discrete manufacturing. In Figure 6, the logic relationship between the "parts supplier," the "machine" and the "factory" is shown in the form of models. All models include, in particular, commissioning and production processes.

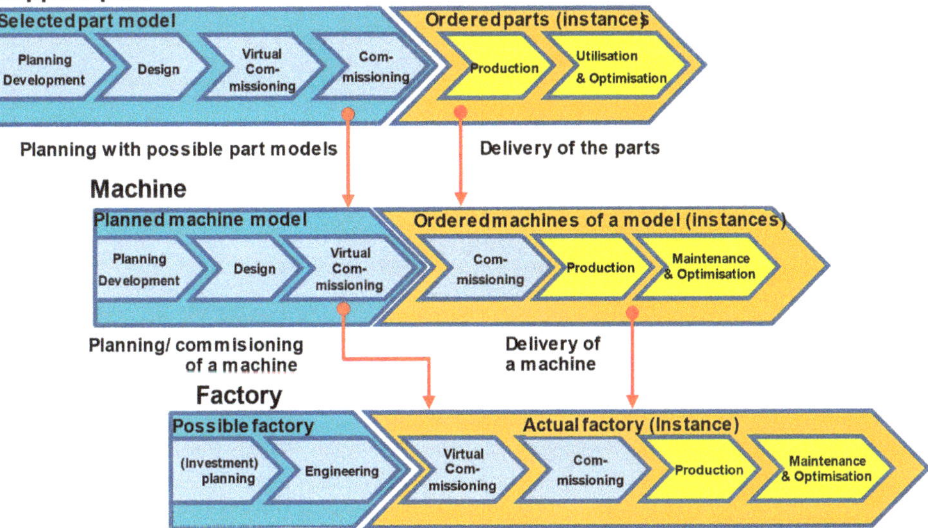

Figure 6. Hierarchy of the different process models in the context of a factory, similar to [31] and adapted.

For these processes, calibrated measurement components are required. However, different pieces of information on the calibration are currently only available as calibration certificates in paper form (Figure 7).

When discrete or continuous manufacturing processes are automated, this offers considerable potential for using a DCC. Whereas in conventional automation with static hierarchies, it is sufficient to manage the calibration data of process sensors analogously and/or decentrally (Figure 8). Due to the fact that there are only a few clearly defined input points for this information, the situation in an Industry 4.0 architecture is different.

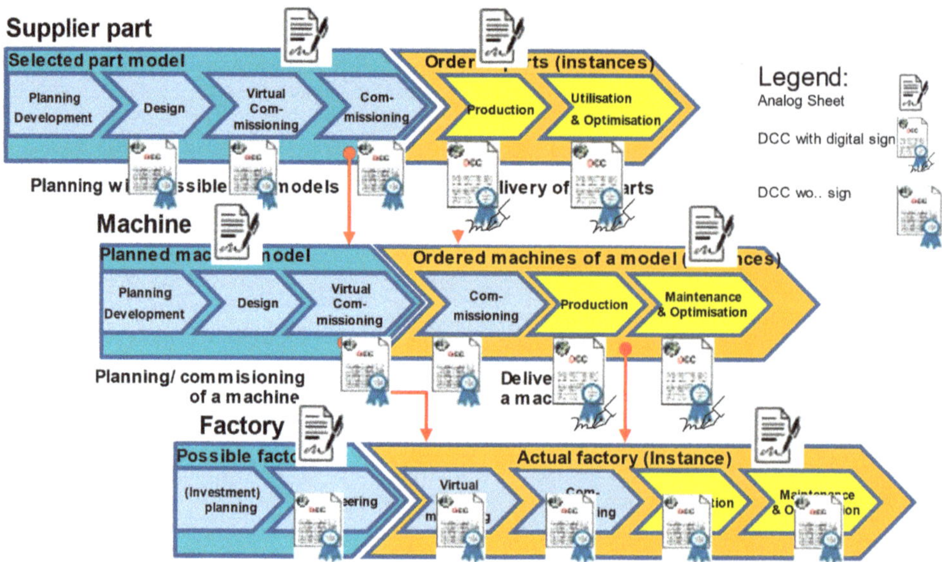

Figure 7. Comparison of how far the calibration information is available in an analogue calibration certificate (in the diagram in Figure 7, this is each time shown above the respective model) and in a DCC.

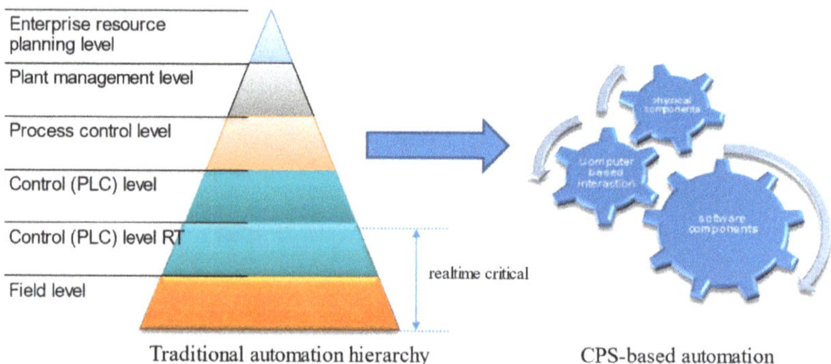

Figure 8. Conventional automation architecture as compared to Industry 4.0 or IIoT structures based on a Cyber Physical System (CPS) based on [32].

CPS-based architectures [33] react adaptively and continuously if it becomes necessary to change manufacturing processes, and this is exactly what is stipulated for Industry 4.0 or IIoT networks. The consequence of this is as follows: The process sensors, which are also used in automation, must not be able any longer to only make the generated measurement values but also—as an independent element—the sensor information (such as calibration certificates) available to other process components independently and in changing architectures. The massive sensor networks, which have already been presented above, might be an option for this and—together with the DCC—fulfil the requirements which have been discussed here.

Figures 7 and 8 clearly show that the calibration information that is contained in a DCC is not only available in the immediate manufacturing processes in which the calibrated measuring components are used. As the DCCs can be seamlessly integrated into the digital infrastructure of a factory, the calibration information can also be used in other processes such as in enterprise planning, commissioning or quality management. The VDI/VDE has already developed a very broad approach for data exchange [28] and—although this is a national standard for Germany—it is already used in various other countries. In this approach, the workflow in industry, in which the DCC can be integrated without any problems, is taken into account.

Within the scope of the GEMIMEG-II project, we are currently working on a software library in the programming language Python (PyDCC). The aim of PyDCC is to facilitate access to the contents of the DCC and their processing so that the advantages of the DCC can be prepared for digital manufacturing in a simple and easy way. As soon as the GEMIMEG-II project has been concluded, this software package, which will be an open-source software, will become generally available. Figure 9 shows the current development and planning status of this software project as well as the focal areas of the future development tasks.

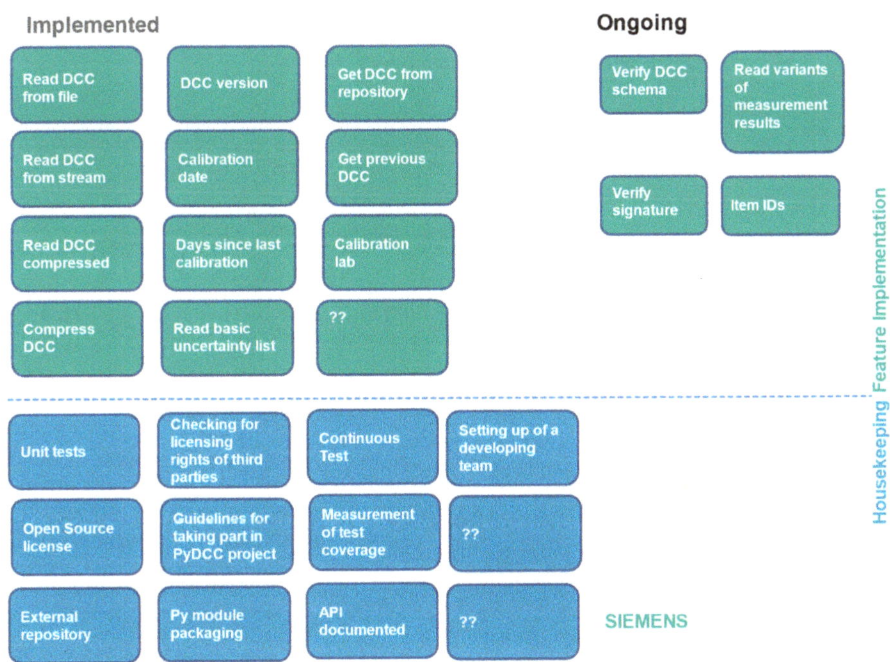

Figure 9. PyDCC Software Management (currently unreleased, see Tobola, Andreas "Introducing PyDCC—a Python module for the DCC" [34], p. 101).

In this way, a basic functionality is also available for using the DCC on edge and smart field devices. The idea is to grant all users easy access to the information contained in the DCC by making this software functionality (which is non-competitive) available as an open-source version that includes the standardized DCC, which is provided with an unambiguous version identification and with the respective XML schema. Publishing the approach as an open-source version, which will reach many users, is supposed to minimize the effort that is needed to implement the basic functionalities in such a way that the DCC can be used by a specific company and/or can be connected with the Enterprise Resource Planning/PLM (Product Lifecycle Management).

6. Conclusions & Outlook

In this article, we showed that all aspects of calibration are of great economic interest. The end user often does not realize how many components have contributed to the quality assurance for a product. Establishing the use of DCCs is a task for both the industrial and trade partners and should be achieved all over the world. When the approach of the digital calibration certificate is recognized and used internationally, the DCC will play a central role. For example, language barriers which still exist today can be eliminated by means of the digital document as the contents will be standardized and can automatically be transferred to the respective language. In this way, the DCC will make an important contribution to the internationalization of measurement technology and the calibration system.

In principle, the DCC system suggested here can also be used for other applications in measurement technology. These range from legal metrology to qualification measurements that are carried out on a system in order to prove its functionality or after it has undergone maintenance work. For this purpose, only the type or level of the issuing authority must be mentioned in the DCC document. Here, the end-to-end use of a DCC would bring about great advantages for all process partners.

Currently, establishing the DCC as a component in an asset administration shell (AAS [35]) in a sub-model is envisaged.

Author Contributions: Conceptualization, methodology, draft preparation and visualization, S.H., S.S., L.D., T.E. and R.B.; supervision, validation, writing—reviewing and editing, L.D. and T.E. All authors have read and agreed to the published version of the manuscript.

Funding: The GEMIMEG II project is funded by the German Federal Ministry for Economic Affairs and Climate Action (BMWK), grant reference GEMIMEG 01 MT20001A.

Institutional Review Board Statement: Not applicable.

Informed Consent Statement: Not applicable.

Data Availability Statement: See 'DCC—GoodPractice · GitLab', GitLab. Available online: https://gitlab.com/ptb/dcc/dcc-goodpractice (accessed on 16 January 2023).

References

1. DCC_PTB. Available online: https://www.ptb.de/dcc/ (accessed on 27 June 2022).
2. IdiS—Initiative Digitale Standards. Available online: https://www.dke.de/idis (accessed on 20 January 2022).
3. BIPM, VIM_JCGM_200_2012. Available online: https://www.bipm.org/documents/20126/2071204/JCGM_200_2012.pdf/f0e1ad45-d337-bbeb-53a6-15fe649d0ff1 (accessed on 23 April 2021).
4. [VIM3] 2.39 Calibration. Available online: https://jcgm.bipm.org/vim/en/2.39.html (accessed on 2 May 2021).
5. DIN EN ISO/IEC 17025:2018-03, Allgemeine Anforderungen an Die Kompetenz von Prüf- Und Kalibrierlaboratorien (ISO/IEC_17025:2017); Deutsche und Englische Fassung EN_ISO/IEC_17025:2017, Beuth Verlag GmbH. Available online: https://www.beuth.de/de/-/-/278030106 (accessed on 24 September 2021).
6. Guasch, J.L.; Racine, J.-L.; Sanchez, I.; Diop, M. *Quality Systems and Standards for a Competitive Edge*; The World Bank: Washington, DC, USA, 2007.
7. Blind, K. Standardization and Standards as Science and Innovation Indicators. In *Springer Handbook of Science and Technology Indicators*; Glänzel, W., Moed, H.F., Schmoch, U., Thelwall, M., Eds.; Springer International Publishing: Cham, Germany, 2019; pp. 1057–1068. ISBN 978-3-030-02511-3.
8. Eichstädt, S. PTB Digitalization Strategy. *Metrol. Für Die Digit. Von Wirtsch. Und Ges.* **2017**, *127*, 40–67. [CrossRef]
9. Bucci, G.; Ciancetta, F.; Fiorucci, E.; Fioravanti, A.; Prudenzi, A.; Mari, S. Challenge and future trends of distributed measurement systems based on Blockchain technology in the European context. In Proceedings of the 2019 IEEE 10th International Workshop on Applied Measurements for Power Systems (AMPS), Aachen, Germany, 25 September 2019; pp. 1–6. [CrossRef]
10. Eichstädt, S. Metrologie für die Digitalisierung von Wirtschaft und Gesellschaft I 2020, p. 1899184 bytes, 13 pages. 2021. Available online: https://oar.ptb.de/resources/show/10.7795/120.20210223 (accessed on 15 June 2021).
11. BIPM—SI Broschüre. Available online: https://www.bipm.org/en/publications/si-brochure/ (accessed on 15 April 2020).
12. Gemimeg-Team. Gemimeg Startseite, 29 June 2022. Available online: https://www.gemimeg.ptb.de/gemimeg-startseite/ (accessed on 19 July 2022).

13. Physikalisch-Technische Bundesanstalt (PTB). DCC Schema Version 3.1.0, DCC Version 3.1.0. Available online: https://www.ptb.de/dcc/v3.1.0/dcc.xsd (accessed on 13 April 2022).
14. Release and Gitlab Repository of DCC Schema/DCC/xsd-dcc, GitLab. Available online: https://gitlab.com/ptb/dcc/xsd-dcc (accessed on 19 April 2022).
15. Industrial Internet Consortium. OpenFog_Reference_Architecture_2_09_17.pdf. Available online: https://www.iiconsortium.org/pdf/OpenFog_Reference_Architecture_2_09_17.pdf (accessed on 20 November 2019).
16. Cloud Computing—Fog Computing & Edge Computing: WinSystems. Available online: https://www.winsystems.com/cloud-fog-and-edge-computing-whats-the-difference/ (accessed on 21 November 2019).
17. Hutzschenreuter, D.; Härtig, F.; Heeren, W.; Wiedenhöfer, T.; Forbes, A.; Brown, C.; Smith, I.; Rhodes, S.; Linkeová, I.; Sýkora, J.; et al. SmartCom Digital System of Units (D-SI) Guide for the Use of the Metadata-Format Used in Metrology for the Easy-to-Use, Safe, Harmonised and Unambiguous Digital Transfer of Metrological Data—Second Edition, July 2020. Available online: https://zenodo.org/record/3816686 (accessed on 1 March 2022).
18. Daniel, H.; Shan, L.; Jan, H.L.; Alexander, S.; Rok, K.; Bojan, A.; Bernd, M.; Lukas, H. SmartCom Digital-SI (D-SI) XML Exchange Format for metrological Data Version 2.0.0, July 2021. Available online: https://zenodo.org/record/4709001 (accessed on 10 March 2022).
19. Yergeau, F.; UTF-8, a Transformation Format of ISO 10646. RFC Editor, RFC3629, November 2003. Available online: https://www.rfc-editor.org/info/rfc3629 (accessed on 9 September 2019).
20. Digital Calibration Certificate—DCC. 14 April 2021. Available online: https://www.ptb.de/cms/en/research-development/into-the-future-with-metrology/the-challenges-of-digital-transformation/kernziel1einheitlichkeitim/digital-calibration-certificate-dcc.html (accessed on 23 April 2021).
21. ISO/IEC 8859-1-1998-04-Beuth.de. Available online: https://www.beuth.de/de/norm/iso-iec-8859-1/7657747 (accessed on 27 June 2021).
22. CSA ISO/IEC 10646-2020-Beuth.de. Available online: https://www.beuth.de/de/norm/csa-iso-iec-10646/320432205 (accessed on 27 June 2021).
23. Johannes, P.C.; Potthoff, J.; Roßnagel, A.; Neumair, B.; Madiesh, M.; Hackel, S. *Beweissicheres Elektronisches Laborbuch: Anforderungen, Konzepte und Umsetzung zur langfristigen, Beweiswerterhaltenden Archivierung Elektronischer Forschungsdaten Und-Dokumentation*; Nomos Verlagsgesellschaft: Baden-Baden, Germany, 2013.
24. Nestor—Publikationen. Available online: https://www.langzeitarchivierung.de/Webs/nestor/DE/Publikationen/publikationen_node.html (accessed on 16 April 2020).
25. DigiSeal Reader—Gratis-Software zum Prüfen von E-Signaturen & E-Siegel. Available online: https://www.secrypt.de/produkte/digiseal-reader/ (accessed on 27 July 2022).
26. PTB, ArchiSafe, 3 May 2016. Available online: https://www.ptb.de/cms/archisafe/startseite.html (accessed on 16 April 2020).
27. Vonhoegen, H. *Einstieg in XML: Grundlagen, Praxis, Referenz, 7*; Aktualisierte und erw. Aufl.; Galileo Press: Bonn, Germany, 2013; p. 655.
28. 'VDI/VDE 2623 Datenaustausch im Prüfmittelmanagement', VDI, 19 November 2020. Available online: https://www.vdi.de/news/detail/datenaustausch-im-pruefmittelmanagement (accessed on 8 February 2023).
29. Schmeh, K. *Kryptografie: Verfahren, Protokolle, Infrastrukturen*; Dpunkt: Heidelberg, Germany, 2016; Volume 6.
30. Pfennig, S. *Sichere und Effiziente Netzwerkcodierung*; Technische Universität Dresden: Dresden, Germany, 2019.
31. Implementation Strategy Industrie 4.0—Report on the Results of Industrie 4.0 Platform, ZVEI. Available online: https://www.zvei.org/en/press-media/publications/implementation-strategy-industrie-40-report-on-the-results-of-industrie-40-platform (accessed on 21 November 2022).
32. Redelinghuys, A.; Basson, A.; Kruger, K. A Six-Layer Digital Twin Architecture for a Manufacturing Cell. In *Service Orientation in Holonic and Multi-Agent Manufacturing*; Springer International Publishing: Cham, Germany, 2019; pp. 412–423. [CrossRef]
33. Cyber-Physical System. Wikipedia. 21 August 2022. Available online: https://en.wikipedia.org/w/index.php?title=Cyber-physical_system&oldid=1105606575 (accessed on 18 October 2022).
34. Laiz, H.; Hackel, S.; Härtig, F.; Schrader, T.; Schönhals, S.; Loewe, J.; Doering, L.; Gloger, B.; Jagieniak, J.; Hutzschenreuter, D. 2nd International DCC-Conference 1–3 March 2022 Proceedings. 2022, p. 318. Available online: https://oar.ptb.de/resources/show/10.7795/820.20220411 (accessed on 28 June 2022).
35. ECLASS, Asset Administration Shell, ECLASS. Available online: https://eclass.eu/en/application/asset-administration-shell (accessed on 20 July 2022).

Article

Industry 4.0 from An Entrepreneurial Transformation and Financing Perspective

Kai Lucks

Bundesverband Mergers & Acquisitions e.V., 82131 Gauting, Germany; kai.lucks@merger-mi.de

Abstract: This paper addresses the management of digital–informational transformation of industrial enterprises. Any transformation requires the coordinated action of several independent actors. Similarly, the digital–informational transformation required for the fourth industrial revolution (i.e., Industry 4.0) requires the involvement of multiple actors from the public and private sectors. This applies to an individual company as well as to the entire sector, regardless of the desired level of transformation. The increasing dissolution of boundaries between industrial and non-industrial actors is therefore essential for Industry 4.0. This paper addresses the above dissolution activities, focusing on cross-company networks and management issues. The management aspects of the following factors are examined: culture change, strategies, degree of digitalization, degree of networking, Internet of Things, digital ecosystems, human resources, organizational development, hierarchies, cross-functional collaboration, cost drivers, innovation pressures, supply chains, enterprise resource planning systems and corporate acquisitions/mergers. Based on the findings on the above factors, a management-driven model of the "transformation to Industry 4.0" for manufacturing companies is presented and discussed. This work thus complements the existing literature on Industry 4.0, as the majority of the literature on Industry 4.0 deals with technical problem solving at the field level.

Keywords: digitalization; business transformation; Industry 4.0; industrial implementation; mergers and acquisitions; knowledge management; networking; process management; informational change; digital ecosystems

Citation: Lucks, K. Industry 4.0 from An Entrepreneurial Transformation and Financing Perspective. *Sci* **2022**, *4*, 47. https://doi.org/10.3390/sci4040047

Academic Editors: Sharifu Ura and Johannes Winter

Received: 19 April 2022
Accepted: 8 November 2022
Published: 1 December 2022

Publisher's Note: MDPI stays neutral with regard to jurisdictional claims in published maps and institutional affiliations.

1. Introduction

1.1. General Remarks

This essay provides a complementary view to the analyses focused on technologies and sciences. The perspective presented here is that of the industrial practitioner and thus focuses on the challenges to managing (!) industrial transformation.

The guiding thesis underlying this paper is that the successful, on-target and on-time implementation of Industry 4.0 is less about the availability of necessary technologies than it is about management competencies, the use of adequate processes, appropriate organizational structures, capabilities for profound cultural and structural change, as well as the involvement of the diverse competence bearers required for this.

1.2. Nomenclature

The term "transformation" is understood here as a fundamental reorganization of structures and processes. This sets it apart from the more or less constantly ongoing and mostly marginal reorganizations.

In the broad public discussion, the term "Industry 4.0" is equated with "digitization". However, this is conceptually incorrect, because according to the German Academy of Science and Engineering (acatech), digitization can be attributed to the "Industry 3.0" development phase. The development spurt, referred to as the "fourth" industrial revolution of modern times after acatech, is characterized by all-encompassing networking. The corporate transformations taking place today include both the complete implementation of digitization and all-encompassing networking. In order to avoid any misunderstandings, the somewhat cumbersome term "digital–informational transformation" is used in this essay as a keyword-like explanation for the implementations according to "Industry 4.0".

1.3. Logic of the Sequence of Steps in This Paper

In view of the variety of major changes that our society is currently undergoing, this paper starts with a classification of the digital–informational transformation in the transformation landscape that is dominant today (Section 2). Section 3 deepens this consideration through a more precise analysis and presentation of the overarching processes and the industrial participation that takes place in them. Building on this, Section 4 deals with the actual implementation management, i.e., the industrial transformation as such, and explains how this is based on the preceding technology and knowledge management. Section 5 can thus conceptually focus on the purely industrial transformation of "Industry 4.0". Section 6 builds on this, dedicated to the experiences with digital–informational transformation in industry. For this purpose, a generic model is presented that has its origins in the systematic management of corporate functions. In addition to the management of transformations, their costs, financing and risks must be dealt with, as announced in the title of the article. Section 7 is dedicated to this aspect. This highly exploratory essay implies that a prospective outlook on the further development of the digital–informational transformation is necessary, especially since the present presentation can also demonstrate the previously insufficient performance in corporate restructuring with the help of individual available data. This is covered in Section 8 under the title "Outlook". The final Section 9 provides summarizing results and their evaluation. The most important findings are that (1) compared to the technical–scientific knowledge about the digital–informational transformation, the empirical knowledge for implementation is far behind; (2) the previous performance in the digital–informational transformation is not satisfactory on average; (3) there are a number of generic models for corporate transformation towards Industry 4.0, which, through combined use, make it possible to develop specific models for implementing the transformation as required; and (4) in view of the unsatisfactory data and study situation, a statistically valid study of failures and success factors in the operational implementation of "Industry 4.0" in companies is recommended. This could support the further orientation of the transformation practice and contribute to a sustainable improvement in performance.

1.4. Radical Change in Management Systems

The change from "classic" corporate management to entirely new management concepts that correspond with the vision of Industry 4.0 is one of the most profound changes that a company can undergo. Thus, the depth of the changes, the all-encompassing readiness of the measures, is most comparable to the fundamental restructuring measures familiar from mergers and major corporate takeovers. As remains to be shown, the degree of verticalization, for example, and the associated concentration of power play a major role. This must be contrasted with new approaches to more horizontally structured management models, which also imply decentralization of power. This primarily involves questions about the degree of autonomy of national organizations and subsidiaries which, according to the "classic model", tend to be "controlled" by the CEO or corporate headquarters. IT-based structures and networks open revolutionary opportunities through rapid data dialogs based on the "countercurrent principle"—i.e., no longer just "direct top-down" but also bottom-up on an equal footing: based on decentralized market and customer proximity, with their specific requirements. The associated innovation potential for new networked management systems cannot be overstated, because ultimately, a company that was previously managed "as a general staff" can be transformed into an "internal digital entrepreneurial ecosystem".

1.5. Professionalization through Transfer of Experience

From large M&A projects, especially after their professionalization phase since the turn of the century, we have a lot of experience to draw on [1,2]. Thus, for the field of Industry 4.0 transformation, for which there is significantly less documented and validated experience, the opportunity to draw on M&A experience is given. After a critical view of their "fit", basic M&A leadership elements can be transferred to digital transformation. However, due to many specifics concerning the reported cases on "digital transformation", we are not yet able to offer generally applicable generic "transformation management models" despite borrowing from the M&A world of today. In this respect, the approaches presented in this paper are to be understood as attempts and working theses, combined with the invitation to develop them further.

Accordingly, the methods underlying this essay deviate from the deductive technological–scientific approach underlying the essays on the fundamentals of Industry 4.0. As was the case with M&A in the 1990s, today's implementation on "managing Industry 4.0" is based less on scientific management research and more on management experience that is strongly aligned with the success and value of business outcomes. To substantiate this with a buzzword, we would have to speak of an "art" of transformation management rather than a science, i.e., heuristics [2,3].

2. Localization of the Digital–Informational Transformation within the Current Transformation Areas

As established in the introduction, the thematic treatment starts at this point with a classification of the digital–informational transformation ("Industry 4.0") in the transformation landscape to which our society is exposed today.

In the following, particular attention is paid to the fact that industrial companies not only represent the "owners" of entire processes, but are also participants in processes in which other partners such as the public sector and private individuals participate. In terms of process optimization for society as a whole, the "non-industrial participants" mentioned would have to be involved in the digital–informational transformation. Industry is to be assigned the role of a pioneer and driver in the digital–informational transformation of our industrial society.

Digital transformations now subsumed under the buzzword "Industry 4.0" involve the most urgent and immediate tasks that industry must address today to secure its future position in international competition. However, the problems facing today's entrepreneurs go far beyond digital-driven transformation. We are in a phase of the greatest upheaval in the history of modern times. The challenges facing business in particular, in its role as a value-generating force in social society, are manifold and severe. Catastrophes, tensions, wars and mega-accidents of all kinds are accumulating and becoming ever more serious, followed by social and economic crises, some of which are moving around the globe in waves [1–3].

The countermeasures required in each case must be taken promptly and, if necessary, simultaneously. Thus, the digital transformation must be integrated into a larger field of transformations.

In Figure 1, the more detailed relationships are shown using examples. We distinguish between influencing factors, (mediating) levers, fields of action ("sectors") and effects. The factors that are occurring simultaneously today include climate change with environmental consequences, social problems (poverty ...), finances (inflation ...), world economy (inequalities, economic cycles ...), refugees (from crisis and war zones, economic refugees ...), as well as crises (and disasters of different kinds). The influencing factors can overlap regionally or worldwide and build on each other. All influencing factors can have an immediate effect on companies, and they can also change their mechanisms of action. For example, environmental disasters can result in a national economic crisis. Thus, there are "intermediary" forces between the influencing factors, referred to here as "levers". The catastrophe theory has a wide variety of explanatory models ready, which are not discussed in detail here.

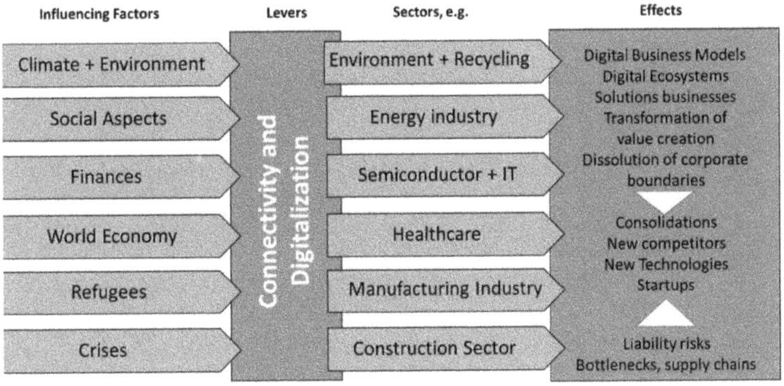

Figure 1. The world in transformation mode.

For the purposes of application and explanation in this essay, the terms "connectivity" (in the sense of all-encompassing networking) and digitalization, which are associated with "Industry 4.0", were used. These factors have a certain role at this point in the sense that there must be a connection ("connectivity") for switching and that every form of data and information must be available digitally in order to be able to transmit it via our networks. This is how they are found in the sectors presented here as examples. The most important ones for the industry are currently the management of the energy transition and the supply chain problem, which are addressed in the present picture with the keywords "Semiconductor + IT" (information technology). The effects alone from the combination of energy change (including energy shortages as a result of the Ukraine war) with breaks in supply chains and (!) the claim to also manage the transformation to Industry 4.0 at the same time shows the dramatic situation industrial companies in particular are in, and how high their existential threat is. A single number may illustrate this: the share of intermediate consumption in the production value in the German industrial sector is "very high" at 63 percent. A large part of this is integrated into the supply chains at various points around the world. This size also shows that the necessary restructuring of supply chains is of unprecedented importance in corporate change. According to our definition, this area alone is a matter of real entrepreneurial transformations. These relationships and the variety of transformation fields to be mastered must be taken into account when we implicitly demand the improved implementation of Industry 4.0 models at this point.

These are undoubtedly enormous challenges for entrepreneurs, who are probably facing the greatest pressure since the Second World War. Digitalization and informational restructuring have cross-cutting functions, without which overarching crises cannot be overcome [4]. The digital transformation is not just a matter for the economy, but a task for the whole of society.

The entrepreneurial answers of our time are shown in the right column of Figure 1, such as digital business models. Special liabilities and bottlenecks have a restrictive effect, which, when combined, bring about changes in entire sectors, such as consolidation movements.

With major transformations and downright disruptive changes, the world's economies are trying to take control over existing ecological, social and economic problems and to counteract escalations that seriously endanger our future.

It is also a favorable circumstance that in the phase of greatest upheavals and radical changes of the modern era, we have those instruments at our disposal, without which none of the pending transformations can be mastered: digitalization and the all-encompassing networking of people, organizations and machines.

Timely deployments of technical solutions are not coincidences but results of extensive research and development. In this paper, we address the challenges, hopes and setbacks to which such processes are fundamentally subject.

In this essay, it is to be shown that "Industry 4.0" not only offers tools for manufacturing industries and specifically their manufacturing processes, but that the so-called fourth industrial revolution encompasses all processes and all stakeholders because all-encompassing communication goes beyond companies. The Internet and data centers, as the informational backbone, connect everyone and everything.

At its core, this paper is dedicated to the drivers of the industry and explores questions about change, deployment factors and structures—but the processes at issue here do not end at virtual "perimeter lines" or "outer boundaries" around commercial enterprises. They extend far beyond that and encompass all organizations, administration, consumption and private citizens, utilities and infrastructures. Thus, if we consider the processes holistically ("end-to-end"), we would have to call them "total societal". However, even this characteristic does not adequately describe our interconnectedness and process landscape unless the aforementioned infrastructure is also explicitly included—or, more expansively, "all things that surround us and are capable of interconnection." To express the totality of society and "its networked things", we choose as working terms the "Integral Processes" that run on the "integral networks of people and things" (i.e., primarily the Internet). This definition also takes into account the basic consideration that needs to maximize benefits while minimizing the use of resources which are not only concerns of the economy but also apply to any organization, the administration, the private citizen, utilities and infrastructure. Integral processes running on integrated digital infrastructure (networks and data processing) promise the greatest benefit for all. Experiences and rules from "Industry 4.0" can be transferred to the aforementioned integral processes. In this respect, we must also deal with our time-typical delimitations of digitalization and networking and not view "industry" in isolation [5,6].

3. On the Embedding of Industry 4.0

After classifying the digital–informational transformation in the overall landscape of transformations, with the driving role that industry plays in the digital–informational transformation (Section 2), the following consideration analyzes the cross-society processes with the industrial processes taking place therein in detail. We begin with a presentation of the overall entrepreneurial situation. This aims to clear up the misunderstanding that "Industry 4.0" only affects the product provision area (procurement, production …). Rather, the company's internal processes go beyond this and involve all contributors to overall corporate performance, namely management, strategy, purchasing, administration, finance, accounting and human resources.

Administration: Digitalization and networking within a company encompass all activities and do not stop at administration. In this respect, administrations within companies are inseparable parts of "Enterprise Models 4.0". Nor does the concept of cross-company ecosystems end at industry boundaries or in the form of a perimeter line around industry. Rather, the hallmark of our highly developed industrial society is that the public sector with its offices, administrations and ministries is also part of the "all-encompassing network". It is common practice for companies to forward their relevant data directly to the tax authorities which use automated processes to calculate taxes and send notices—all paperlessly over the Internet. In this respect, it is time to reinterpret the term "Industry 4.0" in the direction of an "Industrial Society 4.0".

Public sector ecosystem: In this way, public administration can act as a pioneer for a data-based digital ecosystem that will bring companies and citizens the hoped-for efficiency, effectiveness and reduction in bureaucracy. A number of concepts and projects have been developed to achieve this goal. If it is possible to link this preliminary work across levels and sectors, end-to-end process chains and innovative services could be created at the interface between administration, business and civil society. Data play a key role in this process, enabling synergy potential to be tapped and innovative services to be developed. Concepts for data-based service platforms for public administration already exist. Approaches from Industry 4.0 are being taken up and developed further, for example, under the slogan "Smart Data for Public Services". Building on this, administrative ecosystems can be developed in which, for example, city halls are connected, whereby a wide variety of services can be harmonized and provided centrally at different locations [7,8]. So much for concepts and potential. Implementation, on the other hand, looks rather critical.

Dangerous backlogs: Compared to the economy, the public sector is still years behind in digitalization, its networking and the standardization and provision of its services. Germany is finding it particularly difficult to innovate because of the diversity of stakeholders (federal states, administrative levels …) due to overregulation, rigidities and fears. All our neighboring countries are further along. Approaches agreed throughout Europe, such as the "once-only principle" agreed 10 years ago, according to which it should be sufficient to give a basic personal information only once to an authority, after which all offices can access this basic information, are not implemented in our country. Optimistic programs for the economy are published at the highest political level. The implementation within their own ranks is often diametrically opposed to this. This also has a knock-on effect on the business community, which must cope with slow administrative procedures and bureaucracies that still largely work with paper and fax [9]. Recent attempts to standardize processes throughout Germany and to get the responsible administrations to work in a network have failed [10].

4. From Knowledge to Implementation

This section describes the decisive step from "advice to action", namely from the technical–scientific treatment of the digital–informational transformation with its technologies and concepts, to operational implementation. As can be shown, the decisive hurdles today lie in the choice of implementation paths and the actual operationalization. The focus here is therefore the implementation management (in contrast to knowledge management) with the further questions, where do we stand, and where are you going? While knowledge and knowledge management about "Industry 4.0" are dealt with extensively in the literature, we are entering largely new scientific territory with the challenges and solution approaches for the operational implementation of corporate transformations. We do not come across representative scientific investigations. Data are sparse and we have to rely on data published in the trade press. In this inadequate situation, demands for scientific work-up must be made, as they are made at the end of this paper. We start with an analysis of the status quo.

4.1. Political Assessment of the Current Situation

The migration of a classically functioning company to an operation with end-to-end intelligent networking of all processes [11–13], with comprehensive integration of people and machines, still poses a major challenge for most players, even if the IT background is already advanced. This applies both to the extent of change to comprehensive information penetration of the processes and to the associated resources required for conversion and risk prevention.

Thanks to the work of acatech, the German government has also recognized the importance of information technology and digital networking as crucial levers for further development of our economy and for safeguarding our prosperity. In its spring 2022 report on "Industry 4.0 for Germany as a business location", the German Federal Ministry for Economic Affairs and Climate Action stated that:

- 95% of companies see Industry 4.0 as an opportunity;
- 6 out of 10 companies already use Industry 4.0 applications;
- 91% of industrial companies see Industry 4.0 as a prerequisite for maintaining the competitiveness of German industry; and
- 75% of industrial companies believe that Industry 4.0 will reduce CO_2 emissions [14].

These opinions raise hopes and ignore implementation hurdles. We may be rich in concepts and technologies—but we are weak when it comes to implementation. Despite all the inventiveness and research funding, the "capitalization" of ideas, concepts and inventions that can be attributed to "Industry 4.0" is taking place primarily and much more quickly among our competitors, especially in the U.S. and China [9,14]. This is particularly critical for the further development of our economy because, as a result of the economic boom and low interest rates, companies were able to remain in the market that would have long since been squeezed out of the market under more difficult boundary conditions. As a result of the COVID-19 waves (at the beginning of the waves, an additional 100,000 company insolvencies were expected), there was also entrepreneurial damage which has not yet taken the form of business closures [15,16].

Projections by economic institutes predict that the pressure to transform and the large wave of insolvencies still to be expected in the medium term, which we are experiencing as a result of a boom in the economy combined with capital costs that are too low for the market, can only be compensated for by new business approaches that are closely linked to high-tech innovations, far-reaching digitalization and all-encompassing communications based on the latest infrastructure technologies (current expansion to 5G; currently, preparatory research and development for 6G in international consortia, mainly from companies in the U.S., Europe and Asia).

4.2. Diagnoses from the Business World

The dimensions of change can be mapped in highly diverse ways. Frequently mentioned are optimization of processes, flexibilization of activities, fundamental changeability, increase in customer value and minimization of the use of resources. To measure change and ensure reproducibility and sustainability, processes and products must be comprehensively mapped and backed up with data.

A more far-reaching concept calls for virtual images of real products and processes. This concept also generates concern among those affected—especially from the older generation—since they mostly come from real, tangible worlds of action and products. Resistance from the ranks of experienced plant foremen against transformation officers only erupts relatively late, when the depth of the change and the personal consequences only really become clear to the representatives of the "old world" after a series of in-depth discussions. By then, however, considerable effort and time have already been invested, which must be practically written off until the disputing parties diplomatically agree on changes in direction that can be supported jointly.

The fact is, consequently, that implementing change in the field of opposing forces will cost much more time, tie up many more resources and involve much greater risks than the preachers of change could have imagined. This also requires coping with setbacks, as is currently being reported by the chemical industry, for example.

Discrepancies: Surveys among larger companies on the international stage reveal discrepancies between high expectations and practical implementation experience in the transformation toward Industry 4.0. In most companies, there is certainly enthusiasm for transformation and ambitious plans for future investments. At the same time, however, gaps in the networking of plans and measures are conspicuous. "While digital transformation is already taking concrete shape in companies, there are lags in terms of strategy, supply chain transformation, workforce preparation and drivers for investment. Inconsistencies between theory and practice are an indication that while there is a pronounced willingness to address digital transformation, organizations are for the most part still struggling to find a way to balance the optimization of their current business with the opportunities created by technologies in the context of Industry 4.0 [17]."

Individual comments corroborate this:

(2019) "Companies tended to focus on steady evolution such as the gradual networking of machinery, a focus on cost reduction and on increasing efficiency—

rather than on a disruptive revolution, for example, in the form of complete networking and the implementation of new business models." [18]

(2020) "Although manufacturing companies recognize the importance of digitalization, there are only a few that have managed a successful digital transformation. Many companies stall in the pilot phase or struggle to achieve enterprise-wide scalability." [19]

(2020) "digital transformation is still weak at four-fifths of companies in the German economy and is thus still in its infancy." [20]

(2021) "Overall, the industries have become only slightly more digital compared to the previous year . . . . The strongest growth is recorded by tourism . . . . The strongest decline is recorded by the basic materials, chemicals and pharmaceuticals industry group. Its index score drops from 100.6 to 94.5 points." [21]

(2021) "German SMEs have so far made only slow progress with the digitalization of their business processes. Even the government-provided support programs have only been comprehensively used by around five percent of companies to date." [22]

(2021) "Even in 2021, processes in small and medium-sized enterprises are still characterized by paper, makeshift solutions, distributed IT tools and a lot of manual intervention . . . . This means that companies not only give away competitive advantages, but sooner or later even jeopardize their own ability to survive and leave their employees intellectually drained". [22]

As the quotes make clear, the discussion essentially focuses on the concept of digitalization, but not on comprehensive networking, which—according to the creators of the term—represents the core of "Industry 4.0" [23–28].

All-encompassing networking, as the defining criterion of "Industry 4.0", was made possible by the Internet, which is capable of connecting all people, all things ("Internet of Things") and all organizations through unique addresses. The basic element is the cyber–physical system (CPS, see Figure 2), which is composed of mechanical components, electronic components and informational components (hardware and software) and includes the data interface to the network. These prerequisites distinguish this element from its predecessors from the previous industrial generation "3.0", which (like the CPS) have the basic interfaces for energy supply as well as sensors and actuators, which therefore characterize an automaton (robot . . .) even without integration into a higher-level communication network—just functioning autonomously [29].

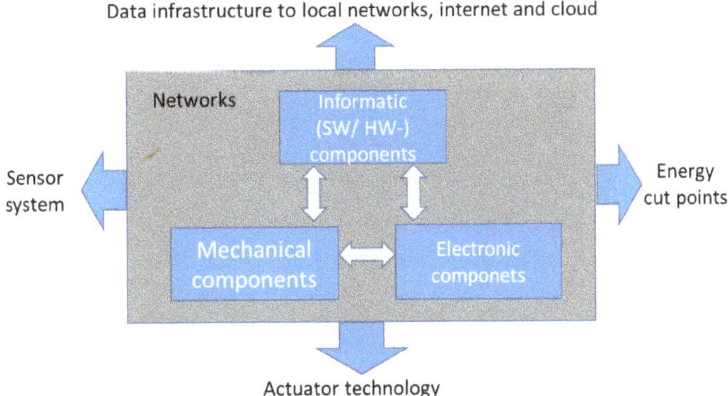

Figure 2. The basic cyber-physical system (CPS). Source: Kai Lucks, Der Wettlauf um die Digitalisierung op. cit. S 202.

5. Industrial Implementation

Building on Section 4, which explains the actual implementation management, i.e., the industrial transformation as such in the context of the preceding technology and knowledge management, Section 5 can now focus conceptually on the purely industrial transformation of "Industry 4.0". We will deepen the definitions given in the introduction to this paper on transformation in general and on "Industry 4.0" in particular. In doing so, we direct our attention to the inter-entrepreneurial networking that is particularly under discussion today, which is closely linked to the dissolution of entrepreneurial boundaries that can be observed today. Then, we go into the main challenges and levers.

5.1. Definition and Basic Understanding: What Does "Industry 4.0" Entail—What Should It Encompass?

"Industry 3.0" characterizes the development thrust of the years from 1950 to 1960 with the resulting implementations [9]. Industry 4.0, on the other hand, refers to the phase toward all-encompassing networking, starting with the spread of the Internet, the foundations of which were laid from the mid-1960s [9]. In today's common linguistic usage, the two development phases are smeared together, and the term "digitalization" is used as a quasi-generic term—as is the case in particular with the German federal government, which mostly speaks of "digitalization" in its presentations, although our current industrial–social upheaval is much more strongly characterized by end-to-end communication encompassing all protagonists. Digital technology in the infrastructure (hardware and software infotech in data centers and networks . . .) provides the technical basis for this. One must be careful here: we may not speak of digital basic technologies if we refer to the informational social transformation.

The networking of companies with each other is to be understood as any form of informational and operational collaboration between companies and thus to be backed up with recommendations drawn from the implementation of "Industry 4.0". This consequently includes:

- Service chains, for example, from the supplier via production and assembly operations to the logistics provider.
- Competitive relationships: opposing in direct competition, cooperation in associations, committees and across organizations such as chambers of industry and commerce.
- Inclusion of service providers at every stage of the value chain and for all processes.
- Service providers and infrastructures for data and communications technology, data hosting and processing (such as cloud, fog and edge computing).
- Dissolution of boundaries between companies, which are increasingly acting as a combination of manufacturers and customers, so-called prosumers or Xsumers (Xsumer stands, for example, for consumers who step in as manufacturers when demand peaks, such as when electricity generators are switched on by the grid operator via photovoltaics), for example.
- Digitalization of industrial projects with their processes, which are simultaneously backed by networks. Example: Company mergers as the most complex project approach in the economy [12,13].
- Emergence of so-called digital ecosystems, where every emerging market niche in supply and demand is occupied in a short time, both by diversifications of existing players and by new entrants such as startups, stationary and online-based founders.
- Permanent, temporary and regional forms of entrepreneurial cooperation such as consortia, project companies, purchasing alliances—mostly without capital backing.
- Capital-backed forms of entrepreneurial mergers such as joint ventures.
- Forms of cooperation between business and the public sector, such as public–private partnerships (PPP).

5.2. Levers and Challenges to Digital–Informational Transformation

Important levers and challenges to digital–informational change are discussed below.

Culture change: The digital transformation of organizations requires significant and often painful behavioral changes from those affected. Entire "operating models" (processes, setup, networking, competence management …) are affected as new forms of collaboration and leadership need to be implemented in organizations to keep up with the competition of change. Sustainable changes as well as new demands on strategies, technologies, people and processes require more dynamic and flexible tools to manage, evaluate and track progress of transformation. This should be carried out step by step, with milestone controlling, and iteratively to practice the new ways of doing things. Indeed, experience shows that changing behaviors that add up to a cultural transformation is particularly time-consuming (time, cost …), and that the risk of falling back into old behavior patterns and of reviving old insider relationships is extremely high.

Strategies: The message of the importance of digital transformation has been received by most companies. Most executives state that digital transformation is one of the most important strategic goals in their organization. However, this does not necessarily mean that they are fully exploring the strategic opportunities that digital transformation offers. Indeed, surveys revealed that around two-thirds of executives see the transformation to "Industry 4.0" merely as a means of increasing profitability [30].

Degree of digitalization: Digitalization is gaining massive importance in companies. More than half of the companies report that responses to the COVID-19 pandemic have brought a significant boost in digitalization. Three-quarters of companies are convinced that companies with a digitally driven business model are in a more stable position and state that companies that have already digitized their business processes will better deal with the COVID-19 pandemic [30]. However, if companies continue to focus only on how digital technologies can accelerate digital transformation, cross-functional collaboration is likely to fall short. Entrepreneurs who have successfully implemented digital transformation, however, rank low-friction inter-functional collaboration as elementally important. Alongside efficiency and productivity, this is becoming an increasingly important barometer of success, especially in economically difficult times.

Degree of networking: Industry 4.0-oriented strategies often do not yet fully target the potential of networking. Capabilities to bundle information from interconnected assets and use it to make informed decisions are critical for full implementation of Industry 4.0, but many organizations are not yet able to fully realize this competency in practice. Referring to the fundamental need to restructure management structures—in a departure from general staff planning from headquarters and enforced by the executive board down to the regional units—reference has already been made to the enormous potential offered by IT-backed digital networking, with its fast bi-directional communication options between headquarters and the periphery. As a result, regional units and special business segments, for example, gain many more opportunities to implement their ideas on management, which, after all, know the needs of their particular customers much better and can consequently deviate from central specifications according to the "fit for all model".

Internet of Things: All-encompassing networking—as mentioned above—includes networking of things. The Internet of Things (IoT) has become a reality in industry and the consumer sector, revolutionizing the entire economy. More and more Internet-enabled "things" such as parts, components, plant areas, finished products and resources are providing automated, efficiency-enhancing control and optimization of manufacturing and logistics processes. "Smart" working factories in Industry 4.0 produce faster and are more resource-efficient, more flexible and scalable.

Digital ecosystems: In a global and highly competitive environment, companies no longer operate autonomously. Instead, they are becoming part of complex, networked and growing ecosystems in which there should be permanent cooperation with the best and most innovative partners. All of this requires new ways of thinking and working: agile and open, flexible and forward-looking. Reference has already been made to the internal digital-backed ecosystems to be formed within the company.

Personnel: Executives are generally confident that suitable personnel are available to shape digital transformation—but they acknowledge that the personnel issue is an ongoing challenge. In fact, only a small minority of top management of internationally active companies see the need to fundamentally change the composition and skills of the workforce in the course of an Industry 4.0 transformation. At the same time, however, the executives rate finding, training and retaining suitable personnel as the greatest organizational and cultural challenge [31].

Organizational development: Traditional inter-functional competition within individual companies inhibits growth and works against the overarching goals of digital transformation. Therefore, digital transformation is also primarily about breaking down functional silos to collaborate better and to be able to drive innovations with higher pressure. Research shows that higher digital transformation investments have seen significantly higher revenue increases than their more conservative competitors. Such "champions" invested 1.1 times more (20.5 percent of their total revenue) than others in digitally transforming their functions. As a result, they achieved twice as much revenue growth as the other companies: 23.7 percent compared with 10.3 percent [31].

Hierarchies: Surveys typically reveal breaks between corporate management and downstream management levels. For example, the management level, which is responsible for managing day-to-day operations, often has little say in the fundamental design of processes. However, during digital transformations, these are quite decisive for success. With increasing informational transformation, hierarchies are becoming less important [32]. Digitalization requires flat hierarchies in which network-like work structures in particular can be implemented. The order of the day is teams with clear role assignments that are not thwarted in their effectiveness, speed of decisions and implementations by hierarchical behaviors [33].

Cross-functional collaboration: It would be a misunderstanding to look for networking under "Industry 4.0" only in the IT infrastructure and the software solutions running on it. The new industrial generation is also based on new forms of interpersonal collaboration as well as on the IT-backed possibilities of human–machine communication. However, cross-functional collaboration is still difficult in many companies. In most companies, different business functions (such as R&D, engineering, production, marketing and sales) are still competing with each other instead of driving the IT transformation forward in a unified and seamless way.

Cost drivers: The problem of functional specialization, with little development of cross-functional careers, which can be observed particularly in Germany (so-called silo structures), is a burden on the informatics transformation. It impairs sales and the success of operational expenditures. This was also reflected in a survey conducted by the management consultancy Accenture for the DACH region:

- Competition between different functions in companies causes superfluous investments in digital projects. In the period studied from 2017 to 2019, actual costs in DACH companies increased by 4.4 percent.
- The digital investments made by function leaders should increase the company's revenue by 12.9 percent annually. In fact, annual revenue increased by an average of 4.2 percent from 2017 to 2019—only one-third of the expected revenue growth in DACH companies.
- According to three out of four DACH companies (74 percent), digital investments do not increase revenue growth [30].

Innovation pressure: With global networking, access to the Internet, and the lowering of market entry costs via online activities, the pressure on national industry players to perform has increased significantly. This is because national market barriers protect them less and less against companies that attack with innovative service offerings from abroad. This is the critical side to competition. However, anyone who makes full use of the opportunities for digitalization and networking in conjunction with smart data technologies, for example, by virtualizing products and processes, can accelerate their product delivery process by a factor of two to four and reduce costs accordingly. This can take place by working on a virtual twin of the future "real" product and can also be reflected in localized, relatively freely "movable" manufacturing value creation. This can be achieved, for example, by having globally distributed design partners and production service providers who have access to the digital twin via the network. Their designs and productions can be used on a demand-driven basis and in compliance with regulations across the globe. Manufacturing modules must be connected with each other and meet regulatory requirements for compliance with technical standards and for localization of value creation [17]. IIoT [34] applications, as the central core of Industry 4.0, play a fundamental role for digital transformation in this context.

Supply chains can be understood as typical extensions of internal company process chains in value generation. Thus, they should also be included as elements of "Industry 4.0"-based management models. The digital transformation provides decisive impetus for the efficiency of the supply chain. As a result, the interaction between the procurement departments of suppliers and manufacturers is gaining momentum. Thus, the implemented ERP systems should be put to the test, for example, whether they still meet the requirements of production plans, quality criteria and budgets of the purchasing departments in the future. The goal is a fully networked supply chain with which the production status can be viewed in real time at any time [35].

Merchandise Management Systems (MMS): control the flow of goods in terms of quantity and value and can thus be designed as complementary solutions for supply chain management. Purchasing, sales and warehousing must be integrated. A digital MMS offers operational improvements and savings potential through links with finance and accounting, human resources and marketing. To this end, all relevant employees must have access to a common MMS data pool, which they can use to track changes in the flow of goods and values in real time. Thus, a digital MMS is a solution approach that can fulfill the requirement criteria according to "Generation Industry 4.0" [36].

6. Approaches to and Experience with Comprehensive Digital–Informational Transformations in Industry

Following the challenges and levers for the operational implementation of the transformation to organizations according to "Industry 4.0", explained in Section 5, we take the next step in this section by investigating the question of what the status of digital–informational implementation is today and how it is about the opportunities and barriers to further implementation. We explore this with individual examples. Above all, these shed light on management models and international contexts, which are crucial for successful implementation. A generic framework model for the transformation of manufacturing companies towards "Industry Generation 4.0" is presented for further discussion. In this context, the typical paths for the implementation of "Industry 4.0" are also presented.

As explained, entrepreneurs' own assessments of the status of their level of digitalization and their ideas about what still needs to be achieved are strongly subjective. Since companies use IT equipment and, as a rule, accounting software, they all believe they are "already somehow in the digital world". The conceptual impurities in everyday language usage encourage this. For example, the blurring between "digitalization" and "all-encompassing networking".

6.1. Chances and Limits for the Reorganization of Entrepreneurial Leadership Models

One of the classic leadership conflicts that a board of management faces is the question of the power with which it wants to and must enforce its fundamental experience and ideas on the management of all business and activities in the business units, right down to the periphery (e.g., regional units, business segments). Or, thinking the other way around, how much autonomy (disobedience?) does it concede to peripheral activities in shaping their businesses? After all, these could have good reasons to deviate from the company's "standard model" because their business partners have different expectations than, say, the "standard customer at the head office". The resulting leadership dilemma must be recognized and endured by a demanding corporate leader, who must then weigh which direction to take, not just according to one of the two polar basic models, but varying it according to the specific needs to be met.

Reference has already been made to the connection with the "Industry 4.0" approach because decisions made by the management board and executives are directly dependent on each other and require the closest possible bidirectional coordination, which can only be achieved with the help of the latest communication and IT systems. These, in turn, must be embedded in the planning and controlling systems of the company concerned, including, for example, the aforementioned merchandise management system.

Regional peculiarities naturally increase the complexity of planning and controlling processes enormously, especially when one considers that regional peculiarities can be found at several locations and each time in a different form. These would also have to be mapped in higher-level IT systems, unless the corporate headquarters decide that deviations are only to be recorded gradually at headquarters and are to be managed from there.

In addition to the need to keep the complexities of management and controlling low, there are other reasons for (gradual) decoupling the periphery. These include economic policy circumstances, such as tendencies toward deglobalization. In concrete terms, this can result in the decoupling of entire regional markets and countries from the international network. This may be due, for example, to local content requirements, political risks, national technology standards or local purchasing regulations.

Despite all such barriers, internationally valid regulations could take effect, such as the Supply Chain Due Diligence Act, according to which a company based in Germany must also ensure responsible management of supply chains for its local subsidiaries abroad and compliance with human rights, safety and protection of employees also in the locally supplying plants. The fact that this also requires the collection of a great deal of data and that the associated processes should obviously be carried out in accordance with standards that apply uniformly to all regional units worldwide is imperative for reasons of the company's legal responsibility for all its activities and for the economic application of the procedures [37].

6.2. Dissolution of Boundaries

As shown, the currently observable dissolution of corporate boundaries must be reflected in the fact that the appropriate "dissolved" corporate models must be included for the respective case.

The most important ones are listed here again in the form of buzzwords: digitally driven ecosystems, data-driven services from procurement to the point of sale, including forecasting from the purchasing market to consumer behavior, online business models (B2B, B2C …), new solution businesses (which can displace product manufacturers from their direct relationship with their customers), role changes (such as manufacturer vs. agency) and function changes, such as between producer and consumer (prosumer, xsumer …).

The involvement of third parties, from whatever organization, plays a special role as a driving force for the innovation capability and flexibilization of the respective company. As shown, stability and change are the two poles between which a company must constantly recalibrate itself. The CEO who makes things easiest for himself is the one who always pursues "business as usual … " and thus falls prey to the assumption that his/her company will remain stable. Especially in times of great change, and when new business models driven by Industry 4.0 are being pushed forward, the effect can go exactly in the opposite direction: the market and competitors change, and the company in question loses touch in its seemingly stable position and goes down.

Studies have shown that in-house strategy analyses whose results recommended "continuing as before" were followed in most cases, while studies that concluded that fundamental changes in direction were recommended were factually questioned in large numbers and consequently failed due to internal resistance. (The author, for example, had carried out such investigations in the context of his assignments at Siemens AG and was able to prove such acceptance and implementation resistances.) Navel-gazing and inside views typically obscure rather than reveal new findings, sometimes dramatically. (An example: the Siemens communications division, which required a strategic market analysis after German reunification, provided a negative example of a misleading national internal view. To this end, the leading market supplier Siemens teamed up with its market-dominant customer, Deutsche Telekom. The "consensus result" of the two partners ignored the emerging Internet-based package switching technology. The U.S. competitor Cisco prevailed on the world market with this technology and Siemens had to withdraw from the communications industry.) In this respect, it is particularly important to install early warning systems, to know early alerters and to listen to them. Such "alerters" are most likely to be found in companies positioned quite differently in the (relevant/related) market and pursue highly innovative concepts. For this purpose, it is advisable to set up appropriate networks staffed with "external" people. These can also be researchers [38].

6.3. Modeling Informational Transformation

In the following, a generic model framework for transformations to "Industry Generation 4.0" for manufacturing companies is presented for discussion. This is an attempt to place the paths to this transformation in an operationally oriented system.

A management cockpit consisting of six instruments is used as a basis for steering and controlling. These instruments are used in parallel:

(I)   Scope and Goals [39]

A project for the transformation to "Industry Generation 4.0" belongs to the category of fundamental corporate restructuring projects that encompass process organization, organizational structure, competence management, leadership structures, (hard) action programs and cultural change.

Before deriving the goals, the perimeter of the company under consideration must be defined, because size, business breadth and depth of value creation define the earnings potential. Uncertainties in this determination and subsequent changes must be eliminated or corrected during the ongoing work process. (Experience shows that there are often serious omissions here because corporate activities are not recorded correctly. For example, non-core business and regional specifics are often not recorded.)

Before the start of the project, the objectives must be derived based on an outside-in analysis, i.e., in concrete terms, on the basis of a dynamic analysis of the following environment:

- Market today (customers, competitors, suppliers …).
- Drivers for change (technologies, behaviors, market definitions, new players, mergers, fundamental power shifts and changes in the environment, e.g., environment, energy, new evaluation criteria such as ESG (Environmental, Social, Governance) and new evaluation methods [40]).
- Extrapolations to a thus "dynamized" future scenario for the period in which the new corporate model is to be applied (i.e., determination of the period, e.g., time horizon 5–10 years, determination of the dynamizing factors with their size and insertion of the same in the calculation of the future scenario).
- Scenario-based determination of the forces and market positions of leading players in the future market.
- Derivation of the own forces and market positions to be achieved in the dynamized future scenario.

(II)   Business Plans

Dynamized top-down objectives in milestones according to time horizon, for the business activities defined according to the perimeter. Nota bene: This can also be used to specifically exclude non-relevant activities. (Hiding is useful where management authority has been relinquished, e.g., outside the scope of consolidation, contract manufacturing based on external specifications, joint ventures in minority positions, support functions performed by third parties.)

Segment plans, where appropriate, each with objectives as above, so that the segment objectives can be consolidated "bottom-up" to the overall objective mentioned above.

(III)   Action Plans

Action plans per segment and as needed below. This includes the action plan, consisting of "hard" (thus also financially assessable) measures.

(IV)   Culture and Communication

So-called "cultural alignment", which means motivating and driving behavioral change at all levels and in all functions so that the action plans to be developed ideally become self-perpetuating. This includes a high degree of identification of the participants with the goals, with company-wide and cross-functional overall goals taking precedence and functional sub-goals and support goals being put behind.

Any cultural change requires preparatory and follow-up communication at all levels and in all sectors. In all cases, this needs to be multi-directional, i.e., top-down versus bottom-up and across the organization in different directions. The respective "communication campaign" is not considered complete until all stakeholders have agreed. Forms of unalterable digital documentation should be found for the approval and person-specific commitment. The priority is to achieve the highest possible quality of the goal and the consensus. The time and effort required for this are secondary. In this respect, the time intervals between the defined milestones should be flexible. Priority is given to quality, consensus and securing commitment (documented voluntary commitments).

(V)   Teams and Terms

In terms of the transformation discussed here, this is the most important instrument of the cockpit presented. Figure 3 presents a highly simplified "generic" picture of a company organization according to its structure, core process, management structure, value streams and data streams.

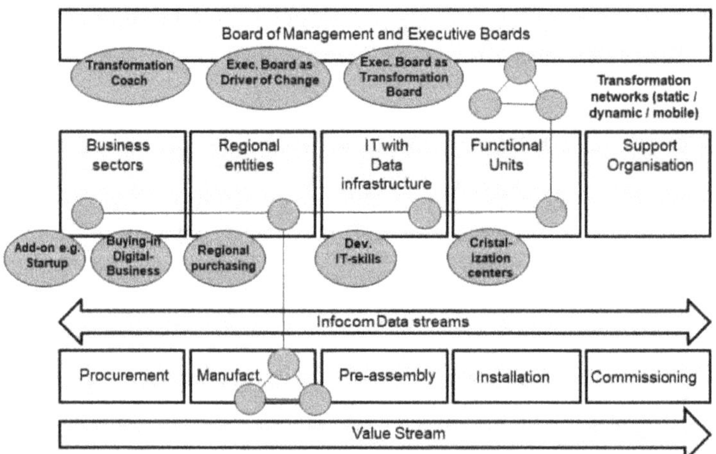

Figure 3. Generic paths for transformations to Industry Generation 4.0.

Preparatory fit of the organization: The circular shapes indicate the levers to be typically applied for transformation. This is to be determined and combined differently from company to company. As already mentioned, companies with a lower degree of verticalization are better suited to implementing transformations. Consequently, organizations that are as flat as possible are to be preferred. If there is a need for preparatory action in this regard, opportunities for horizontalization should be sought, such as the juxtaposition of organizational units that were previously nested. The resulting increase in management span of top management with the corresponding increase in work and presence in the company can be compensated for by hierarchical mergers at the top levels as well, such as harmonization of the executive board and management. Certain differences in rank and function should be regulated in employment contracts, but these should not permit renewed "verticalization through the back door". Rather, the top management team should be emotionally and contractually bound to each other by a common "mission & vision document".

Disruption management: Disruptions can occur at any time, but they should be resolved to the best of their ability in the processes of struggling for the new. In the case of an accumulation of disturbances, however, it can become unavoidable that disruptors and disruptive structures (also insider relationships . . .), which oppose the agreed change, must be eliminated before too much energy is wasted and perhaps even the agreed overall goal is corrupted. For this purpose, rules and instruments must be kept ready in advance to lead such forces out of the "organization of change" without loss of face and injury.

Crystallization cores: The circular shapes shown in this figure localize most of the drivers of change to be found in today's world. These can be used in almost any combination and can also be reconfigured over time as needed. There is a lot of individual experience available on which crystallization cores were used in which cases, in which form and how successfully. The specifics of the reported cases, however, only reveal certain measures of generally valid rules for success. Interestingly, these experiences are similar to those made in completely different types of reorganization projects:

- The more far-reaching the conversion, the more difficult it is.
- The more far-reaching the conversion, the higher the increase in value and the longer the positive prospects.
- Project types based on experience patterns and the more common ones have higher potential for success and involve lower risks.
- The more dissimilar the partners to be brought together, the more difficult and the higher the risk of a later rupture. (These are transferred results after evaluations at Siemens AG from strategy projects at the corporate level and mergers and acquisitions (M&A), for which the author had worldwide responsibility until 2008.)

The forces for change described as "crystallization cores" in the figure are briefly characterized below.

Add-on, e.g., startup

A typical path to change is to merge with a startup. The main reasons for this are found in activity types (overlaps/additions/complementary competencies . . .) for resource bundling (access to funding for startup/access to competence bearers, especially to IT staff for the company to be transformed/strengthened to improve the position in the race for new developments vis-à-vis the competition . . .) and for cultural learning (learning promising behaviors and work structures from each other/exploring stress limits through differences/developing migrations to new and promising work models . . .).

In most cases, this model is applied in the form of test projects and individual projects which, in the best case, turn out to be promising and can be "attached" to the existing organization. In the optimal case, this model can also be transferred to the wider organization of the company in transition and penetrate the entire organization more or less deeply.

Experience shows that the proportion of such pilot projects that lead to sustainable success is relatively low, because working methods, motivations and the forms of incorporation into the organization do not fit. A central reason for failures are "imposed" goals and milestones that correspond to the basic pattern of the parent company but do not meet the expectations of the "young entrepreneurs". (A few years ago, the author was called in to assess M&A projects by a leading German IT group. Out of 10 "classic" acquisitions, all 10 were classified as "successful". Of 10 mergers with startups, all 10 were classified as "not promising". Here it turned out that the success criteria for the startup collaborations were borrowed from classic M&A and were therefore inappropriate. After changing (fitting) the criteria, half of the startup cooperations could be classified as "promising in perspective". At the same time, the time horizons for the necessary migrations had to be extended considerably.)

The "hype" about startup mergers with medium-sized companies has died down in the meantime. In contrast, successful direct transfers of startup founders and employees to medium-sized companies are more common. In addition to the rather weak success rate of this migration model, the time required to penetrate an entire company is high, so this model is only suitable as a pilot and as a supplement to a top-down approach to managing change [41].

Buying-in digital business

This describes the "classic M&A path", namely the acquisition of a digital-driven company by a strategically operating company with a long-term focus (so-called corporate, i.e., less applicable to private equity, which is usually already planning its exit at the time of the acquisition).

Here, the typical M&A processes and work stages come into play. The most important is the entry into the project with the help of a candidate screening, which in the field of digital business approaches is extremely broad, time-consuming and highly updating. Thus, M&A databases can contribute rather little. Most of the data gathering and analytics must be executed in a timely manner by in-house teams. Frequent buyers generate their own target data beacons, which must be constantly updated and adapted to the changing target search. The number of candidates to be captured in an international search can range in size from tens to hundreds of thousands of potential companies.

Integration usually follows the "hang-on" model in the existing organization. Because of cultural and national differences, such integrations are not easy and are "lost" if the hang-on is too low and if management pays too little attention.

A special case is when, due to the low availability of free IT specialists on the labor market, entire IT companies are purchased to then completely dismantle them and assign the individual employees to various organizational units. In terms of costs, this can pay off. However, the risks of dismantling are high, legal hurdles (e.g., § 613a) must be considered and the risk of loss of employees who feel deprived of their colleague network is high.

Regional acquisition

This is another M&A path, directed at a target region where digitally driven business models are more entrenched and where a suitable pool of companies can be found to choose a target. Strategically, this path would be classified as "diversification or business expansion on a regional level" and typically applied in the Far East, for example.

From such a newly attached unit, which can also be launched as a "trial balloon" according to size and orientation, a new core business can be gradually developed—as in an upgrading process—which even has the potential to substitute older business approaches.

Such a path can turn out to be tolerable for the company because initially it is sufficiently far away from potentially threatened established businesses in the parent company's region and because such a business can develop well under special "digital-ecological" circumstances. Once this has reached a certain size and a certain threshold of earning power, the proof of success is also given to critics within the company. Once this model has proven itself, further attempts can follow.

Developments ex IT competencies

Another direction of development is offered by IT competence centers. These can be close to the CIO, more likely to be assigned to IT administration (e.g., for master data), to hardware and software development or to IT service providers. In all of these cases, the key lies with the IT specialists already embedded in the company, who (a) on the one hand, bring specific competencies with them that experience in recent years has shown to be suitable for new management approaches, and (b) are most likely to be able to build bridges between "old" (knowledge of and understanding for outdated business approaches) and "new": through their own visions and ideas, as well as through their connections to developers and founders.

The specific use of this resource enables simultaneous "Industry 4.0" penetrations of companies at various points—including the possibility of using such IT competence carriers specifically as ambassadors across the company.

Crystallization centers

In addition to IT, there are many crystallization centers that are suitable for highly innovative business approaches. These include permanently established specialist and staff departments that are predestined for innovations due to their degree of specialization. Temporary project organizations are also suitable—especially if they bring together different competencies from different organizational units, for example, for business development, investment fields, venture capital and basic research. These are particularly noteworthy when it comes to completely new approaches to solutions. These often hold the key to business innovations. They can be triggered by the overall organization committing to higher innovation frequency and having the appropriate tools in place to do so (e.g., measurement at departmental levels, innovation competitions, publication of benchmarks, bonuses, team building across the organization, linking top-down to bottom-up initiatives …).

Board and management levels

The aforementioned solutions can be used individually or in combination, triggered by senior management, at workshops in factories or through the operational improvement system. Ultimately, the lead-up to an "Industry 4.0" organization is the responsibility of the top management levels. Depending on the size and breadth of the company's setup, this means the board level plus its managing directors as the second and those responsible for the operational business (divisions, regions …). Because of the many special transformation tasks, this team should be of a certain size, conceivably at least 5 to 15 people, to be able to assign the necessary responsibilities personally. The operative management level must ensure access deep into the organization and thus also prevent (!) that detached decisions are made far away from the business.

Ideally, a coach is appointed for this purpose and assigned a corresponding initiative and leadership role. The CEO can "take" this role (in accordance with management regulations). The entire executive board can dedicate itself to this task as a "top team" and also appoint a "transformation coach" who will be actively and continuously involved personally and at the same time serve as a central contact person in the company. Because of his/her professional qualifications, it makes sense to entrust the chief information officer (CIO) with such a task. The CIO should be a member of the top-level transformation committee, whose members work together as colleagues on an equal footing.

Network organization

The top team presented, consisting of the aforementioned management levels, should see itself as the driver of a network for implementing the transformation, which is placed on the company as a flexibly positioned "overlay organization". This network must permeate the entire enterprise by having its named members drive transformation workshops and other initiatives at all levels of the enterprise, in all business segments, at all stages of the value stream. The named (primary) network members ensure that working groups, workshops and other activities occur. The "primary circle" should be allocated a contingent of their working time in which they dedicate themselves to the transformation task. Additional members should be brought in so that larger teams and initiative groups can be formed and meet regularly.

The transformation network must link all transformation activities with each other and can also present individual of the presented paths/models (e.g., startup cooperations) as examples for discussion.

As a brace function and for overarching cohesion, regular town hall meetings for the entire workforce should be scheduled by the management level. A milestone program could provide the clocking for this.

All channels and media of communication should be used; in addition to the face-to-face meetings mentioned above, in-house newspapers, programmatic notices in plant halls, an intranet site, Q&A forums, videos and podcasts are all suitable for this purpose. A special editorial department should be created in the communications and press department for this purpose, in which all productions are coordinated, timed and their content coordinated.

A comprehensive and coordinated stakeholder management system must be developed. This covers not only all internal forces, friends and families but must (!) also be directed to the business partners (customer, suppliers …) and to the general public (press, local representatives, politics, …). Depending on the size and importance of the company, higher levels (district associations, federal state, federal government, international alliances) should also be addressed.

This is not just about informing the immediate or wider environment, but to a much greater extent about a kind of reflection: employees want to know how "their" company is seen from the outside. A positive external image reinforces and confirms their commitment, and they feel appreciated as fellow activists. Pride in one's own company is an important motivator.

Highly differentiated time management also contributes to success. Messages must first (!) be exchanged within the respective innermost circle, then across the breadth of the company, and only then (!) externally—possibly in cascades to meet external expectations. For example, the mayor might want to be informed about important information firsthand before the department heads are involved. All this can be exercised with military precision with hourly cycles and under role allocation (e.g., who contacts who on the board?). For this to be safe and repeatable, the processes should be standardized and backed up with IT.

Evaluation

The model presented here should be understood as a generic framework and adapted to each individual case. A fundamental transformation belongs to the most demanding and highest project category that a company can undertake—on a par, for example, with a full takeover by and merger with another company. The change can lead to a kind of "rebirth". The highest level of attention, the most precise target management, high commitments, the involvement of everyone and the use of a complete set of instruments are called for.

7. Comparison: Empirical Values and Recommendation

This section supplements the operational side of the digital–informational transformation discussed so far with its financing and the management of risks. Because with every fundamental reorientation of a company, special risks arise as a result of temporary uncertainties. After all, the entrepreneurial principle of maintaining the balance between stability and change must always be observed. Stability is often understood as maintaining the status quo. However, in times of great upheaval in the environment, this can pose existential threats to a company. Too radical change can cause losses of corporate identities, employees and established customer relationships. Therefore, even so-called disruptive changes should be approached with caution. The timing is crucial: both when fundamental changes are initiated and at what speed they must be implemented. As shown in Section 6, various generic management models are available.

Potential for improvement: Digitalization in conjunction with networking undoubtedly unlocks enormous potential in terms of cost savings, acceleration and avoidance of work that does not create value. In a recent survey of European companies, most cited increased efficiency as an important benefit resulting from the use of innovative technologies in production. European decision makers also cited potential cost reductions and improved product and service quality as other benefits.

Hurdles: When asked about reasons for the progress of transformation to date, the majority of decision makers in all European survey countries cite the factors of time and money. This assessment cuts across all company segments and is mentioned in every third survey. Around one in four also cite IT security concerns, incompatibility with existing machines and data protection regulations as major challenges in their own company transformation.

Costs and pricing: Motivated by the realization that they cannot bring Industry 4.0 into their own organization on their own due to a lack of expertise, budget or capacity, European company leaders are consciously relying on partnerships. When selecting technology providers, flexible and simple pricing is the most important criterion for one in five companies. The focus is on the ability to flexibly adapt services to specific needs at comprehensible prices. Also particularly important are the aspects of security and the provision of advice and support as well as professional services.

Return on investment expectations: Organizations that focus on innovation or internal transformation report increases in their return on investment. Companies in advanced stages of transformation had invested just over one-fifth of their total revenue in digital projects. The result was profitable growth. Their EBITDA (earnings before interest, taxes, depreciation and amortization) grew by 12.2 percent from 2017 to 2019. By contrast, the EBITDA growth of the other companies in the DACH region shrank by 6.2 percent [41]. Cross-functional collaboration is neither a goal nor a means to an end. Rather, it should be a key organizational and strategic imperative for leaders who want to drive digital transformation. Effective collaboration across functional boundaries not only reduces unnecessary effort and costs—it also leads to measurable improvements in returns [33].

It is almost impossible to put a general figure on the potential savings from the transformation of companies. Nevertheless, individual consultants dare to make overarching statements. According to Accenture, "The optimal technology mix could save large companies up to \$16 billion. And yet, only 13 percent of companies have realized the full impact of their digital investments, achieving cost savings and creating growth [41]."

Rules: Companies that have successfully restructured according to Industry 4.0 criteria differ significantly from their non-transformed competitors in several ways:

1. They make it clear to function holders what digital transformation means for the organization and why all functions should work together.
2. They hold executives accountable for how well they collaborate across organizations in digital transformation projects.
3. They prioritize projects that foster cross-functional collaboration.
4. They invest in platforms that enable seamless collaboration and scale them rapidly. IT island solutions have become a thing of the past.
5. They define clear rules for their information technology and operational processes and make transparent how the two mesh [41].

Risk assessments: The implementation of "Industry 4.0"-based leadership and management concepts to date revealed that the problem lies not in recognizing the benefits and not in the lack of "golden rules" and the lack of advice disseminated via relevant institutions, via ministries and consulting firms. Rather, the crucial problems lie in operational implementation, from the board to the store floor. In addition, it can be observed that mistakes and dead ends that companies run into lead to disappointment and frustration at a variety of levels and across organizational units, so that companies abandon the digital transformation due to exceeding deadlines and budgeted resources, even after part of the implementation process, and even row back when they have to recognize that there are signs of division in the company, that values are being destroyed and that the operational business is being jeopardized.

Two-level approach: For this reason, a fundamental set of rules for implementing industrial transformation is recommended. Progress must be measurable, risks and threats must be perceived at an early stage, and ways and means to counteract them must be constantly identified and, in the event of dead ends, redeveloped in a timely manner as a substitute [42]. Thus, we need ("**Level 1**") a **competence group from industry practice** to advise the federal government specifically on implementation issues. This must be closely interlinked with ("**Level 2**") the companies in industry. For **companies in transformation**, the basic rules can be summarized as follows.

Basic recommendations

I.   Appoint a board of management that acts as a coach at the top management level to drive the digital transformation forward and is responsible for its success.
II. This person also acts as a top decision maker for organizational adjustments, with the aim of getting the most out of the relevant investments.
III. Sustainability takes precedence over short-term effects.
IV. Prioritize projects that promote cross-company collaboration: across organizational units, involving different functions, and along the entire value chain.
V.   The aim should be to harmonize the various technology platforms that are to be introduced and stored in the cloud.
VI. To achieve optimal results, it is necessary to ensure that the various platform protagonists cooperate optimally by practicing close exchange.
VII. Develop cross-activity standards for IT and operational governance.
VIII. Their consistent application must always be checked and thus ensured in the long term.

8. Conclusions

This section provides, in addition to preliminary considerations "where do we come from", a perspective (i.e., limited in its temporal preview) of "where is it going (probably or with the best will in the world) to". We are pursuing an exploratory approach here, because scientifically sound process descriptions require that the processes have been experienced, are completed and can therefore also be scientifically recorded in toto. However, if we find ourselves in the midst of current development processes, as here, then, strictly speaking, these are "open-ended" and cannot be precisely predicted until they are completed. Thus, at least in part, they elude a scientifically verifiable description. This presents itself as follows:

> "The traditional value chain is evolving toward hyper-personalized experiences, products and services, driven by innovative business models and new revenue streams [42]."

Just as the transition from digitalization ("Industry 3.0") to comprehensive networking and automation ("Industry 4.0") was a smooth one, further development will take place in a kind of progressive process, with special thrusts that can probably also have a disruptive character.

New technology generations always include entire waves of fundamental developments and inventions whose appearance is often difficult to predict. Some of them lie within the shorter horizon of expectation, such as in the energy sector with new battery generations. Others we seem to be pushing ahead with a constant lead time, such as economically viable energy conversion via nuclear fusion, which is not expected until beyond the target horizon we have set for implementing the energy transition (2050 . . .). How quantum computing, neural networks, machine learning for higher forms of artificial intelligence will be reflected in the industrial value stream remains unknown.

Nevertheless, the purely technological potential can be predicted quite well on the basis of fundamental knowledge and the statistically expected variety of new solution offerings. On this basis, former Chancellor Merkel was able to say time and again, with some justification, "We'll manage (meant: somehow)".

However, our challenges lie above all in implementation, in the ability to generate value from new concepts, ideas and technologies: cash, business value, job value (through sustainable ability to pay salaries) and value for the state community (in the form of tax payments)—all of which, after all, can be traced back to entrepreneurial activity. This is specifically where our German crux lies: we must compete with the leading industrial nations such as the U.S., China, Japan, South Korea and a few others, while also establishing new global leadership positions. "Industry 4.0" is the main battlefield. If we fail here, we will not be able to catch up with our numerous backlogs. We will not be able to pay for the ecological energy transformation and will not be able to close the COVID-19-induced gaps and the open flanks in security policy. Our affluent society will be at risk.

9. Summary Results and Their Evaluation

Strictly speaking, this essay is only an attempt to bring together and structure the previous and mostly isolated project experiences for the implementation of digital–informational transformations. The quasi-intermediate results that this paper provides are as follows:

- Compared to the technical–scientific knowledge of the digital–informational transformation, the practical knowledge of the implementation is far behind.
- The previous performance in the digital–informational transformation is not satisfactory on average.
- A systematic cross-company, industry-specific or region-specific transfer of experience is not recognizable.
- There are a number of generic models for corporate transformation towards Industry 4.0.
- Through individual or combined use, these allow a wide variety of models for the implementation of the above-mentioned transformation, which also correspond to the different requirements in terms of time and range.
- Basic experiences from comparably profound transformations are available and can be transferred to a certain extent to the digital–transformational transformation.
- The current data situation is unsatisfactory. In order to provide practice with scientific assistance in the implementation of the above-mentioned transformation, further research is recommended.
- This includes a statistically valid study of failures and success factors, ideally broken down by industry, type of company and region.

Funding: This research received no external funding.

Institutional Review Board Statement: Not applicable.

Informed Consent Statement: Not applicable.

Data Availability Statement: Not applicable.

Acknowledgments: The author thanks Johannes Winter and Reinhard Meckl for reviews and advice on issues to be covered, and Johannes Winter and Jan Biehler for technical support for the production of this article.

Conflicts of Interest: The author declares no conflict of interest.

References

1. Hürden für Fusionen. Available online: https://www.dasinvestment.com/huerden-fuer-fusionen/ (accessed on 3 March 2022).
2. Kraft in der Krise. Available online: https://www.dasinvestment.com/kraft-in-der-krise/ (accessed on 3 March 2022).
3. Neue Wege in der Krise. 2022. Available online: https://www.dasinvestment.com/neue-wege-in-der-krise/ (accessed on 3 March 2022).
4. Lucks, K. Introductory lecture. In Proceedings of the 19th Corporate M&A Congress, Munich, Germany, 16 November 2021.
5. Die Entgrenzung von M&A. Available online: https://www.tax-legal-excellence.com/die-entgrenzung-von-ma/ (accessed on 17 February 2022).
6. Wie Fusionen Gelingen. Available online: https://www.dasinvestment.com/kai-lucks-uebernahmen-merger-bundesverband-mergers-und-acquisitions/ (accessed on 14 February 2022).
7. Lucks, K. Forschung und Implementierung von Künstlicher Intelligenz zur Harmonisierung und Repositionierung Europas im globalen Wettbewerb. *Digitale Welt Online-Magazin*, 18 September 2020.
8. Steffen, Andreas. Digitalisierung und Vernetzung der Verwaltung als Basis für ein daten- und dienstebasiertes Ökosystem. In *Marktplätze im Umbruch: Digitale Strategien für Services im Mobilen Internet*; Linnhoff-Popien, C., Zaddach, M., Grahl, A., Eds.; Springer: Berlin/Heidelberg, Germany, 2015; pp. 521–530.

9. Lucks, K. *Der Wettlauf um die Digitalisierung—Potenziale und Hürden in Industrie Gesellschaft und Verwaltung*; Schaeffer-Poeschel: Stuttgart, Germany, 2020.

10. Lucks, K.; (Bundesverband Mergers & Acquisitions e.V., Frankfurt, Germany; Peter Parycek Digital Advisory Council of the Federal Government, Germany). Personal communication, 2021.

11. Lucks, K.; Meckl, R. *Internationale Mergers & Acquisitions—Der prozessorientierte Ansatz*, 2nd ed.; Springer: Berlin/Heidelberg, Germany, 2015.

12. Lucks, K. *M&A-Projekte erfolgreich führen—Instrumente und Best Practices*; Schaeffer-Poeschel: Stuttgart, Germany, 2013.

13. Lucks, K. M&A als Kunst oder Wissenschaft—Brückenfunktion von zeitgemäßem Wissensmanagement. *CFO Aktuell Z. Für Financ. Control.* **2013**, *7*, 129–131.

14. Digitale Transformation in der Industrie. Available online: https://www.bmwi.de/Redaktion/DE/Dossier/industrie-40.html (accessed on 10 February 2022).

15. Aussetzung der Insolvenzantragsfrist endet zum 30. April 2021. Available online: https://www.hk24.de/produktmarken/start/ coronavirus/aussetzung-insolvenzantragspflicht-4870698 (accessed on 17 February 2022).

16. Insolvenzen in Deutschland, Jahr 2021. Available online: https://www.creditreform.de/aachen/aktuelles-wissen/ pressemeldungen-fachbeitraege/news-details/show/insolvenzen-in-deutschland-jahr-2021 (accessed on 17 February 2022).

17. Industrie 4.0 Strategien: Den Wandel Gestalten. Available online: https://www2.deloitte.com/de/de/pages/energy-and-resources/articles/industrie-40-strategien.html (accessed on 11 February 2022).

18. Warum Sich Die Einführung von Industrie 4.0 bei vielen Unternehmen verzögert. Available online: https://www.handelsblatt. com/technik/it-internet/digitalisierung-warum-sich-die-einfuehrung-von-industrie-4-0-bei-vielen-unternehmen-verzoegert/ 25343886.html?ticket=ST-3319173-yRYjXNCMdA9ZXwmHwBym-ap6 (accessed on 29 January 2022).

19. Studie: Atos und Forrester Consulting Untersuchen Herausforderungen für eine Erfolgreiche IoT-Einführung in der Fertigungsin-dustrie. Available online: https://atos.net/de/2020/news-de_2020_09_03/studie-atos-und-forrester-consulting-untersuchen-herausforderungen-fuer-eine-erfolgreiche-iot-einfuehrung-in-der-fertigungsindustrie (accessed on 18 February 2022).

20. Wie Digital ist Deutschlands Wirtschaft? Available online: https://www.de.digital/DIGITAL/Navigation/DE/Lagebild/ lagebild.html (accessed on 29 January 2022).

21. Digitalisierungsindex. Available online: https://www.de.digital/DIGITAL/Navigation/DE/Lagebild/Digitalisierungsindex/ digitalisierungsindex.html (accessed on 29 January 2022).

22. Available online: https://firmen.handelsblatt.com/geschaeftsprozesse-digitalisieren.html (accessed on 19 April 2022).

23. Lucks, K. *Praxishandbuch Industrie 4.0—Branchen -Unternehmen—M&A*; Schaeffer-Poeschel-Verlag: Stuttgart, Germany, 2017.

24. Graning, P.; Hartlieb, E.; Heiden, B. *Mit Innovationsmanagement zu Industrie 4.0*; Springer Gabler: Wiesbaden, Germany, 2018.

25. Huber, W. *Industrie 4.0 kompakt—Wie Technologien unsere Wirtschaft und unsere Unternehmen verändern*; Springer Vieweg: Wiesbaden, Germany, 2018.

26. Blokdyk, G. *Industry 4.0 A Complete Guide—2019 Edition*; 5starcooks: Brendale, Australia, 2019.

27. Reinhart, G. *Handbuch Industrie 4.0*; Hanser-Verlag: München, Germany, 2018.

28. Lucks, K. Der Wettlauf um die Digitalisierung: Deutschland zerrieben zwischen USA und China? *Rotary Magazin Deutschland.* Available online: https://rotary.de/wirtschaft/der-wettlauf-um-die-digitalisierung-a-16261.html (accessed on 10 July 2020).

29. Industrie 4.0—So Digital Sind Deutschlands Fabriken. Available online: https://www.bitkom.org/sites/default/files/2021-04/ bitkom-charts-industrie-4.0-07-04-2021_final.pdf (accessed on 11 February 2022).

30. Vereint zum Ziel—Warum Digitale Transformation Funktionsübergreifende Zusammenarbeit Braucht. Available online: https: //www.accenture.com/de-de/insights/industry-x-0/cross-functional-collaboration (accessed on 11 February 2022).

31. Digitalisierung—Hierarchien in Unternehmen verlieren an Bedeutung. Available online: https://www.bertelsmann-stiftung.de/ de/themen/aktuelle-meldungen/2015/juni/digitalisierung-hierarchien-in-unternehmen-verlieren-an-bedeutung/ (accessed on 26 February 2022).

32. Die Digitalisierung Erfordert Zunehmend Flache Hierarchien in Teams. Available online: https://kitoko-people.ch/die-digitalisierung-erfordert-zunehmend-flache-hierarchien-in-teams/ (accessed on 26 February 2022).

33. Von der Hierarchie zum Netzwerk: Warum Digitalisierung neue Strukturen Braucht. Available online: https://www.tandemploy. com/de/knowledge-transfer/von-der-hierarchie-zum-netzwerk-warum-digitalisierung-neue-strukturen-braucht/ (accessed on 26 February 2022).

34. Industrial Internet of Things. Available online: https://en.wikipedia.org/wiki/Industrial_internet_of_things (accessed on 26 February 2022).

35. Industrie 4.0 und Supply Chain Management: Wie KI und Blockchain die Lieferkette Revolutionieren. Available online: https://www.sage.com/de-de/blog/industrie-4-0-und-supply-chain-management-fy21/ (accessed on 26 February 2022).

36. Digitale Warenwirtschaftssysteme. Available online: https://www.hk24.de/produktmarken/digitalportal/online-marketing-vertrieb/ecommerce/digitale-warenwirtschaft-4319206 (accessed on 26 February 2022).

37. Lieferkettensorgfaltspflichtengesetz. Available online: https://www.bmwi.de/Redaktion/DE/Gesetze/Wirtschaft/ lieferkettensorgfaltspflichtengesetz.html (accessed on 30 March 2022).

38. Picot, A. Available online: https://www.iom.bwl.uni-muenchen.de/personen/leitung/picot/index.html (accessed on 30 March 2022).

39. Lucks, K. Die Organisation von M&A in internationalen Konzernen. *Schweiz. Z. Für Betr. Forsch. Und Prax.* **2002**, *56*, 197–213.

40. Lucks, K. Digitale Transformation: Implikationen für das Controlling. *Controller Magazin*, January 2021; pp. 66–72.
41. Kooperationen Zwischen Startups und Mittelstand. Available online: https://www.hiig.de/wp-content/uploads/2017/11/Kooperationen_Startups_Mittelstand_small.pdf (accessed on 28 February 2022).
42. Industry X. Available online: https://www.accenture.com/de-de/insights/industry-x-0-index?c=acn_glb_brandexpressiongoogle_12315426&n=psgs_0721&gclid=Cj0KCQiAr5iQBhCsARIsAPcwROOtRe5_qy90Suu8cDxy5XU1aRwXyOWNbl7Fr6QQTy5dThMyYqaWS5caAo-gEALw_wcB (accessed on 11 February 2022).

Article

Increasing Firm Performance through Industry 4.0—A Method to Define and Reach Meaningful Goals

Christian Koldewey [1,*], Daniela Hobscheidt [2], Christoph Pierenkemper [3], Arno Kühn [2] and Roman Dumitrescu [1,2]

1 Advanced Systems Engineering, Heinz Nixdorf Institute, Paderborn University, 33102 Paderborn, Germany
2 Product Creation, Fraunhofer Institute for Mechatronic Systems Design, 33102 Paderborn, Germany
3 CP Contech Electronic GmbH, 33818 Leopoldshöhe, Germany
* Correspondence: christian.koldewey@hni.upb.de

Abstract: Industry 4.0 is one of the most influential trends in manufacturing as of now. Coined as the fourth industrial revolution it promises to overthrow entrenched structures opening new pathways for innovation and value creation. Like all revolutions, it is accompanied by disruption and uncertainty. Consequently, many manufacturing companies struggle to adopt an Industry 4.0 perspective that benefits their performance. Hence, our goal was to develop a method for increasing firm performance through Industry 4.0. A key factor was to focus on the entire company as a socio-technical system to depict the numerous interactions between people, technology, and business/organization. To realize the method, we combined consortium research, design science, and method engineering. We gathered comprehensive data from workshops, interviews, and five case studies, which we used to develop the method. It consists of four phases: a maturity model to determine the status quo, a procedure to derive a target position, a pattern-based approach to design the socio-technical system, and a procedure to define a transformation setup. Our approach is the first to combine maturity models with foresight and extensive prescriptive knowledge. For practitioners, the method gives orientation for the future-oriented planning of their transformation processes.

Keywords: Industry 4.0; maturity model; transformation; methodology; Industry 4.0 strategy; socio-technical system

Citation: Koldewey, C.; Hobscheidt, D.; Pierenkemper, C.; Kühn, A.; Dumitrescu, R. Increasing Firm Performance through Industry 4.0—A Method to Define and Reach Meaningful Goals. *Sci* **2022**, *4*, 39. https://doi.org/10.3390/sci4040039

Academic Editor: Johannes Winter

Received: 3 June 2022
Accepted: 12 October 2022
Published: 18 October 2022

Publisher's Note: MDPI stays neutral with regard to jurisdictional claims in published maps and institutional affiliations.

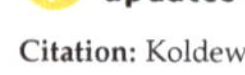

1. Introduction

The root of Industry 4.0 is a German strategic initiative; the term itself was first mentioned during the Hannover Fair in 2011 [1]. Based upon the emergence of the internet of things, data, and services, it is anticipated to describe a fourth industrial revolution [2]. Fundamental concepts of Industry 4.0 include smart factories, cyber-physical systems, self-organization, new distribution and procurement systems, new systems for product and service development, adaptation to human needs, and corporate social responsibility [3]. Industry 4.0, hence, fosters radically new, highly dynamic, ad hoc networked, and real-time modes of collaboration within and between companies [4]. This is associated with a wide range of advantages, e.g., the production of custom products with minimal use of time and resources [5]. Consequently, Industry 4.0 will transform value creation, value capture, and value offer of manufacturing companies, and thus the whole business model [6].

While technology represents the origin, it is critical for the implementation of Industry 4.0 to recognize it as a socio-technical endeavor [7]. Socio-technical systems are instantiations of social and technical elements engaged in purposeful goal-directed behavior, where both, social and technical factors, are responsible for successful performance [8]. This thinking is derived from the insight, that any production system requires technology and a social structure linking human operators with the technology and to each other [9]. Ulich (2013) introduces the so-called "MTO" concept, which considers the dimensions "people", "technology" and "organization", hence explicating the relevance of organizational structures [10]. In this paper, we follow a further extension of the concept and replace organization with business to further emphasize the value orientation. Simplified, business represents organization and earning money [11]. Following the initial understanding of socio-technical systems, we consider business, technology, and people as responsible for the successful performance of Industry 4.0 approaches.

These linkages can be illustrated by a simplified example: machines and end devices are increasingly equipped with sensors in the context of Industry 4.0. As intelligent cyber-physical systems (technology), they collect and process data, which is made available to networked systems via web technologies (e.g., via a digital platform). The analysis of the provided data leads to new value creation opportunities such as complementary services or new revenue models, e.g., pay-per-use (business). However, these possibilities demand reflections on the work organization within the company (business) and the necessary competencies (people).

While the overall potential of Industry 4.0 has been recognized by companies, the actual implementation imposes major, heterogenous challenges (c.f., [12,13]). Given the multitude of possibilities and their associated challenges, companies should not strive for the introduction of the highest possible Industry 4.0 level, but rather concentrate on what is beneficial in their current position. Hence, the first step for the beneficial introduction of Industry 4.0 should be the individual determination of the company's current performance [14]. This should be contextualized compared to the competitors [15], to strive for a competitive advantage. Therefore, the performance level of the company needs to be systematically determined and expressed in an objectively measurable way.

Next, companies need to determine the direction they want to take. This is not trivial and has to be chosen conscientiously taking the company's contingencies into account [16]. However, one study finds in its sample that although 73 percent of companies specify that digital transformation has a major influence on their corporate strategy, more than 50 percent do not yet have a comprehensive and, above all, holistic implementation strategy [17].

Based on the initial situation and the target position, a coherent set of solutions that address the gap between both is needed. Implementation projects for Industry 4.0 often involve an uncertain and non-sequential strategic approach [15]. According to a McKinsey study, 70 percent of the companies have already introduced a pilot project for Industry 4.0, but at the same time, only 29 percent have been able to generate added value as a result [18]. Comparing large companies to small and medium-sized enterprises (SMEs), it seems to be a particular challenge for the latter to identify and select individually suitable solutions. According to another study, 65 percent of SMEs (compared with 19 percent of large companies) indicated the selection of solutions that meet their needs as a major obstacle concerning Industry 4.0 [19].

Concluding the challenges above, it becomes clear that companies need a method to support the implementation of Industry 4.0 to increase their performance. Hereby, a socio-technical approach should be adopted. The method should answer the research question *"How can companies identify and attain an individual and performance increasing Industry 4.0 target position?"*. Based on the considerations above and supported by our practice partners, the method should cover four fields of action:

1. **Maturity Check:** To evaluate the performance of the considered company, Industry 4.0 has to be depicted in detail as a socio-technical system in the areas of technology, business, and people. An objective evaluation scheme, i.e., a maturity model, is required to make the company's initial position measurable.

2. **Target definition:** Based on the individual initial situation, an appropriate target maturity level is to be determined for the company, taking into account the actual and future contingencies of the market, technology, and environment.

3. **Implementation planning:** To select suitable solutions to reach the target position, it is promising to draw on the experience of Industry 4.0 pioneers. The target is to enable

companies to participate in the dynamic development of Industry 4.0 with relatively little effort. For that, implementation patterns need to be identified. They allow to tackle typical Industry 4.0 tasks. The selection and combination of implementation patterns must be supported, taking into account the effects on people, technology, and business.

4. **Transformation setup:** For the concrete implementation of Industry 4.0 the status quo, target position, and implementation solutions must be integrated into a coherent view, which can be communicated to the employees. Measures must be defined that translate the plan into action.

In this paper, we reflect on one of our recent research projects concerning this problem. The paper unfolds as follows. Next, we investigate the relevant body of knowledge in the literature. After that, we describe our research setting and methodology. Then, the results of the project are described according to an abridged and revised excerpt from our final project report [20]. A discussion of the results follows before we close the paper with a short conclusion.

2. Literature Review

Chapter 1 has introduced the four major fields of action in the course of our research. To further clarify the research gap, we address with our method, the literature for the fields of action is analyzed. While Industry 4.0 is a manufacturing-centric concept, in general, the term digital transformation is also used quite often [21]. Hence, in the following first the main concepts of the field of action and then both frameworks and methods for Industry 4.0 and digital transformation are considered.

2.1. Maturity Check for Industry 4.0

To investigate the maturity (or competency, capability, level of sophistication) of companies regarding certain domain so-called maturity models are used, which comprise a set of more or less comprehensive criteria [22]. Maturity models consist of sequences of maturity levels ranging from an initial state to an optimal stage following an evolution path of discrete steps. To this end, criteria and corresponding characteristics that need to be fulfilled for a certain maturity level are provided [23]. Using maturity models managers can determine (1) the actual performance of the company today, (2) the current status of the industry (benchmark), (3) targets for improvement, and (4) required changes between the status quo and the target [24]. There are numerous maturity models in literature today. In terms of Industry 4.0/Digitalization new maturity models keep appearing. Wagire and colleagues for example recently introduced an Industry 4.0 maturity model emphasizing organizational awareness and emerging technologies like cobots [25]. Gökalp and Martinez propose a digital transformation maturity model (DX-CMM) that allows for a process-centric, holistic, and integrated view for companies across all sectors [26]. Other approaches proved to be very successful in practice. The Industry 4.0 Maturity Index for example is continuously used in different industries by the Industrie 4.0 Maturity Center consultancy. It is structured along the four capabilities for industry 4.0 resources, information systems, organizational structure, and culture [14]. From our point of view existing Industry 4.0 maturity models are already very sophisticated but lack depth regarding the business perspective of Industry 4.0. Furthermore, they lack methodological support to define meaningful goals for Industry 4.0. Deriving concrete measures from the gap between two maturity levels is also associated with high cognitive demands.

2.2. Defining Target Maturity Levels for Industry 4.0

Defining a target position is one of the major activities during strategy development [27]. It is also an essential component of many methods for the Industry 4.0 transformation, whether they are based on a maturity model or not. Oleff and Malessa for example distinguish between visionary goals and operational goals. Visionary goals are derived from a maturity model investigating which maximum maturity level might be reached within a limited project. Operative goals are derived iteratively matching current problems and potentials [28]. Tüllmann and colleagues on the other hand recommend defining the targets by consolidating the expectations of the involved stakeholders [29]. Other approaches simply recommend conducting expert workshops to define targets within a maturity model [30–32]. Another approach that utilizes expert workshops is by Jodlbauer and Schagerl, who recommend defining the target position considering strategy, (company) goals, maturity, and economic and technical restrictions [33]. The Industrie 4.0 Maturity Index defines two consecutive generic targets: first, equalize the maturity levels, then raise the maturity levels [14]. Summarizing the findings from the literature, it is clear, that most methods use only rudimentary tools to determine the target position for Industry 4.0. This does not seem to do justice to the importance and the necessary investments in Industry 4.0. Moreover, no approach even considers thinking ahead the future contingencies in which the company has to flourish in the future, which is paramount since the transformation towards Industry 4.0 is a long-term project. A future-oriented target definition for Industry 4.0 is therefore needed.

2.3. Solution Pattern as a Means to Close the Gap between Maturity Levels

Even when meaningful goals for industry 4.0 as defined by a target maturity level are formulated, closing the gap between as-is and should-be situations is not trivial. Suitable measures must be defined and enacted. Since Industry 4.0 is a widespread phenomenon, there are already vast amounts of solutions for typical problems of companies. However, these so-called "Best Practices" are difficult to identify and structure for individual needs [15]. Solution patterns describe recurring problems and the core to their solutions so that they can be used over and over again, without ever leading to the same result [34]. Hence, solution patterns for Industry 4.0 implementation are a suitable approach to overcome the gaps. Each solution pattern has a name and is described by context, problem, and solution [35]: (1) The name is a descriptive representation of the solution contained in the pattern. (2) The context classifies the underlying problem into the situation in which it occurs. (3) The problem describes the challenges or issues addressed by the pattern. (4) The solution provides appropriate ways and means to solve the problem. Besides that, the notation can be adapted to the application context.

Anacker synthesizes six major benefits using solution patterns from literature [36]: (1) transferability across disciplinary boundaries, (2) improvement of communication through explicit knowledge representation, (3) long-term documentation of solution knowledge, (4) reduced complexity by breaking down extensive problems, (5) increased efficiency through targeted reuse, and (6) promotion of creativity.

Solution patterns have been discussed for almost 50 years, starting with Alexander's book "*A Pattern Language—Towns, Buildings, Construction*" [34]. Since then, their application can be investigated in numerous disciplines like software engineering, product engineering, and business model development [37]. This shows the extensive uses for solution patterns. Compared to other forms of knowledge representation solution patterns seem to be especially promising for Industry 4.0 because they (1) are focused on problem-solution-combinations and hence allow for an easy transfer into practice, (2) externalize and generalize knowledge to close knowledge gaps, (3) allow for the creation of individual solutions, (4) are comparable to each other, and (5) can be continuously extended by new patterns, which is necessary given the rapid evolution of Industry 4.0.

There are only a few solution pattern approaches for Industry 4.0. Weking and colleagues for example propose a business model pattern framework [38], while Gausemeier and colleagues describe a pattern system for Industry 4.0 business models [39]. Many other approaches can be seen as enablers for Industry 4.0, e.g., Dumitrescu's design patterns for cognitive functions [40]. However, patterns that take into account the socio-technical transformation of the enterprise are not existing in literature as of now.

2.4. Transformation Setup

Adopting Industry 4.0 in a firm requires a deep transformation. A transformation involves the "redefinition of mission and purpose, and a substantial shift in goals, to reflect a new direction and therefore encompassing a fundamental shift in the business model of the organization, touching all cultural, structural, and processual aspects" [41]. This is reflected in many approaches for the introduction of or the transformation towards Industry 4.0. Merz for example includes the management of processes, technologies, organization, and employees in her method [42]. Hennegriff and colleagues define concrete projects and responsibilities and include a controlling phase for the transformation [43]. The acatech Industry 4.0 maturity index recommends defining measures, which are then clustered into action streams and are planned in a factual-logic sequence [14]. However, despite a transformation being a high-risk endeavor, no approach considers a holistic, sociotechnical management of risks within the transformation towards Industry 4.0, leaving a research gap from both theory and practice [44].

2.5. Summary

The analysis of the existing literature in Industry 4.0 leads to the conclusion, that besides numerous existing maturity models, there are still many open questions regarding the utilization of the maturity models to define and reach meaningful goals. Solutions to define the target maturity are needed as well as approaches to transfer a maturity level towards concrete measures. Furthermore, looking at the transformation itself, managing the risks for Industry 4.0 transformation seems to be a significant research gap.

3. Materials and Methods

Our research approach is inspired by Otto and colleagues (2015) [45], who combine consortium research and method engineering to develop an approach for digital business model design. This can be considered as an analogous research endeavor and, thus, it provides a first indication of what our research approach might look like [45]. Next, we describe the research process in general before we elaborate on data collection and artifact design in detail.

3.1. Research Process

The research need implies that the desired resulting artifact is a method. Methods can be considered design artifacts [46]. Developing a method to increase firm performance in the context of Industry 4.0 is a complex task that requires comprehensive data from the field. When we planned the research project, Industry 4.0 was still in its infancy. Hence, we chose a strong focus on case study research, which is suitable for investigating new phenomena that cannot be separated from their organizational context [47]. To this end, suitable cases had to be identified and organized.

Given these boundary conditions, we chose consortium research (CR) as an overarching research process. It especially suits research projects, where the desired result is an artifact designed to solve practical problems (e.g., a method) and where close long-term cooperation of researchers and different companies (i.e., the case companies) is necessary for data collection, artifact design, and artifact evaluation in real business settings [48]. CR gives guidelines to organize the collaboration and allowed us to create a stimulating research setting for theoretical and practical insights. A CR project follows four phases: analysis, design, evaluation, and diffusion [48]. The concrete activities and methods used in each phase during our research project are shown in Table 1. The resulting research process, hence, comprises a multilayered approach: The general research setting follows the CR approach, while the method engineering approach, which is a subdiscipline of design science research, was utilized to develop the concrete artifact.

Table 1. Research process with activities and methods.

Consortium Research Phase	Activities and Methods
Analysis	Literature review Interviews and consortium workshops Best practice analysis (e.g., platform Industry 4.0, it's owl)
Design	Rigor: Review of Industry 4.0 and Business Transformation (in Manufacturing) literature Relevance: Interviews and workshops with consortium partners Method Engineering as design paradigm for the development of the method Action research to solve real-world problems within the consortium, check the relevance, and iterate toward the solution
Evaluation	Case studies Pilot application Review via workshops
Diffusion	Knowledge transfer workshops Homepage, Online-Tool Research papers

It becomes clear, that a strong involvement of practitioners was the condensation point of our research process. We integrated their knowledge through case studies, interviews, and cross-case workshops. The next chapter describes how we collected the necessary data in detail.

3.2. Data Collection

The nature of the five main case studies was predominantly participative. Hence, the researchers were actively involved in the solving of the concrete problems within the companies [47]. In addition to the researchers, a consultancy for digital transformation and innovation worked on the cases. Data was collected through hundreds of informal interviews and talks, internal and external documents, site visits, and workshops. Table 2 shows the case studies considered in our research. Overall, we conducted 50 workshops and 69 jour fixes within the CR project.

Table 2. Case Study Overview.

Case	Industry	Size (Empl.)	Collection Period and Setting	Key Experts	Type of Case Study
A	White Goods	800	July 2016–June 2019, 12 workshops	Industrial engineers, strategic planning, and project management	Explorative, participatory, application company
B	Electronics and intelligent technical systems	40	July 2016–June 2019, 10 workshops	Business management	Explorative, participatory, application company
C	Engineering for printing machines	60	July 2016–June 2019, 6 workshops	Technical director, technical engineering team	Explorative, participatory, application company
D	Engineering for HVAC	130	July 2016–June 2019, 9 workshops	Research and development team	Explorative, participatory, application company

Table 2. *Cont.*

Case	Industry	Size (Empl.)	Collection Period and Setting	Key Experts	Type of Case Study
E	Engineering for security and access solutions	7.000	July 2016–June 2017, 6 workshops	Senior manager strategic innovation	Explorative, participatory, application company
X	Cross-case Workshops	-	July 2016–June 2019, 4 workshops	Project leads from the consortium	Explorative, interviews
Z	Individual third-party transfer	-	July 2016–June 2019, 3 workshops	Diverse	Explorative, participatory, partial application

In addition to the case studies, we conducted focus group workshops to challenge our findings with third parties from outside the consortium (see Table 3). The participants of the workshops and interviews were predominantly from the German manufacturing industry, whereby the companies themselves served a wide variety of industries (e.g., mechanical engineering, food, advertising, preliminary products, etc.). Many experts were located within the state of North Rhine Westphalia. Company sizes ranged from SMEs to international enterprises with tens of thousands of employees. In general, one expert per company attended. Only in a few cases, two or more experts from one company participated.

Table 3. Focus Groups.

Date	Topic	Method	Participants
22 June 2017, 13:00–17:15	Change of the company as a socio-technical system (technology, business, people)	World Café	28 experts from industry and research
26 June 2018, 13:00–17:00	Increasing firm performance: prototypes of Industry 4.0, patterns for Industry 4.0, use cases with the firm	World Café	34 experts from industry and research
27 June 2019, 14:00–18:00	Industry 4.0 Expert Group: From digitalization strategy to implementation	Presentation and Workshop	85 experts from industry and research

3.3. Method Engineering

The envisioned result of our research was a method to empower companies to adapt an individually beneficial maturity regarding Industry 4.0. A method in this context can be understood as guidance for projects employing a certain way of thinking through directions and rules structured in a systematic way applying activities and techniques to realize certain deliverables [49]. Such a method must meet three quality criteria: it has to fit the situation, consist of sufficient components to deliver the results, and its components themselves have to be proven to work [50]. To ensure this, we followed the guiding principles of method engineering (ME), which is defined as the engineering discipline to design, construct and adapt methods [51]. ME considers five concepts: metamodel, results, activities, techniques, and role. The metamodel describes relevant concepts and relationships of the application domain (e.g., organizational structure). Results are the artifacts to be delivered by the application of the method (e.g., transformation setup). Activities (or phases) describe how results are created, e.g., determine the degree of maturity. All activities as a whole form the procedure model. Techniques describe in detail, how a result is created within an activity, e.g., using a maturity model. Furthermore, activities are carried out by roles within the project team, e.g., software developer [45,52]. Such a method should always follow a distinct method rationale explicating the values and goals behind the method. The method rationale structures the requirements for the method derived from practical needs (relevance) and an existing knowledge base (rigor) [53].

According to the method rational, the method was developed in close collaboration with the practitioners. First, the key results were defined. Then, the procedure model was drafted, roughly defining the method phases (i.e., general activities). Following, the concrete activities and techniques for each phase were chosen or developed when no suitable technique was available. Activities and techniques were discussed within the consortium and tested within the case studies. Learning from the experiences during the application, they were further improved until they fit the requirements regarding usability and usefulness.

4. Results

The result is a method for the improvement of firm performance through the adoption of Industry 4.0. In the following, we will first introduce the method rationale as the basis for our method. Then, an overview of the method components is given before the components are described in detail. As stated in the introduction, the results shown here are a shortened and revised excerpt from our final project report to which we refer for further information [20].

4.1. Method Rationale

The research process is guided by the so-called method rationale, which includes the arguments complementing the method to be developed. In that way, the method rational gathers requirements from theory and practice for the research [53]. The requirements fall into four categories: maturity check, target definition, implementation planning, and transformation setup. Table 4 gives an overview of the relevant requirements identified through literature review and workshops with the consortium.

Table 4. Method requirements.

Cat.	Req.	Description	Supporting Literature
Maturity Check	R1	Consideration of the relevant aspects of Industry 4.0	[54,55]
	R2	Objective evaluation criteria for Industry 4.0	[56]
	R3	Benchmarking with similar companies	[15,57]
Target defini-tion	R4	Integration of foresight into target definition	[58]
	R5	Internal and external consistency of the target	[59]
Implemen-tation planning	R6	Inductive development of Industry 4.0 implementation patterns	[35,60]
	R7	Interdisciplinary notation scheme	[61,62]
	R8	Identification of consistent pattern paths	[63,64]
Transfor-mation setup	R9	Socio-technical view on transformation	[10,65]
	R10	Transparent and holistic transformation set-up	[66–68]

4.2. Method Overview

An overview of the resulting method is shown in Figure 1. The method is structured into four major phases: (1) Industry 4.0 Maturity Check, (2) Industry 4.0 Target Definition, (3) Industry 4.0 Implementation Planning, and (4) Industry 4.0 Transformation Setup.

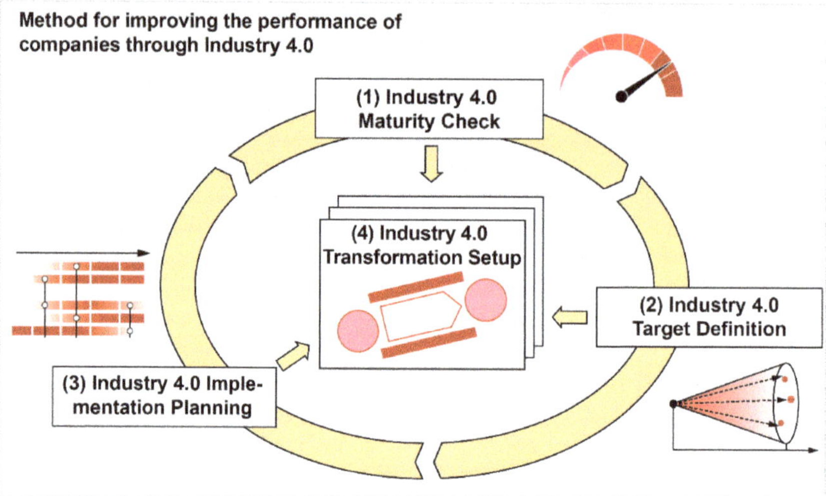

Figure 1. Method for improving firm performance through Industry 4.0 adapted from [20].

(1) Industry 4.0 Maturity Check: The first phase aims to determine the actual state of Industry 4.0 in the focal company. For that, an Industry 4.0 quick check is used. It allows assessing the maturity utilizing 59 criteria, spanning the categories technology, business, and people. The 59 criteria are then evaluated to determine the most relevant fields of action for the company at hand. Analyzing those criteria in-depth with a tool kit leads to concrete improvement potentials.

(2) Industry 4.0 Target Definition: This phase aims to define adequate goals for Industry 4.0 maturity of the firm. Using foresight methods, the future of markets, technologies, and business environments is anticipated. Based on these insights, the target maturity is determined.

(3) Industry 4.0 Implementation Planning: The third phase leads to generic solutions for the implementation of the envisioned target position. To that, solution patterns for Industry 4.0 are used. The solution patterns characterize the socio-technical system, that realizes the target position of the firm. Fields of action are prioritized. Then, pattern combinations are built and arranged.

(4) Industry 4.0 Transformation Setup: The implementation of Industry 4.0 is a complex task, that must be set up comprehensively. Socio-technical risks of each implementation project are determined and managed. At last, the insights are consolidated within a master plan of action, that summarizes the transformation setup.

Table 5 gives a detailed overview of the corresponding method components and their concrete goals.

Table 5. Method components and goals for each phase.

Phase	Method Components	Goal
Maturity Check	Quick check Industry 4.0	Determine the current maturity level
	Relevance ranking	Find the most important maturity criteria
	Derivation of fields of action	Find fields of action for further investigation
	In-depth analysis	Identify concrete potentials for improvement within fields of action and rank them
Target definition	Anticipation of the future	Gain an idea of the future environmental conditions
	Impact analysis	Find out how the future environment influences the different maturity criteria and levels
	Target position definition	Define medium- and long-term target maturity levels
Implementation planning	Implementation patterns for Industry 4.0	Provide generic solutions for recurring problems/potentials within Industry 4.0, that can be concretized for the specific company
	Identification of relevant implementation fields	Narrow down the solution space according to the concrete transformation needs
	Assessment of implementation patterns	Find established solutions that contribute to reaching the target position
	Combination analysis	Build a set of solutions that support each other and sort them in a meaningful way
Transformation setup	Definition of measures	Break down general solutions into concrete work packages
	Risk assessment	Identify risks associated with the implementation and measures to mitigate them
	Masterplan of action	Condense the previous results into a document, that can be used for communication
	Transformation controlling	Continuously check if the assumptions are still correct and if the transformation is going according to plan

4.3. Method Phases

Next, the four method phases (1) Industry 4.0 Maturity Check, (2) Industry 4.0 Target Definition, (3) Industry 4.0 Implementation Planning, and (4) Industry 4.0 Transformation Setup are described in detail.

4.3.1. Phase 1: Industry 4.0 Maturity Check

For the successful introduction of Industry 4.0, it is first imperative to determine the current performance level regarding objectively measurable criteria. Only then, a realistic transformation setup may be developed. When assessing the current Industry 4.0 level, it is important to understand Industry 4.0 as a socio-technical endeavor [8]. Therefore, the dimensions people, technology, and business must be considered integratively.

A socio-technical maturity model is employed for this task. Its' application and analysis are conducted in four steps. First, the maturity model "Quick Check Industry 4.0" is used to determine the current performance profile (status quo). Then, the criteria of the maturity model are prioritized. Within the most relevant criteria, fields of action for improving performance are identified. At last, an in-depth analysis leads to concrete improvement potentials. Below, the individual activities are explained in more detail.

Quick Check Industry 4.0: The basis for deriving customized fields of action is the determination of current maturity in the context of Industry 4.0. This results from using a so-called "Quick Check Industry 4.0", which is structured according to the socio-technical dimensions technology, business, and people. It allows the assessment of a company using 59 criteria. For each criterion, one of four performance levels has to be chosen, with the fourth performance level reflecting the ideal vision of Industry 4.0. They are based on established literature, existing maturity models, and empirical values. The final criteria were selected in workshops within the consortium, taking into account the perspectives of researchers and practitioners. The criteria selected were those that both parties agreed would have the greatest impact. Within the technology dimension it is sometimes necessary to differentiate criteria regarding product and production (e.g., data storage for products and data storage for production). Table 6 shows an overview of the criteria of the Quick Check Industry 4.0.

Table 6. Criteria of the Quick Check Industry 4.0.

Technology	Business	People
T1 Horizontal Integration	B1 Industry 4.0 Strategy	P1 Scope of activity and autonomy
T2 Vertical Integration	B2 Strategy Controlling	P2 Variety of requirements
T3 IT Process Support	B3 IT Security Concept	P3 Flexibility of Working Hours
T4 Tool Landscape	B4 Value-creation Cooperation	P4 Co-dependency
T5 Systems Engineering	B5 Access to capital	P5 Performance Feedback
T6 Sensor Technology (production)	B6 Approach to New Product Development	P6 Collaboration and Social Interaction
T7 Actuator Technology (production)	B7 Customer Integration	P7 Ergonomics
T8 Information Processing (production)	B8 Pioneering Spirit	P8 Continuing Education
T9 Human Machine Interface (production)	B9 Technology Transfer	P9 Documentation of Experiential Knowledge
T10 Data Storage (production)	B10 Participation in Innovation Networks	P10 Availability of Support
T11 Data Usage (production)	B11 Innovation Organization	P11 Leadership Transparency
T12 External Data Integration (production)	B12 Approach to Business Model Development	P12 Employee Participation
T13 Digitalization of production processes	B13 Product-Service-Systems	P13 Strategy for Change
T14 Connectivity (production)	B14 Penetration of Digital Services	P14 Software Usability
T15 Intralogistics	B15 Data Collection and Analysis	P15 Assistance Systems

Table 6. *Cont.*

Technology	Business	People
T16 Organization of Production Planning and Steering	B16 Data Exploitation	P16 Human-technology Dependency
T17 Production Flexibility	B17 Digital Customer Channels	
T18 Assistance Systems in Assembly		
T19 Sensor Technology (product)		
T20 Actuator Technology (product)		
T21 Information Processing (product)		
T22 Human Machine Interface (product)		
T23 Data Storage (product)		
T24 Data Usage (product)		
T25 External Data Integration (product)		
T26 Connectivity (product)		

For the assessment, a workshop with representatives from various disciplines is conducted. To avoid misinterpretation, the assessment is supported by a question for each criterion and an explanatory text for each performance level. After the completion of the Quick Check Industry 4.0, a database with results from over 250 companies allows for a benchmark with a comparative collective. Based on this and the organizational framework conditions, a first estimation of the target position in five years is requested.

Relevance ranking: To reduce the number of criteria to investigate in detail to a manageable extent, a relevance ranking of the criteria is conducted. The relevance of the criteria can depend on the difference between the actual and target position, resources, organizational structure, customer requirements, or further aspects. Hence, an individual evaluation of different stakeholders is necessary to integrate different perspectives. For that, a workshop is held in which each participant is asked to identify the five most relevant criteria from their point of view. The quantity of five has proven itself to be manageable in practice and was chosen for this reason. It is recommended to select at least one criterion from each one of the dimensions people, technology, and business. The resulting criteria are those, that the experts agree to present the greatest need for action.

Derivation of the fields of action: Based on the relevance assessment, the prioritized criteria are to be checked for synergies and dependencies. The goal is to determine relevant fields of action that support the selection of suitable methods for Industry 4.0 in-depth analyses. In total, 17 fields of action are available—five fields of action in the dimension technology (*digitalization of processes, human-technology-interaction, process organization, value chain, self-optimization*), six fields of action in the dimension business (*I4.0 strategy, innovation culture, data management, digital services, strategy controlling, business models*), and six fields of action in the dimension people (*work design, communication and change, usability, qualification, human–machine-interaction, ergonomics*). Selecting the fields of action is facilitated for the company via guiding questions and investigating the assigned Quick Check criteria. An example is shown in Figure 2.

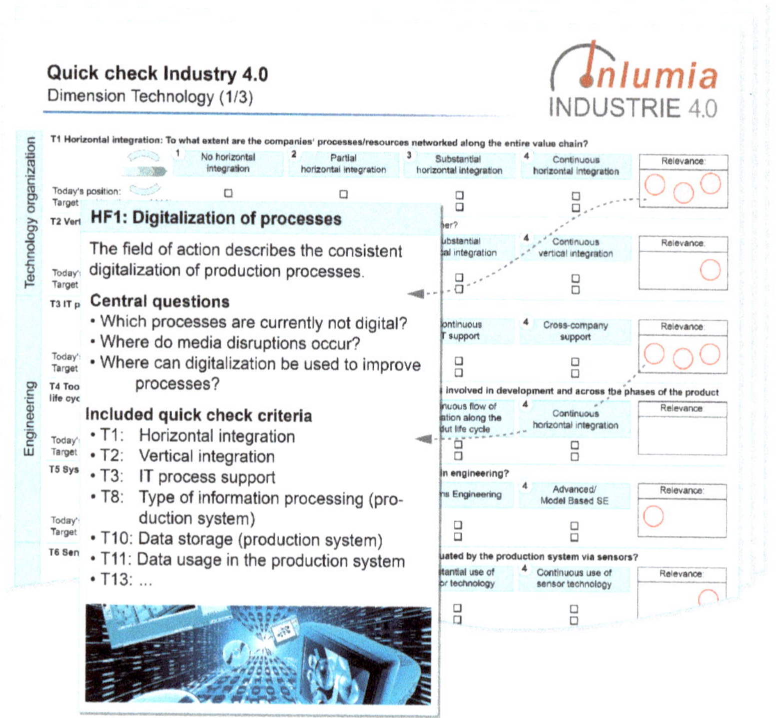

Figure 2. Example of a derived field of action from the Quick Check criteria [20].

Industry 4.0 in-depth analysis: A toolbox of methods is provided for the in-depth analysis of the selected fields of action. It includes both, already established methods and specifically developed methods, which are used during workshops. Experts from different disciplines and management levels should be involved to obtain the most comprehensive discussions possible. The toolbox includes methods like OMEGA (method for business process modeling and analysis) [69] or a modeling language for value creation systems [39]. Applying the methods allows to identify concrete potentials for improvement in the context of Industry 4.0 (e.g., complicated resource planning due to lacking predictions for upcoming orders). This serves as the starting point for improving the performance. Generally, speaking from our experience, companies should invest three to six months for in-depth analysis. This results in many potential improvements, which are then prioritized with the help of a bubble chart (Figure 3). The chart considers the two evaluation dimensions benefit and development effort. Improvement potentials at the bottom left of the portfolio have a low benefit and at the same time require a high development effort. Hence, they should be neglected at first. Potentials in the middle of the portfolio should be reviewed on an individual basis to determine whether the potential should be exploited immediately or put on hold for the time being. Potentials at the top right of the portfolio should be given priority as they offer considerable benefits while requiring little effort. These so-called "low-hanging fruits" help to significantly boost the Industry 4.0 performance within a short period.

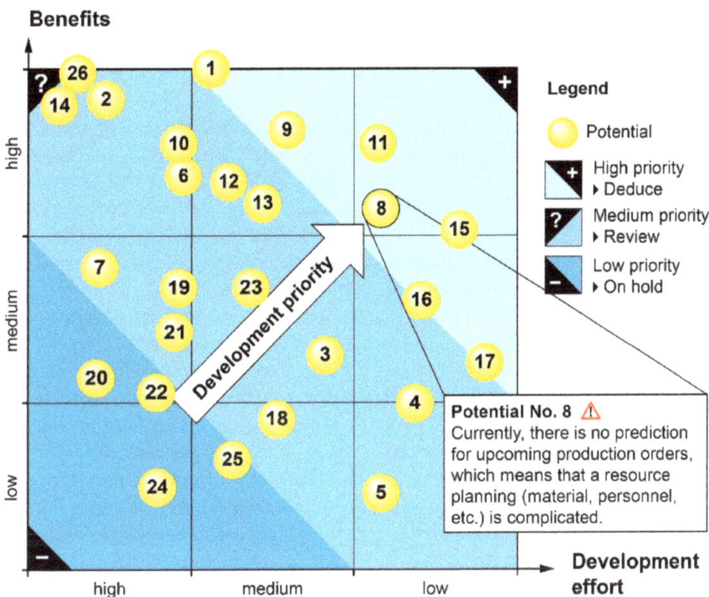

Figure 3. Evaluation portfolio with exemplary potentials derived from one case [20].

4.3.2. Phase 2: Industry 4.0 Target Position

The aim of the second phase is the Industry 4.0 target position for the considered company. The target position is a strategic long-term commitment to change and, hence, must consider forging a fit [70]. The term fit in this case means the consistency of multiple contingencies and structural aspects [71], i.e., the fit of the target position to the future business environment. To ensure that, this phase involves three activities: (1) anticipate the future, (2) analyze the impact, and (3) set the target position.

Anticipation of the future: To anticipate the future of markets, technologies, and the business environment for Industry 4.0 two foresight methods are combined: trend analysis and scenario technique in combination allow for the envisioning of a medium and a long-term view.

The trend analysis (c.f., [58]) is suitable to anticipate medium-term developments (i.e., approx. five years into the future). A trend is a possible trajectory into the future that can be observed to a certain degree today. Trends can be identified by scanning relevant business-specific and global sources like studies and publications or conducting expert interviews. The trend identification should consider trends that influence the technology, business, and people dimensions of the firm. Identified trends are documented using trend profiles that include a first firm-specific analysis of the trend (i.e., chances and risks) as well as further information (e.g., drivers of the trend). This allows assessing the trends regarding their probability of occurrence and foreseeable impact strength. To communicate the results of the trend analysis, a trend radar is a suitable tool. Each trend is represented by a bubble on the radar. Trends with a high probability of occurrence are placed in the center and the bubble size indicates the impact strength of the trend. Furthermore, the trends can be sorted into one of the three socio-technical dimensions. Trends with a high impact and a high probability of occurrence should be considered in developing the Industry 4.0 strategy.

Scenarios describe possible situations in the future which are based upon a complex network of influence factors and a plausible explanation of the progress from today to that situation. The scenario technique is a suitable tool to develop these kinds of scenarios (c.f., [58]). The results of the scenario technique are multiple, internally consistent scenarios. For the development of the Industry 4.0 target position, one must be chosen as a reference scenario that serves as an orientation. Usually, the scenario with the highest perceived combination of the probability of occurrence and relevance is chosen.

Impact analysis: The impact analysis delivers insights on the effect of future developments on the Industry 4.0 maturity. This allows for an assessment if a higher maturity might be achieved. For that, an influence matrix and an impact matrix are used. This is done on the one hand for the trends and on the other hand for the reference scenario. In the following, the procedure for the analysis of the trends is described. First, the influence matrix is filled out. It answers the question of how a trend (column) influences a criterion of the maturity model (row). For the assessment, five evaluations are possible ranging from −2 (trend hinders the performance improvement significantly) to +2 (trend benefits the performance improvement significantly). For example, the trend "market penetration of cyber-physical systems" benefits a performance improvement of the criterion "horizontal integration" and is, hence, evaluated with +2.

Building upon the influence matrix, the impact matrix is filled out. Here, the overall impact of a trend (row) on a maturity criterion (column) is assessed. That means, in addition to the influence matrix, the individual assessment of trends from the firms' point of view is included. The impact of each trend is calculated by multiplying influence strength (InS), probability of occurrence (PO), and impact strength (ImS). Now, for each maturity criterion, the line sum is calculated. This allows the estimation of the influence of all identified trends on the considered criterion. Figure 4 shows the matrices and their relations.

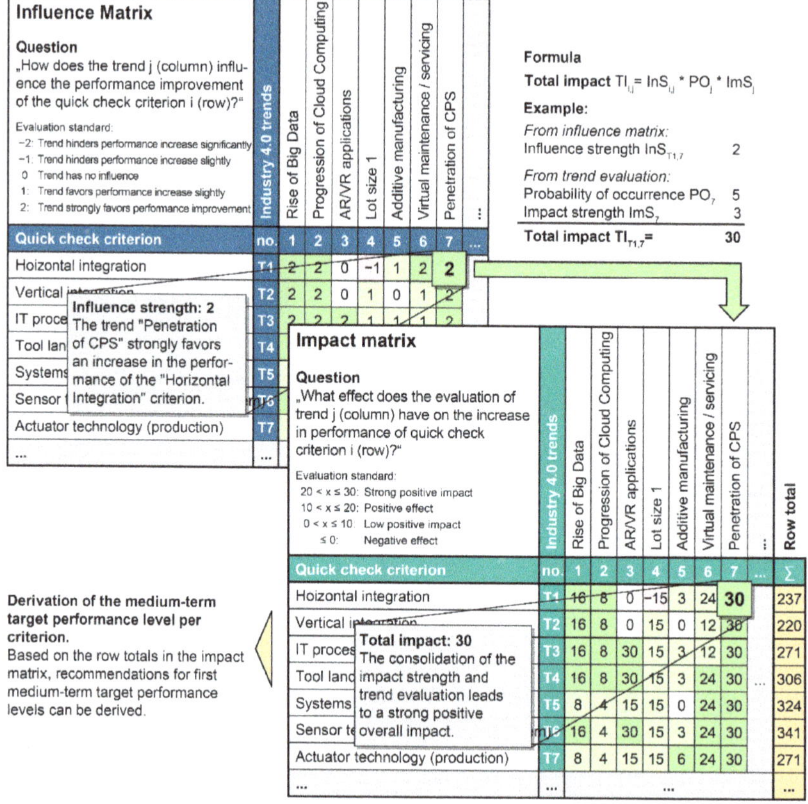

Influence Matrix:

Quick check criterion	no.	1	2	3	4	5	6	7	...
Hoizontal integration	T1	2	2	0	−1	1	2	2	
Vertical integration	T2	2	2	0	1	0	1	2	
IT proce...	T3	2	2	2	1	1	1	2	
Tool lan...	T4								
Systems	T5								
Sensor ...	T6								
Actuator technology (production)	T7								
...	...								

Impact matrix:

Quick check criterion	no.	1	2	3	4	5	6	7	...	Σ
Hoizontal integration	T1	16	8	0	−15	3	24	30		237
Vertical integration	T2	16	8	0	15	0	12	30		220
IT proces...	T3	16	8	30	15	3	12	30		271
Tool land...	T4	16	8	30	15	3	24	30	...	306
Systems	T5	8	4	15	15	0	24	30		324
Sensor te...	T6	16	4	30	15	3	24	30		341
Actuator technology (production)	T7	8	4	15	15	6	24	30		271
...	...				...					...

Figure 4. Influence and impact matrix [20,72].

Building on these insights, a recommendation for the medium-term target maturity level can be derived. This is done using an interval scale, which must be created depending on the considered trends. One possible solution might be to divide the criteria into quartiles according to the row total. A criterion that falls into the first quartile would then allow for an improvement of three levels, a criterion from the second quartile would allow for two levels, and so on. The medium-term target maturity level (e.g., level 2) can then be calculated by adding the current maturity level (e.g., level 1) for a criterion and the performance improvement possible through the Industry 4.0 trends (e.g., one level).

Finding the long-term target maturity level is conducted similarly. Instead of trends, here, the future projections of the descriptors within the reference scenario are considered in the influence and impact matrices.

Target position definition: The results from the impact analysis allow the creation of mid and long-term target profiles. The profiles summarize the results and show the necessity to act. Figure 5 shows the profiles from one of the case studies within the project. First measures and projects can be derived and structured, e.g., using a high-level road map. To detail the necessary transformation process, the target position has to be translated into an implementation plan.

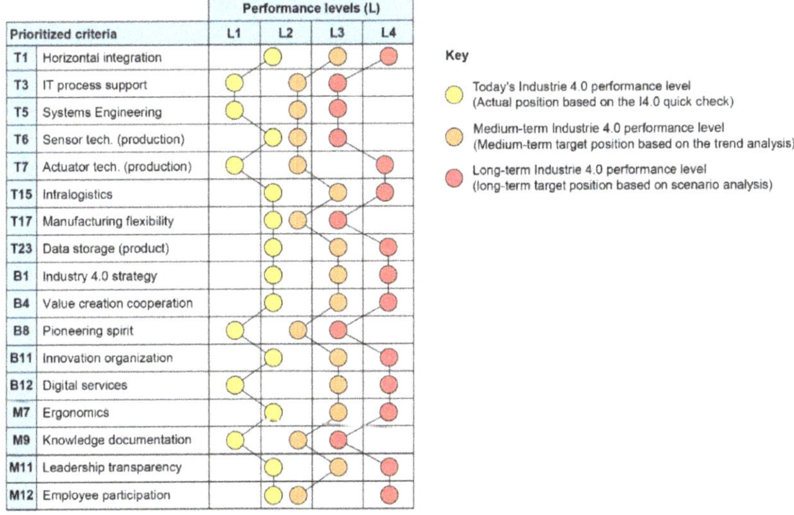

Figure 5. Mid- and long-term Industry 4.0 target profile [16,20].

4.3.3. Phase 3: Industry 4.0 Implementation Planning

When the status quo and target position are known, the implementation of Industry 4.0 within the company can be planned. The goal of this phase is, hence, a suitable implementation path. Since this is a complex and challenging task, solution patterns are utilized to integrate existing knowledge about Industry 4.0 into the method. First, the patterns are described, then it is explained how they can be assessed, combined, and structured to drive Industry 4.0.

Solution patterns for Industry 4.0 Implementation Planning: There is already a large body of knowledge regarding options for action in the context of Industry 4.0 through pioneers. As argued, this knowledge can be made accessible for companies following the solution pattern approach. In our case, the term Industry 4.0 implementation pattern is used. The use and proficient combination of several Industry 4.0 implementation patterns enable the successive transformation of the company's status quo (maturity level) today to its desired target position (target maturity level) in the future (Figure 6).

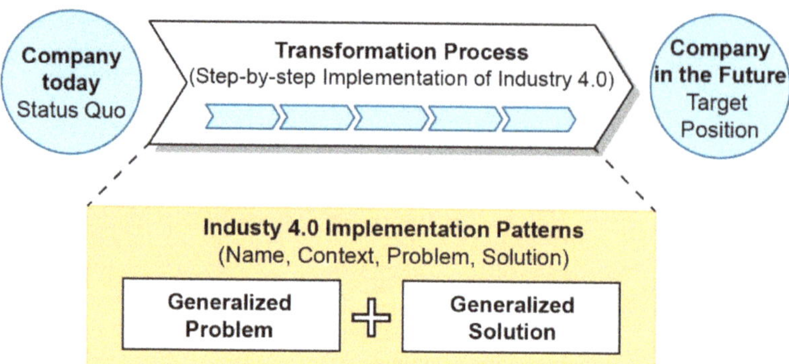

Figure 6. Pattern-based transformation planning [20].

The patterns stem from different so-called implementation fields. An implementation field is composed of the criteria of the quick check and includes criteria from up to all three socio-technical dimensions. In this context, an implementation field represents a thematic alignment of options for action and company-specific goals. A short profile with a concise description and criteria for performance evaluation relevant to the implementation field supports the subsequent selection of relevant implementation fields. The number of criteria per dimension determines the direction of impact of the implementation field. A total of 12 implementation fields have been identified (Figure 7).

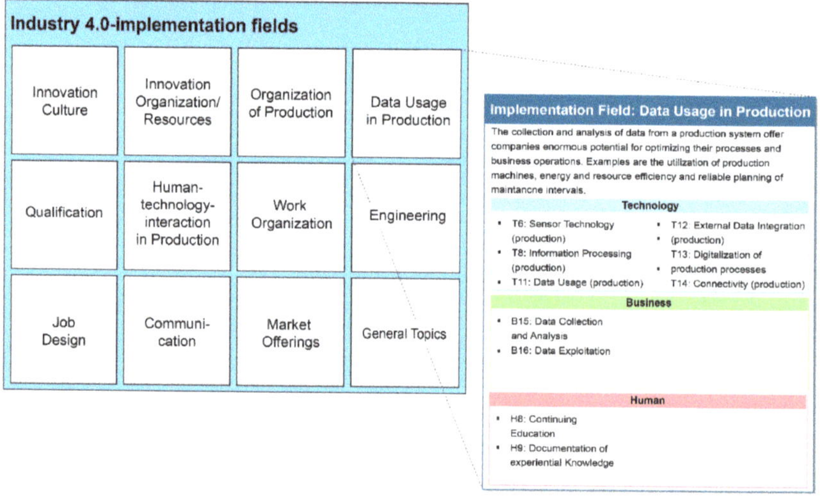

Figure 7. Industry 4.0 implementation fields [20].

Within 12 implementation fields, 83 Industry 4.0 implementation patterns were identified. This was done analyzing successful examples, i.e., best practices. In addition to this approach, there are two other widely acknowledged ways to identify patterns: observation and analysis of unsuccessful examples and the derivation of patterns based on abstract arguments [35]. For Industry 4.0, the best practice approach is best suitable since the field is still young, but old enough that many interesting solutions have been discovered. The analyzed knowledge base consisted of the following four sources:

1. **Practical project examples:** In the context of this study, best practices represent Industry 4.0 projects that have been successfully implemented by pioneers—mostly large companies. This allows identifying both, problems and associated solutions. A comprehensive list of projects is for example provided by the Industry 4.0 platform initiated by the German Federal Ministry of Education and Research as well as by acatech (www.plattform-i40.de (accessed on 14 July 2020)).

2. **Studies:** Publications that deal with challenges and successfully implemented solutions or with future developments provide direct or indirect indications of potential implementation patterns. An example of this is the accompanying research for AUTONOMIK, an Industry 4.0 technology program carried out by the German Federal Ministry for Economic Affairs and Energy [19].

3. **It's OWL transfer projects:** Within the cluster of excellence it's OWL 171 so-called transfer projects were successfully carried out. Solutions for Industry 4.0 problems were developed in 8-to-10-month project collaborations between research institutes and SMEs on various cross-sectional topics such as self-optimization or systems engineering. Both the problems as well as the associated solutions represent a valuable source of knowledge [55].

4. **Industry 4.0 demonstrators:** Representations of prototypical solutions in smart factories (e.g., SmartFactoryOWL) internal, and external exhibition demonstrators also represent suitable sources for implementation patterns.

134 Best Practices were analyzed regarding repetitive problems and associated solutions to identify patterns. The resulting 83 implementation patterns were then visualized in the form of so-called pattern cards. They comprise a detailed description of the established components of a solution pattern—name, problem, solution, and context (Figure 8). Based on the implementation field, each implementation pattern is assigned to a direction of impact.

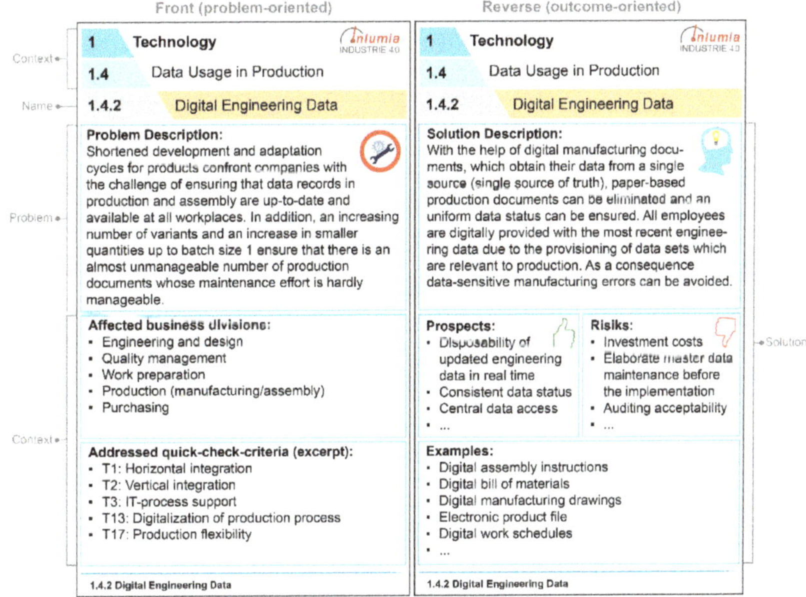

Figure 8. Example of a documented implementation pattern [20].

Identification of relevant implementation fields: To narrow the solution space, first, the relevant implementation fields are identified. This is done based on the current and target maturity levels. Because every implementation field is characterized by a bundle of criteria of the maturity model, the implementation fields can be ranked based on their contribution to bridging the gaps. In most of the considered case studies, this allowed to focus on three or four implementation fields, drastically reducing the number of patterns to be considered (to around 20), and, hence, allowing us to master the complexity.

Assessment of Industry 4.0 implementation patterns: The implementation patterns within the chosen implementation fields are then evaluated in a two-stage process. First, restrictions are examined regarding the feasibility of implementing the patterns in the respective company. Such restrictions are, e.g., the type of production (pure assembly), the market offerings, or the average age structure of the employees of a company. For example, a purely mechanical product is difficult to use for the introduction of data-based services. Second, the feasible implementation patterns are systematically evaluated. For this purpose, an utility analysis is applied. Five evaluation criteria are used for assessing the implementation patterns, they can be weighed individually for each company (Figure 9).

Figure 9. Evaluation criteria for the assessment of implementation patterns [20,73].

Following the assessment, the highest-rated implementation patterns are selected for further consideration.

Combination analysis: In the context of socio-technical systems theory, the three dimensions people, technology, and organization are to be considered as independent subsystems, but their mutual dependence and interaction always have to be taken into account [10]. In addition to the socio-technical dependencies, there are also dependencies between implementation patterns themselves. The application of a pattern usually leads to other patterns that have to be considered upstream or downstream. Therefore, when the Industry 4.0 implementation patterns are used, they must form a consistent whole and be put into a meaningful sequence. For this purpose, the implementation patterns are investigated with dependency and combination matrices. The matrices show, which patterns require other patterns as prerequisites, and which patterns can be combined for mutual benefit. Depending on the evaluations within the matrices and starting from an individual initial pattern (which resulted from the pattern assessment), an algorithm for topological sorting is applied and an individual pattern system is derived (Figure 10).

The pattern system maps the dependencies between the patterns and puts them in order. This results in multiple implementation phases, from which possible socio-technical implementation paths are generated.

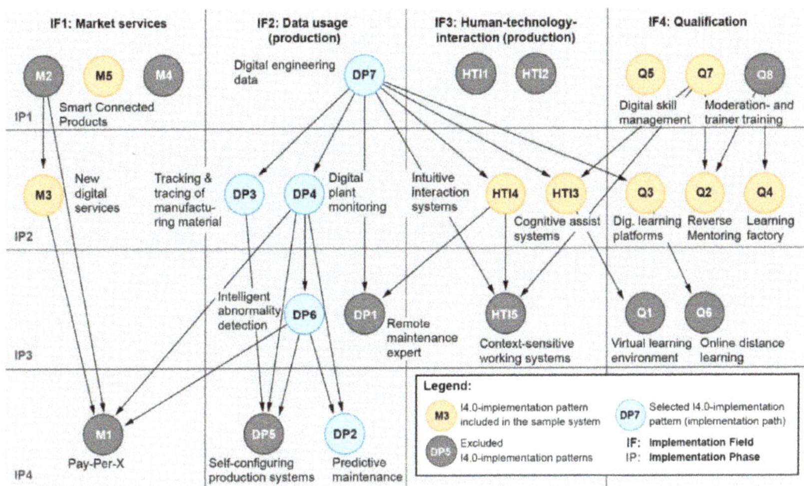

Figure 10. Pattern system for Industry 4.0 Implementation [20].

4.3.4. Industry 4.0 Transformation Setup

After the status quo is known, a target position is determined, and a desirable implementation path is derived, the concrete transformation must be planned. For that, measures are defined. Since the transformation is associated with many demanding changes, the risks of the transformation have to be taken into consideration as well. Only then, the transformation setup can be specified within a master plan of action.

Definition of measures: The general prescriptions documented in the chosen patterns must be concretized. Individual measures have to be derived, that are necessary to realize the patterns. Usually, this is done within a workshop setting. Here, the insights and documents from the in-depth analysis should be considered. The workshop results in a first concrete implementation roadmap.

Risk assessment: Various issues arise for companies, especially regarding the smooth introduction of the use cases. These are often less technical, but rather risks that are difficult to assess concerning the organization, e.g., economic viability, and in terms of people. Even if the installation of new technologies is successful (e.g., assembly assistance system), the beneficial operation depends to a large extent on investment costs, accurately fitting processes, adequate competencies, and employee acceptance. The resulting socio-technical risks and their interdependencies are difficult to manage and should be considered systematically at an early stage. For that, workshops for risk identification, analysis, and mitigation should be held. This results in a prioritized list of measures for the mitigation of risks.

Masterplan of action: Current maturity, target maturity, implementation paths, and measures and risk assessment are combined in a master plan of action. This is elaborated on and discussed in an interdisciplinary workshop. On the one side the master plan comprises the company's initial situation described by the fields of action from the maturity check. On the other side, the target situation is included in the form of the identified target profile. The iterative transformation process is documented connecting both. For this purpose, the intended use cases and critical risks are listed in accordance with the implementation path. Then, both, implementation activities and risk mitigations, are defined in the form of measures, after being differentiated into the socio-technical dimensions people, business, and technology. Existing activities should be considered, and synergy potentials should be exploited. The transformation setup summarizes all the results of the method and translates them into a gradual plan of action for the company. Furthermore, it serves as a transparent means of communication within the entire company regarding digitalization efforts and as a basis for the capacity planning of subsequent pilot projects.

Transformation controlling: Since a transformation process is dynamic in nature, the master plan of action should be reviewed regularly in terms of the underlying premises and the progress made. For this purpose, it is first necessary to monitor the trends and scenarios to ensure the plausibility of the perception of the future. This review is necessary because changes in the assumptions about the future may lead to changes in the target maturity levels. Depending on the identified effects, the implementation controlling then checks whether the implementation is going according to plan or needs to be adjusted.

5. Discussion

This chapter will reflect on the results mentioned before. First, we shortly emphasize the main results and highlight peculiarities compared to literature. Then, we will provide insights from the application, show limitations, and elaborate on further research paths.

5.1. Main Results

In summary, our research delivers a comprehensive method for socio-technical performance improvement in the course of Industry 4.0. It combines a maturity model, future-oriented target definition, pattern-based implementation planning, and transformation setup to enable companies to plan their transformation from a socio-technical viewpoint and to realize their intended performance improvement. The method components provide consistent support for the companies, hence answering the research question *"How can companies identify and attain an individual and performance increasing Industry 4.0 target position?"*. The application within five use cases showed that the method is both useful and usable for companies.

5.2. Peculiarities Compared to Literature

Maturity models are nothing new as per se. According to research by De Bruin, for example, more than 150 different maturity models were already in place by 2005 [22]. Established Industry 4.0 specific approaches in practice include the acatech Maturity Index [14] or the VDMA Guideline [74]. Knowing the degree of one's digitalization in comparison to the competition is an essential prerequisite for the successful digital transformation of a company [15]. However, the numerous existing maturity models are often very generic reference systems with only descriptive characters. Therefore, it is difficult to adapt them to the individual needs of companies, especially SMEs. Furthermore, the focus of most maturity models is on the technical perspective. The maturity level of digitalization is assessed based on the use of technology in the company. Especially, aspects of business in the sense of socio-technical perspective are often neglected. The method presented in this paper solves these challenges and hence extends the existing knowledge base. In addition to this, our method is—to the best of our knowledge—the first to combine a maturity model with foresight approaches. Since the transformation towards Industry 4.0 is a long-term endeavor, this seems appropriate and promising. Additionally, there is a lack of prescriptive solution proposals for achieving a targeted maturity level. Deriving knowledge about concrete solutions associated with a problem is therefore very difficult [75]. Hence, our method offers significant added value compared to the state of the art. At last, our method integrates first risk management aspects into the transformation towards Industry 4.0 extending the perspective.

5.3. Insights from Application

In the course of validating the instrument, it was possible to gain further insights into Industry 4.0. Besides the application within the case studies, the quick check was made accessible online, which allowed identifying commonalities between companies, particularly with regard to the performance assessment. The online quick check resulted in a database that includes a sample of over 250 companies that can be used for comparisons between companies. A cross-company evaluation yields the following exemplary findings in the socio-technical dimensions as of 2020 [20]:

In the dimension *technology*, it can be recorded that digitalization is not yet far advanced. For most companies (54%), the exchange of information is aggravated by media breaks. Only a small amount of information is digitally available. Horizontal integration across different IT systems is a major challenge for companies. A total of 75% of companies state that there is no or only partial networking of the IT systems of individual value-creation steps within the company (e.g., production and logistics). After all, 51% of respondents state that so far only selected data is stored. However, in 39% of the cases, there is no further processing and use of the data by upstream or downstream systems.

Investigating the *business* dimension, 59% of companies pay attention to Industry 4.0 during the strategy process, but a concrete strategy has not yet been formulated. Challenges lie in the transformation of the entire organization to form more flexible and open structures. Most of the companies surveyed maintain long-term, contractually bound ("rigid") relationships (26%) or short-term adaptable ("flexible") business relationships with a few selected partners (45%). In most companies, there are no clear (24%) or clear but rigid responsibilities within the company for incremental and radical innovations, as well as rigid innovation processes that are independent of projects (41%). In addition, business model development within the company is also often unstructured (38%).

For the *people* dimension, rigid structures still prevail among employees. Only 8% decide completely independently on the design of their activities or the planning of their work schedule, and only 10% have irregular daily schedules with ever-changing requirements. This impression is also confirmed regarding the flexibility of working hours, even though this might have significantly improved in the course of the Corona pandemic. Most respondents have clearly defined attendance and break times for employees (31%) or flextime with flexibly scheduled breaks (35%). Overall, digitalization and people are not yet ideally synced. Still, 26% of the companies surveyed state that the disruptions regularly lead to delays because people and technology are waiting for each other.

A further fuzzy-set qualitative comparative analysis of the data from the quick check by Schneider and colleagues (2022) revealed two success paths for technological maturity: intensive training and strong worker participation combined with strong entrepreneurial culture or with strong customer-oriented innovation in larger firms [76].

The results presented here show the commonalities among companies in Germany about the status quo of Industry 4.0. However, even if many share the same problems, the respective target positions and the solutions depend on their individual requirements and contingencies. To validate that our method is suitable to help companies define and reach those, the method was conducted with the five early adopters from our case studies.

In our case studies, the initial situations and the relevant problems regarding Industry 4.0 were quite similar. Topics included eliminating analog documents, reducing media breaks, and increasing the competencies of employees. However, while the topics were similar, the concrete solutions created were quite different since the contingencies of the companies had to be taken into account. The strive to eliminate analog documents, for example, led one company to digitalize their order documents in the production, while another company focused on machine documents and the use of data from machines. This shows that companies must focus on what is specifically necessary and not on what is fundamentally possible. Despite the generalized content of the solution patterns, it was possible to tailor them individually to the different companies.

5.4. Limitations

Although our results stem from a comprehensive consortium research project, there remain some limitations. First, qualitative research per se is limited in terms of generalizability. We applied the method in multiple cases in practice, but further studies on the application of the method are necessary to further prove the validity. Furthermore, in our work, we focus on the intellectual perspective of the transformation process without addressing the social dimension. The people participating in the application of the method and their relations are not considered in detail (c.f., [77]). The third limitation is a temporal one: Our results reflect the insights on Industry 4.0 as of today. Hence, our method should be continuously reflected on, and new insights must be integrated. At last, regarding the insights from the quick check, it must be stated, that the companies participating through the online tool are not a representative sample because they were not randomly selected but participated because of their interest.

5.5. Future Research

Research on Industry 4.0 is not exhausted yet; there are still myriads of open questions. Especially driven by the socio-political discourse, sustainability is becoming more important for Industry 4.0. Research should emphasize an integrative view. The socio-technical dimensions technology, people, and organization/business of Industry 4.0 should be aligned to the social, ecological, and economic views of sustainability. Implementing Industry 4.0 in a company is also associated with risks. Our method delivers a first approach to deal with these risks that suits the scope of the method, but further research into this topic seems promising and necessary (c.f., [44,78]). Future research should focus on methods and approaches, that support these tasks.

6. Conclusions

Our study investigates how companies can adopt an individual, suitable approach to Industry 4.0 to increase their firm performance. The resulting method was developed and validated within a consortium research project of three years employing design science research and method engineering. The method comprises four phases: (1) Industry 4.0 maturity check, (2) Industry 4.0 target definition, (3) Industry 4.0 implementation planning, and (4) Industry 4.0 transformation setup. The first phase helps companies to determine their current performance level regarding Industry 4.0 using a socio-technical maturity model. Relevant fields of action are analyzed using a set of distinct methods. The second phase allows companies to determine their target position utilizing corporate foresight techniques. During the third phase, the implementation of the target position is planned by building combinations of established solution patterns. The last phase transfers the implementation path into a transformation setup comprising a gradual plan of action. The findings contribute both to theory and practice. For practitioners, we deliver a method they can use for the transformation of their own company. At the same time, however, we advise using the support of consultants in this regard. Comprehensive data regarding the performance level of other companies, that participated through the online tool, give them orientation regarding their peer group. For the scientific community, we provide one of the few socio-technical approaches to Industry 4.0 maturity models and the first approach to combine maturity models with foresight and extensive prescriptive knowledge for the transformation process.

Author Contributions: Conceptualization, C.K. and D.H.; methodology, C.K.; validation, C.P. and D.H.; investigation, C.P. and D.H.; writing—original draft preparation, C.K. and D.H.; writing—review and editing, C.K. and D.H.; visualization, D.H., C.P. and C.K.; supervision, R.D. and A.K.; project administration, R.D. and A.K. All authors have read and agreed to the published version of the manuscript.

Funding: This article is based on the results of the consortium research project INLUMIA—Instruments for Improving the Performance of Companies through Industry 4.0. The joint project consisting of 11 partners was funded by the European Regional Development Fund NRW (EFDF.NRW). Further insights stem from the research project SORISMA—Sociotechnical Risk Management for the Introduction of Industry 4.0, which is conducted by three research institutes and six companies. The project is funded by the European Regional Development Fund NRW (ERDF.NRW) as well.

Institutional Review Board Statement: Not applicable.

Data Availability Statement: Not applicable.

Acknowledgments: We want to acknowledge the valuable contributions from the research partners, project members, and all students who participated in the research. The results shown are a shortened and revised excerpt from our final project report [20]. We also thank the reviewers for their valuable comments.

Conflicts of Interest: The authors declare no conflict of interest. The funders had no role in the design of the study; in the collection, analyses, or interpretation of data; in the writing of the manuscript, or in the decision to publish the results.

References

1. Da Xu, L.; Xu, E.L.; Li, L. Industry 4.0: State of the art and future trends. *Int. J. Prod. Res.* **2018**, *56*, 2941–2962. [CrossRef]
2. Kagermann, H. Change Through Digitization—Value Creation in the Age of Industry 4.0. In *Management of Permanent Change*; Albach, H., Meffert, H., Pinkwart, A., Reichwald, R., Eds.; Springer Gabler: Wiesbaden, Germany, 2015; pp. 23–45. ISBN 978-3-658-05013-9.
3. Lasi, H.; Fettke, P.; Kemper, H.-G.; Feld, T.; Hoffmann, M. Industry 4.0. *Bus. Inf. Syst. Eng.* **2014**, *6*, 239–242. [CrossRef]
4. Kagermann, H.; Wahlster, W.; Helbig, J. *Recommendations for Implementing the Strategic Initiative INDUSTRIE 4.0: Final Report of the Industrie 4.0 Working Group*; Acatech—Deutsche Akademie der Technikwissenschaften e.V.: Berlin, Germany, 2013.
5. Gausemeier, J.; Klocke, F.; Dülme, C.; Eckelt, D.; Kabasci, P.; Kohlhuber, M.; Schön, N.; Schröder, S.; Wellensiek, M. *Industrie 4.0: International Benchmark, Future Options and Recommendations for Action in Production Research*; Heinz Nixdorf Institut, Universität Paderborn, Werkzeugmaschinenlabor WZL der Rheinisch-Westfälischen Technischen Hochschule Aachen: Paderborn, Aachen, 2016.
6. Müller, J.M.; Buliga, O.; Voigt, K.-I. Fortune favors the prepared: How SMEs approach business model innovations in Industry 4.0. *Technol. Forecast. Soc. Chang.* **2018**, *132*, 2–17. [CrossRef]
7. Davies, R.; Coole, T.; Smith, A. Review of Socio-technical Considerations to Ensure Successful Implementation of Industry 4.0. *Procedia Manuf.* **2017**, *11*, 1288–1295. [CrossRef]
8. Walker, G.H.; Stanton, N.A.; Salmon, P.M.; Jenkins, D.P. A review of sociotechnical systems theory: A classic concept for new command and control paradigms. *Theor. Issues Ergon. Sci.* **2008**, *9*, 479–499. [CrossRef]
9. Rousseau, D.M. Technological differences in job characteristics, employee satisfaction, and motivation: A synthesis of job design research and sociotechnical systems theory. *Organ. Behav. Hum. Perform.* **1977**, *19*, 18–42. [CrossRef]
10. Ulich, E. Arbeitssysteme als Soziotechnische Systeme—Eine Erinnerung [Work systems as sociotechnical systems—A memory]. *J. Psychol. Des Alltagshandelns* **2013**, *6*, 4–12.
11. Pierenkemper, C.; Drewel, M.; Gausemeier, J. Zukunftsorientierte Leistungssteigerung von Unternehmen durch Industrie 4.0 [Future-oriented performance enhancement of companies through Industry 4.0]. In *Tagungsband AALE 2018, Proceedings of the AALE 2018, Cologne, Germany, 1–2 March 2018*; Smajic, H., Ed.; VDE Verlag: Berlin, Germany, 2018; pp. 19–27.
12. Bajic, B.; Rikalovic, A.; Suzic, N.; Piuri, V. Industry 4.0 Implementation Challenges and Opportunities: A Managerial Perspective. *IEEE Syst. J.* **2021**, *15*, 546–559. [CrossRef]
13. Masood, T.; Sonntag, P. Industry 4.0: Adoption challenges and benefits for SMEs. *Comput. Ind.* **2020**, *121*, 103261. [CrossRef]
14. Schuh, G.; Anderl, R.; Dumitrescu, R.; Krüger, A.; Hompel, M.t. *Industrie 4.0 Maturity Index: Managing the Digital Transformation of Companies—UPDATE 2020*; Acatech STUDY: Munich, Germany, 2020. Available online: https://www.acatech.de/publikation/industrie-4-0-maturity-index-update-2020/ (accessed on 2 July 2022).
15. Ennemann, M.; Linden, S. *Survival of the Smartest 2016: Studie zur Digitalen Transformation [Study Regarding Digital Transformation]*; KPMG: Amstelveen, The Netherlands, 2016.
16. Pierenkemper, C.; Gausemeier, J. Developing Strategies for Digital Transformation in SMEs with Maturity Models. In *Digitalization: Approaches, Case Studies, and Tools for Strategy, Transformation and Implementation*; Schallmo, D., Tidd, J., Eds.; Springer: Cham, Switzerland, 2021; pp. 103–124. ISBN 978-3-030-69379-4.
17. Büst, R.; Hille, M.; Schestakow, J. *Digital Business Readiness*; Crisp Research: Kassel, Germany, 2015.
18. Schmitz, C.; Tschiesner, A.; Jansen, C.; Hallerstede, S.; Garms, F. *Industry 4.0: Capturing Value at Scale in Discrete Manufacturing*; McKinsey & Company: Chicago, IL, USA, 2019.
19. Federal Ministry for Economic Affairs and Energy. *Studie Industrie 4.0—Volkswirtschaftliche Faktoren für den Standort Deutschland [Study Industry 4.0—Economic Factors for Germany as a Business Location]*; Federal Ministry for Economic Affairs and Energy: Berlin, Germany, 2015.
20. Dumitrescu, R.; Maier, G.; Pierenkemper, C.; Reinhold, J.; Hobscheidt, D.; Lipsmeier, A.; Gundlach, T.; Schlicher, K.; Heppner, H.; Seifert, L. *Soziotechnische Leistungssteigerung von Unternehmen durch Industrie 4.0 [Socio-Technical Performance Enhancement of Companies through Industry 4.0]*; Heinz Nixdorf Institute: Paderborn, Germany, 2022; in press.
21. Culot, G.; Nassimbeni, G.; Orzes, G.; Sartor, M. Behind the definition of Industry 4.0: Analysis and open questions. *Int. J. Prod. Econ.* **2020**, *226*, 107617. [CrossRef]
22. de Bruin, T.; Freeze, R.; Kulkarni, U.; Rosemann, M. Understanding the Main Phases of Developing a Maturity Assessment Model. In Proceedings of the 16th Australasian Conference on Information Systems, Sydney, Australia, 29 November–2 December 2005.

23. Becker, J.; Knackstedt, R.; Pöppelbuß, J. Developing Maturity Models for IT Management. *Bus. Inf. Syst. Eng.* **2009**, *1*, 213–222. [CrossRef]
24. IT Governance Institute. *COBIT 4.1: Framework, Control Objectives, Management Guidelines, Maturity Models*; IT Governance Institute: Rolling Meadows, IL, USA, 2007. ISBN 1-933284-72-2.
25. Wagire, A.A.; Joshi, R.; Rathore, A.P.S.; Jain, R. Development of maturity model for assessing the implementation of Industry 4.0: Learning from theory and practice. *Prod. Plan. Control.* **2021**, *32*, 603–622. [CrossRef]
26. Gökalp, E.; Martinez, V. Digital transformation capability maturity model enabling the assessment of industrial manufacturers. *Comput. Ind.* **2021**, *132*, 103522. [CrossRef]
27. Chandler, A.D. *Strategy and Structure: Chapters in the History of the Industrial Enterprise*; M.I.T. Press: Cambridge, MA, USA, 1962; ISBN 9780262530095.
28. Oleff, A.; Malessa, N. Strategischer Ansatz zur Industrie 4.0-Transformation [Strategic Approach for Industry 4.0 Transformation]. *ZWF* **2018**, *113*, 173–177. [CrossRef]
29. Tüllmann, C.; Sagner, D.; Prasse, C.; Piastowski, H. Vernetzt denken—Vernetzt handeln [Networked Thinking—Networked Acting]. In *Disruption und Transformation Management: Digital Leadership—Digitales Mindset—Digitale Strategie [Disruption and Transformation Management: Digital Leadership—Digital Mindset—Digital Strategy]*; Keuper, F., Schomann, M., Sikora, L.I., Wassef, R., Eds.; Springer Gabler: Wiesbaden/Heidelberg, Germany, 2018; pp. 165–186. ISBN 978-3-658-19130-6.
30. Schumacher, A.; Nemeth, T.; Sihn, W. Roadmapping towards industrial digitalization based on an Industry 4.0 maturity model for manufacturing enterprises. In *Procedia CIRP, Proceedings of the 12th CIRP Conference on Intelligent Computation in Manufacturing Engineering, Gulf of Naples, Italy, 18–20 July 2018*; Teti, R., D'Adonna, D.M., Eds.; Elsevier B.V.: Amsterdam, The Netherlands, 2018; pp. 409–414.
31. Pessl, E.; Sorko, S.R.; Mayer, B. Roadmap Industry 4.0—Implementation Guideline for Enterprises. *Int. J. Sci. Technol. Soc.* **2017**, *5*, 193–202. [CrossRef]
32. Morlock, F.; Wienbruch, T.; Leineweber, S.; Kreimeier, D.; Kuhlenkötter, B. Industrie 4.0-Transformation für produzierende Unternehmen [Industrie 4.0-transformation of manufacturing companies]. *ZWF* **2016**, *111*, 306–309. [CrossRef]
33. Jodlbauer, H.; Schagerl, M. Reifegradmodell Industrie 4.0—Ein Vorgehensmodell zur Identifikation von Industrie 4.0 Potentialen [Industrie 4.0 maturity model—A process model for identifying Industrie 4.0 potentials]. In *Informatik 2016*; Mayr, H.C., Pinzger, M., Eds.; Gesellschaft für Informatik: Bonn, Germany, 2016; ISBN 9783885796534.
34. Alexander, C.; Ishikawa, S.; Silverstein, M.; Jacobson, M.; Fiksdahl-King, I.; Angel, S. *A Pattern Language: Towns Buildings Construction*; Oxford University Press: New York, NY, USA, 1977; ISBN 13 978-0-19-501919-3.
35. Alexander, C. *The Timeless Way of Building*; Oxford University Press: New York, NY, USA, 1979; ISBN 0195022483.
36. Anacker, H.; Dumitrescu, R.; Kharatyan, A.; Lipsmeier, A. Pattern Based Systems Engineering—Application of Solution Patterns in the Design of Intelligent Technical Systems. *Proc. Des. Soc. Des. Conf.* **2020**, *1*, 1195–1204. [CrossRef]
37. Echterfeld, J.; Gausemeier, J. Digitising Product Portfolios. *Int. J. Innov. Mgt.* **2018**, *22*, 1840003. [CrossRef]
38. Weking, J.; Stöcker, M.; Kowalkiewicz, M.; Böhm, M.; Krcmar, H. Leveraging industry 4.0—A business model pattern framework. *Int. J. Prod. Econ.* **2020**, *225*, 107588. [CrossRef]
39. Gausemeier, J.; Wieseke, J.; Echterhoff, B.; Koldewey, C.; Mittag, T.; Schneider, M.; Isenberg, L. *Mit Industrie 4.0 zum Unternehmenserfolg [With Industry 4.0 to Business Success]: Integrative Planung von Geschäftsmodellen und Wertschöpfungssystemen [Integrative Planning of Business Models and Value Creation Systems]*; Heinz Nixdorf Institute: Paderborn, Germany, 2017.
40. Dumitrescu, R.; Anacker, H.; Gausemeier, J. Design framework for the integration of cognitive functions into intelligent technical systems. *Prod. Eng. Res. Devel.* **2013**, *7*, 111–121. [CrossRef]
41. Balogun, J.; Hope Hailey, V.; Gustafsson, S. *Exploring Strategic Change*, 4th ed.; Pearson Education Limited: Harlow, UK, 2016; ISBN 978-0-273-77891-2.
42. Merz, S.L. Industrie 4.0—Vorgehensmodell für die Einführung [Industry 4.0—Procedure model for the implementation]. In *Einführung und Umsetzung von Industrie 4.0 [Introduction and Implementation of Industry 4.0]*; Roth, A., Ed.; Springer: Berlin, Heidelberg, 2016; pp. 83–132. ISBN 978-3-662-48504-0.
43. Hennegriff, S.; Terstegen, S.; Stowasser, S.; Dander, H.; Adler, P. Vorgehensmodell für die Industrie 4.0: Strukturierte Einführung und Umsetzung von Digitalisierungsmaßnahmen in der produzierenden Industrie [Procedure model for Industry 4.0: Structured introduction and implementation of digitization measures in the manufacturing industry]. *Ind. 4.0 Manag.* **2019**, *3*, 47–50.
44. Menzefricke, J.S.; Wiederkehr, I.; Koldewey, C.; Dumitrescu, R. Socio-technical risk management in the age of digital transformation-identification and analysis of existing approaches. *Procedia CIRP* **2021**, *100*, 708–713. [CrossRef]
45. Otto, B.; Bärenfänger, R.; Steinbuß, S. Digital Business Engineering: Methodological Foundations and First Experiences from the Field. In *BLED 2015, Proceedings of the 28th Bled eConference "#eWellbeing", Bled, Slovenia, 7–10 June 2015*; University of Maribor, University Press: Maribor, Slovenia, 2015; pp. 58–76.
46. March, S.T.; Smith, G.F. Design and natural science research on information technology. *Decis. Support Syst.* **1995**, *15*, 251–266. [CrossRef]
47. Yin, R.K. *Case Study Research and Applications: Design and Methods*, 6th ed.; Sage: Los Angeles, CA, USA, 2018; ISBN 9781544308388.
48. Österle, H.; Otto, B. Consortium Research. *Bus. Inf. Syst. Eng.* **2010**, *2*, 283–293. [CrossRef]
49. Brinkkemper, S. Method engineering: Engineering of information systems development methods and tools. *Inf. Softw. Technol.* **1996**, *38*, 275–280. [CrossRef]

50. Kumar, K.; Welke, R.J. Methodology Engineering: A Proposal for situation-specific methodology construction. In *Challenges and Strategies for Research in Systems Development*; Cotterman, W.W., Senn, J.A., Eds.; John Wiley & Sons, Ltd.: Hoboken, NJ, USA, 1992; pp. 257–269.
51. Harmsen, F.; Brinkkemper, S.; Oei, H. Situational method engineering for information system project approaches. In *Methods and Associated Tools for the Information Systems Life Cycle, Proceedings of the IFIP WG 8.1 Working Conference on Methods and Associated Tools for the Information Systems Life Cycle, Maastricht, The Netherlands, 26–28 September 1994*; Verrijn-Stuart, A.A., Ed.; Elsevier: Amsterdam, The Netherlands, 1994; ISBN 0-444-82074-4.
52. Gutzwiller, T.A. *Das CC RIM-Referenzmodell für den Entwurf von Betrieblichen, Transaktionsorientierten Informationssystemen [The CC RIM Reference Model for the Design of Operational, Transaction-Oriented Information Systems]*; Physica-Verlag HD: Heidelberg, Germany, 1994; ISBN 978-3-642-52406-6.
53. Goldkuhl, G.; Karlsson, F. Method Engineering as Design Science. *JAIS* **2020**, *21*, 1237–1278. [CrossRef]
54. Deflorin, P.; Scherrer, M.; Eberhardt, N. Digitale Intensität und Management der Transformation [Digital Intensity and Transformation Management]. In *Digitale Transformation von Geschäftsmodellen [Digital Transformation of Business Models]*; Schallmo, D., Rusnjak, A., Anzengruber, J., Werani, T., Jünger, M., Eds.; Springer Fachmedien: Wiesbaden, Germany, 2017; pp. 265–282. ISBN 978-3-658-12387-1.
55. Dumitrescu, R.; Ebbesmeyer, P.; Fechtelpeter, C.; Gausemeier, J.; Hobscheidt, D.; Kühn, A. *On the Road to Industry 4.0—Technology Transfer in the SME Sector*; It's OWL Clustermanagement GmbH: Paderborn, Germany, 2017.
56. Westermann, T. *Systematik zur Reifegradmodell-Basierten Planung von Cyber-Physical Systems des Maschinen-und Anlagenbaus [Systematic for Maturity Model-Based Planning of Cyber-Physical Systems in Mechanical and Plant Engineering]*; Paderborn University: Paderborn, Germany, 2017.
57. Rauter, R. Interorganisationaler Wissenstransfer [Interorganizational Knowledge Transfer]: Zusammenarbeit Zwischen Forschungseinrichtungen und KMU [Cooperation between Research Institutions and SMEs]. Ph.D. Dissertation, Graz University, Wiesbaden, Germany, 2013.
58. Gausemeier, J.; Plass, C. *Zukunftsorientierte Unternehmensgestaltung [Future-Oriented Company Design]: Strategien, Geschäftsprozesse und IT-Systeme für die Produktion von Morgen [Strategies, Business Processes and IT Systems for Tomorrow's Production]*, 2nd ed.; Carl Hanser Verlag: Munich, Germany, 2014; ISBN 978-3446436312.
59. Müller-Stewens, G.; Lechner, C. *Strategisches Management [Strategic Management]: Wie Strategische Initiativen zum Wandel Führen: Der Strategic Management Navigator [How Strategic Initiatives Lead to Change: The Strategic Management Navigator]*, 5th ed.; Schäffer-Poeschel Verlag: Stuttgart, Germany, 2016; ISBN 978-3-7910-3439-3.
60. Kerth, N.L.; Cunningham, W. Using patterns to improve our architectural vision. *IEEE Softw.* **1997**, *14*, 53–59. [CrossRef]
61. Cloutier, R.J.; Dinesh, V. Applying the concept of patterns to systems architecture. *Syst. Eng.* **2007**, *10*, 138–154. [CrossRef]
62. Deigendesch, T. Kreativität in der Produktentwicklung und Muster als methodisches Hilfsmittel [Creativity in Product Development and Patterns as a Methodological Tool]. Ph.D. Dissertation, Karlsruhe Institute for Technology, Karlsruhe, Germany, 2009.
63. Schumacher, M. *Security Engineering with Patterns: Origins, Theoretical Model, and New Applications*; Springer: Berlin/Heidelberg, Germany, 2003; ISBN 978-3-540-45180-8.
64. Zimmermann, B. Pattern-basierte Prozessbeschreibung und unterstützung—Ein Werkzeug zur Unterstützung von Prozessen zur Anpassung von E-Learning Materialien [Pattern-Based Process Description and Support—A Tool to Support Processes for Adapting E-Learning Materials]. Ph.D. Dissertation, Technical University of Darmstadt, Darmstadt, Germany, 2008.
65. Hirsch-Kreinsen, H.; Hompel, M.t.; Ittermann, P.; Niehaus, J.; Dregger, J.; Mättig, B.; Kirks, T. *Digitalisierung von Industriearbeit: Forschungsstand und Entwicklungsperspektiven—Ein Leitbild der Technologischen, Organisatorischen und Sozialen Herausforderungen der Industrie 4.0 [Digitization of Industrial Work: State of Research and Development Prospects—A Guide to the Technological, Organizational and Social Challenges of Industry 4.0]*; Technical University of Dortmund: Dortmund, Germany, 2015.
66. Kollmann, T.; Schmidt, H. *Deutschland 1.0 [Germany 4.0]: Wie die Digitale Transformation Gelingt [How the Digital Transformation Can Succeed]*; Springer Gabler: Wiesbaden, Germany, 2016; ISBN 978-3-658-13145 6.
67. Fujitsu. *Walking the Digital Tightrope*; Fujitsu: Munich, Germany, 2016. Available online: https://www.fujitsu.com/global/imagesgig5/br-fujitsu-digital-tightrope-report-em-en.pdf (accessed on 2 June 2022).
68. Cole, T. *Digitale Transformation [Digital Transformation]: Warum die Deutsche Wirtschaft Gerade Die Digitale Zukunft Verschläft und Was Jetzt Getan Werden Muss! [Why the German Economy is Currently Sleeping through the Digital Future and What Needs to Be Done Now!]*, 2nd ed.; Verlag Franz Vahlen: Munich, Germany, 2017; ISBN 9783800653997.
69. Gausemeier, J.; Lewandowski, A.; Siebe, A.; Soetebeer, M. OMEGA: Object-Oriented Method Strategic Redesign of Business Processes. In *Changing the Ways We Work, Proceedings of the Shaping the ICT-Solutions for the Next Century, Conference on Integration in Manufacturing, Göteborg, Sweden, 6–8 October 1998*; Mårtensson, N., Mackay, R., Björgvinsson, S., Eds.; IOS Press: Amsterdam, The Netherlands, 1998; pp. 381–391. ISBN 4274902528.
70. Porter, M.E. What is strategy? *Harv. Bus. Rev.* **1996**, *74*, 61–78.
71. Drazin, R.; de van Ven, A.H. Alternative Forms of Fit in Contingency Theory. *Adm. Sci. Q.* **1985**, *30*, 514–539. [CrossRef]
72. Pierenkemper, C.; Reinhold, J.; Dumitrescu, R.; Gausemeier, J. Erfolg versprechende Industrie 4.0-Zielposition [Promising Industry 4.0 target position]. *I40M* **2019**, *2019*, 30–34. [CrossRef]
73. Hobscheidt, D.; Kühn, A.; Dumitrescu, R. Development of risk-optimized implementation paths for Industry 4.0 based on socio-technical pattern. *Procedia CIRP* **2020**, *91*, 832–837. [CrossRef]

74. Anderl, R.; Fleischer, J. *Guideline Industrie 4.0: Guiding Principles for the Implementation of Industrie 4.0 in Small and Medium Sized Businesses*; VDMA Verlag: Frankfurt am Main, Germany, 2015; ISBN 978-3-8163-0677-1.
75. Berghaus, S.; Back, A. Gestaltungsbereiche der Digitalen Transformation von Unternehmen: Entwicklung eines Reifegradmodells [Design areas of the digital transformation of companies: Development of a maturity model]. *Die Unternehm.* **2016**, *70*, 98–123. [CrossRef]
76. Schneider, M.; Hellweg, T.; Menzefricke, J.S. Identification of Human and Organizational Key Design Factors for Digital Maturity—A Fuzzy-Set Qualitative Comparative Analysis. *Proc. Des. Soc. Des. Conf.* **2022**, *2*, 791–800. [CrossRef]
77. Horovitz, J. New Perspectives on Strategic Management. *J. Bus. Strategy* **1984**, *4*, 19–33. [CrossRef]
78. Gabriel, S.; Grauthoff, T.; Joppen, R.; Kühn, A.; Dumitrescu, R. Analyzing socio-technical risks in implementation of Industry 4.0-use cases. *Procedia CIRP* **2021**, *100*, 241–246. [CrossRef]

Concept Paper

Digital Factory Transformation from a Servitization Perspective: Fields of Action for Developing Internal Smart Services

Jens Neuhüttler [1,2,*], Maximilian Feike [1,3,*], Janika Kutz [1,4], Christian Blümel [1] and Bernd Bienzeisler [1]

1   Fraunhofer IAO, Fraunhofer Institute for Industrial Engineering IAO, 74076 Heilbronn, Germany
2   Institute of Human Factors and Technology Management (IAT), University of Stuttgart, Nobelstr. 12, 70569 Stuttgart, Germany
3   Faculty of Media, Bauhaus-University Weimar, 99423 Weimar, Germany
4   Center for Cognitive Science, University of Kaiserslautern-Landau, 67663 Kaiserslautern, Germany
*   Correspondence: jens.neuhuettler@iao.fraunhofer.de (J.N.); maximilian.feike@iao.fraunhofer.de (M.F.)

Abstract: In recent years, a complex set of dynamic developments driven by both the economy and the emergence of digital technologies has put pressure on manufacturing companies to adapt. The concept of servitization, i.e., the shift from a product-centric to a service-centric value creation logic, can help manufacturing companies stabilize their business in such volatile times. Existing academic literature investigates the potential and challenges of servitization and the associated development of data-based services, so-called smart services, with a view to external market performance. However, with the increasing use of digital technologies in manufacturing and the development of internal smart services based on them, we argue that the existing insights on external servitization are also of interest for internal transformation. In this paper, we identify key findings from service literature, apply them to digital factory transformation, and structure them into six fields of action along the dimensions of people, technology, and organization. As a result, recommendations for designing digital factory transformation in manufacturing companies are derived from the perspective of servitization and developing internal smart services.

Keywords: servitization; digital factory transformation; smart services; IoT; AI; internal services

Citation: Neuhüttler, J.; Feike, M.; Kutz, J.; Blümel, C.; Bienzeisler, B. Digital Factory Transformation from a Servitization Perspective: Fields of Action for Developing Internal Smart Services. *Sci* **2023**, *5*, 22. https://doi.org/10.3390/sci5020022

Academic Editor: Johannes Winter

Received: 5 April 2023
Revised: 29 April 2023
Accepted: 11 May 2023
Published: 16 May 2023

1. Introduction

In recent years, the manufacturing sector is facing profound changes [1]. In addition to the changes posed by technological advances themselves, primarily by digitalization, there are several other challenges for the manufacturing industry to overcome [2]. These difficulties could be an effect of the digital transformation itself. Still, they could also be the outcome of regulation or recent international events that are fundamentally altering the market situation. This applies, in particular, to manufacturing companies' production and logistics areas as their effectiveness and efficiency are crucial when they want to remain competitive, even if the production plants are located in high-wage countries, such as Germany. Selected challenges that can be observed in different companies and sectors are listed below:

- **Difficulties with international supply chains:** Recently, partly due to the COVID-19 pandemic, global supply chains have faltered. As a consequence, for example, a seamless supply of spare parts and components is not always guaranteed. This has disrupted even particularly well-optimized JIT and JIS production processes. Sourcing and supply chain issues dominate the planning and management of manufacturing processes, making it difficult to fully assemble products and requiring improvisation, as well as ad hoc decision-making. Supply constraints in some industrial sectors have become so severe that they call into question the flexible mass production's prior achievements [3]. Energy supply and price have recently forced companies to adjust their manufacturing processes. Overall, this will lead to a re-evaluation of their supply chain strategies [4].

- **Sustainability**: Sustainability is a challenge with several origins, such as the need to stay cost competitive while energy costs are increasing. Additionally, there is an increasing focus of the customers on the impact of their own consumption. The most obvious and most controversially discussed cause are regulations, for example, the European Green Deal or the restriction of internal combustion engine vehicles. As a result, the manufacturing industry has to rethink the whole life cycle of its products, starting from the design to how they are produced or even their whole business models [5,6].

- **New global players:** In addition to the growing international pressure of existing competitors and the advance of new technologies, new companies are increasingly entering the markets, putting pressure on established manufacturers. These companies do not have a production history stretching back decades but instead, build their manufacturing processes closely based on procedures and processes previously only known to the digital economy. The most prominent example is Tesla, which is focusing everything on software and digitalization and whose manufacturing plants are literally being built as greenfield projects [7].

- **Digitalization of brown-field factories:** Although traditional companies have advantages due to decades of experience, efficient product design, and deep knowledge of production processes, they must face challenges originating in the organic growth of factory layouts, processes, and technology. Most notably, this organic growth also took place in IT systems and led to a fast number of different systems that are not well-connected, hard to maintain, and difficult to replace. This brown-field burden prevents companies from benefiting easily from adopting and scaling new technological advances and, therefore, they are not able to react flexibly to changing conditions [8].

- **Skilled labor shortage:** In recent years, the labor market in industrialized countries is favoring the employee's side. In addition to the rapidly expanded demand for specialized skills in IT and engineering, the obvious factor is the ever-increasing age of the average worker. This leads to a competitive environment where, in addition to the financial incentives, the working conditions in the form of work–life balance play an increasingly more prominent role [9]. These changes will have the added benefit that older workers can extend their working life [10]. The downside is that, especially in manufacturing, predominant shift models adapting to more flexible working conditions is difficult.

Even though some generally recognizable challenges have been listed here, the actual set of barriers is always multi-faceted and can vary depending on the particular company. This is accompanied by major economic risks, but these challenges also point to significant opportunities. These include creating intelligent and adaptive production structures that make it possible to manufacture high-quality products in series under competitive conditions, even under massive uncertainty. However, this will only be possible if production processes are fully digitized, from capacity and demand planning to delivery. The associated change process can also be described as digital transformation [9]. Therefore, the ability to implement digital transformation will be a key factor for the future competitiveness of companies and for overcoming the challenges described above. In the context of industrial manufacturing, this fundamental change can also be referred to as digital factory transformation.

The increasing use of digital technologies, such as artificial intelligence (AI) and the Internet of Things (IoT), is accompanied by high investment costs. In order to be able to amortize these, it is necessary to develop services that use digital technologies and the resulting data for concrete added value, such as automated and flexible production steps [10]. The resulting increase in the importance of services and a service-oriented value creation logic in manufacturing companies is discussed under the term servitization. In recent years, scientific research has produced a wide range of findings on the concept of servitization but is particularly concerned with the question of how products can be enriched with services and marketed to external customers in service-oriented business models [11]. It has been shown, for example, that companies can increase their resilience and flexibility in turbulent times by switching from a product-oriented to a service-oriented business model [12].

Against the backdrop of the challenges and need for change outlined above, as well as the availability of digital technologies, the importance of developing internal services for manufacturing companies is increasing. We argue that the approach of servitization and its dimensions are worth transferring to production and other internal processes. In our paper, we, therefore, want to explore how existing insights from the servitization literature and a service-specific perspective can be applied to key action areas of digital factory transformation and provide a new impetus.

2. Servitization and the Development of Internal Smart Services

2.1. Digital Servitization: The Changing Character of Value Creation

Changes in industrial value creation structures have been observed at the global level, increasing the importance of services in the industrial sector, not only in an end customers' context but also between partners of an industrial value chain [13,14]. With the ongoing digitalization, seizing data-related opportunities gains significance [15]. As a result of technological developments, the increasing relevance of value co-creation and customer centricity can be observed in the manufacturing environment [16]. Research shows that the implementation of data-driven services in a manufacturing environment can provide opportunities for a firm's long-term competitive advantage [15]. Industrial firms that consequently seize the existing opportunities for digitally enabled service growth could be more resilient to global crises [12]. Moreover, following a structured servitization strategy can improve a firm's internationalization and increase its competitiveness [17]. Improving the quality of internal services can result in higher external customer satisfaction [18] and improvement in financial performance [19]. Most of the work to date has a focus on external offerings when considering servitization in the context of digital transformation. We argue that due to the increasing use of digital technologies and the changing demands on processes and employees, the approach of servitization and its dimensions are worth transferring to internal production processes.

Taking the path of servitization in a traditional, product-driven industry comes with various business opportunities but also poses several challenges. To unveil the manifold potentials of digital servitization, firms need to undertake a series of transformational steps, including processual, organizational, and ecosystem changes [15,20]. Companies that want to achieve digital servitization are faced with a variety of tension within the organizational boundaries inherent in their business relationships, including performance priorities, organizational identity, and data utilization [20]. Other challenges include inappropriate culture, a lack of customer focus and resources, and poor processes [21]. Digital servitization also demands a change in business logic. To become digital, a firm must adjust its organizational identity and culture [15]. To develop resilient businesses, it is crucial to build service-led strategies and design product-service offerings while maintaining the existing expertise in the engineering domain [12]. At the center of digital servitization stands the development of data-based services that create added value and allow new business models. The following section, therefore, describes the special features and characteristics of so-called smart services.

2.2. Characterization of Smart Services

Being researched in several domains (e.g., marketing, management, and information systems) and from different perspectives (e.g., customer perspective, company perspective), the phenomenon of smart services emerges from the fields of service science and service engineering [22]. In general, smart services can be defined as data-based, individually configurable bundles of intelligent products, digital services, or personally delivered services that are organized and performed via integrated service platforms [23]. One technological core component of smart services is the IoT. Data collected by sensors is combined with other data on digital platforms and analyzed to gain insights into the condition, use, or application-specific environment of networked physical objects. Based on these insights, digital and personally delivered service modules are combined and adapted to situational needs in a specific context. In addition, machine learning methods represent the second core technological component of smart services [24]. Trained with appropriate data, machine learning processes are able to independently perform tasks described in advance without having to reprogram each step. Smart services are, therefore, characterized by their ability to deliver individualized value propositions in a highly automated and scalable manner [25]. In addition to these special characteristics, which result from their "smartness", smart services also have traditional characteristics of services that distinguish them from physical products. To understand the shift in the offerings' nature during servitization, the following three dimensions can be considered [26]:

- First, services exhibit a high degree of immateriality, which means they are also represented by intangible elements. This can refer, on the one hand, to the resources used by the provider and, on the other hand, to the service outcome, for example, in the form of generated knowledge or customer experiences. Looking at smart services, data as a core characterizing element that is of an immaterial nature in the first place enables interaction between different actors, setting the fundament for value co-creation activities.

- The second dimension, "interactivity", describes the integration of the customer and its resources into the value-creation process, in which an intensive exchange of resources occurs between the supplier and the customer. This can mean either the exchange of data, ideas, and information or the integration of physical resources of the customer (e.g., a physical machine component) that is processed by the supplier.

- The third dimension addresses the degree of individualization, i.e., the adaptation of service offerings to the needs of the customer in a specific situation. In the context of smart services, individualization is based on the availability of data on individual actors or individual activities. Allowing a case-by-case distinction, the execution of a service can be adapted to its specific requirements in each case.

Along the dimensions, different value propositions can be classified and characterized according to the degree of proficiency. While simple physical products have a low proficiency in the three dimensions, it is relatively high for many service offerings. Against the background of the three dimensions, it is clear that successful internal servitization requires the development and management of capabilities for dealing with immateriality, interactivity and individualization, as well as a high degree of automation and scalability. This holds true, in particular, for the development of internal smart services. In the following section, an example of such a service is provided, in order to make the features easier to grasp.

2.3. AI-Based Quality Control as an Example of an Internal Smart Service

As already described, digitalization is an essential driver for service transformation in the industry. AI technologies are particularly highlighted here, as they contribute to optimizing production processes and factories in a wide range of areas. Industrial AI has the potential to improve productivity and quality throughout the entire value stream, as it can help manufacturing companies improve efficiency, reduce errors, and optimize processes [27]. The application for industrial AI-based solutions can be found, for example, in the maintenance of systems, capacity control, and internal quality assurance [28]. Industrial AI-based solutions may initially seem like completely automated IT systems, but they have a stronger service component than one might initially think. The three core characteristics of services, (1) immateriality, (2) interactivity, and (3) individualization, are also reflected in AI applications in the industrial environment. From a service-specific perspective, AI-enabled services can be understood as intelligent and, thus, "smart" digital service products [29,30]. This can be due to their varied application potential views as an archetype of smart services.

A particularly instructive example of a smart service using industrial AI is image recognition (industrial computer vision) for quality assurance. This is one of the key areas where industrial computer vision (ICV) is used to ensure that products meet the required quality standards. ICV systems typically consist of a camera or multiple cameras that capture images or videos of the product or process, software algorithms that analyze the images or videos, and a user interface that displays the results of the analysis. Algorithms can be trained using machine learning techniques to recognize specific features and defects in images or videos. For example, sheet metal components for machine housings or car bodies can be inspected for damage, such as cracks. ICV applications are not merely technical products but can be understood as smart services. This becomes clear once the three characteristics of services are considered in the context of the development and use of ICV. A key component of an ICV application is the output delivered through software algorithms that analyze images or videos of products, which are then interpreted by humans. These outputs are intangible and cannot be physically touched or experienced. This means that the value of an ICV solution is not in the physical product but rather in the insights and information provided by the computer vision system. Therefore, they can be considered immaterial.

Furthermore, ICV solutions are characterized by interactivity, as their value only arises in the collaboration between the ICV system and the user. The ICV system can detect potential damage and flag it to alert employees to a possible defect. They can then make informed decisions about whether to reject or refinish a product based on the information provided by the system. In addition, ICV-based services are individual because they can train and evolve on their own. Thus, they adapt to the specific needs of a particular manufacturing department in an application context and meet their unique requirements and processes. Thus, if the input changes permanently, this leads to different output results. Second, system decisions are dependent on environmental factors. For example, in image recognition systems, minimal changes in hardware setup or the environment of the system (light, dust, etc.) can affect the results.

If AI-based smart services are provided internally, for example, by the IT department, they are referred to as internal smart services. To successfully shape the digital service transformation in companies internally, it is essential to adopt a service-oriented perspective in the development, implementation, and roll-out of industrial AI-based applications. This service-oriented understanding is leading to changes in how business units within a company work together and between partner companies in ecosystems. However, the fact that AI services are frequently delivered through digital platforms, cloud frameworks, and edge computing capabilities also results in structural adjustments within an organization. Against the background of the three dimensions, successful servitization requires, in particular, the development and management of capabilities for dealing with immateriality, interactivity, and individualization.

3. Taking a Service Perspective on Digital Factory Transformation

The development of internal smart services is increasingly important in digital factory transformation to generate added value from data. Against this background, digital factory transformation in manufacturing can also be understood as internal servitization. By digitally linking all manufacturing processes in conjunction with data-driven service products, the production process itself is organized as a service system. Production sites develop, productize, and market their own manufacturing know-how in the form of internal smart services and thus create added value for upstream and downstream areas, other company sites, or even external cooperation partners who obtain and use these digital service products. It is important to take a holistic view of the service system in order to move from a local optimum for individual services to a global optimum for the entire production. A concerted strategy and management of intra-factory services is thus essential.

To provide new insights, we look at fields for digital factory transformation from a service perspective. Therefore, the framework of socio-technical systems is adopted, which is well-established when it comes to the analysis of production systems [31]. The framework distinguishes three levels of consideration: *people, technology,* and *organization.*

It is widely accepted that *people* play a decisive role in technological change processes. Technologies are used by people, and people work together in organizations. The growing importance of digital technologies further enhances the significance of the human factor in that the technologies influence human interaction. *Technology* is the basis for new production methods and manufacturing processes. What is new is the speed at which technology is developing, which is also a consequence of the fact that digital technologies have significantly shorter development times and life cycles. On the *organizational* level, a holistic and systemic view of rules, processes, and decisions with data and data analytic systems is increasingly providing the basis for these decisions.

A systemic view implies that levels are not considered separately. It is often at interfaces and overlaps the levels that are particularly relevant as design fields arise. The following simplified analytical grid with six design fields emerges for considering digital factory transformation from the internal smart service perspective (see Figure 1). The six fields of action are described in the following sections.

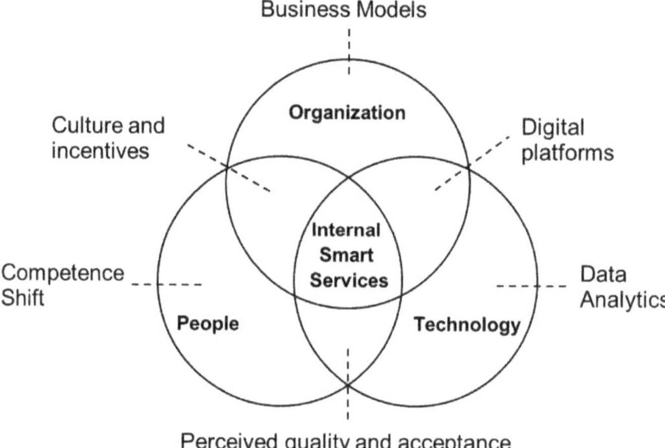

Figure 1. Fields of action for internal smart services (source: own representation).

3.1. Developing Business Models for Internal Smart Services

The development of new, service-oriented business models is at the center of the discussion about digital servitization with a perspective on external customers. The increasing use of data and the application of machine learning methods support everything-as-a-service (XaaS) business models and, in particular, use- or outcome-based revenue models in which products are not sold but rather their use or result of use are evaluated and billed on the basis of empirical data. The use of digital technologies, such as AI and the IoT, can increase the number of parameters considered, automate the evaluation of product performance, and thus create mutual transparency, which reduces the economic risk of these business models [32].

The concept of thinking and acting in business models also provides potential for manufacturing companies regarding internal value creation. On the one hand, manufacturing companies can benefit from equipment-as-a-service business models of suppliers, in which not machines and equipment parts but their usage or results are purchased. The advantages here arise from transferring risks from the factory to the equipment provider. These include, for example, the investment risk since the manufacturer assumes the financing and the costs of production, the availability risk since the supplier keeps the machine available and guarantees its performance, or even the market risk, which arises from potential fluctuations in demand and the associated low-capacity utilization [33]. Especially in times of high volatility, this can create an advantage. However, potential risk transfer is also countered by potentially negative effects. These include, in particular, the necessary sharing of data from the production process with equipment providers, the outflow of production-relevant knowledge, and the associated dependence on external suppliers. Therefore, from the factory operator's point of view, it is essential to identify the non-critical process steps and to critically reflect on the advantages and disadvantages of such business models for each case and production step individually. On the other hand, considering business models and especially evaluating profitability also plays a central role in developing internal smart services. Establishing digital infrastructures for collecting and using data must be translated into added value for the company via these services, which, at best, can be expressed in monetary terms.

However, assessing the economic benefits of smart services is not always easy. For one thing, the costs of acquiring data of appropriate quality and the resulting benefits are difficult to estimate before implementing and training appropriate algorithms. In addition, many digital technologies have marginal costs close to zero due to their scalability, which makes a precise cost analysis at the level of individual services in the production process even more difficult. On the other hand, the result is also not always easy to evaluate in monetary terms. It is easy, for example, in the case of quantifiable targets, such as productivity gains or energy savings, which are measured using the data. In other cases, the outcome is less easily quantifiable, such as generating new insights or increasing process transparency, the value of which depends on the recipient and the application context. Here, indirect benefit measurements, such as conjoint analysis, can help to supplement a purely cost-driven view with indirectly unfolding potential benefits. A problem that often exists is that the necessary data for developing smart services is collected at one point in the supply chain, but the added value from the data unfolds at another point in the process, or one department focuses on maximizing its own value from the service instead of considering its effects on the whole service system. One answer to this challenge could be the collaborative business and operating model development methodology presented in the service literature. Based on a common value proposition (e.g., saving resources by a certain percentage), the various players can jointly classify their expenditures for this and the revenues from it and develop rules and compensation mechanisms.

In order to monetize data through additional revenues, in addition to cost savings, the developed services can also be offered to other production sites within the company or even to external partners or competitors with similar production processes. The advantage of innovative companies is that they collect data early on and train algorithms and analysis models that are needed to offer smart services. For other production sites or external companies, the question later arises as to whether they want to carry out this effort themselves or use the existing "as-a-service" offerings. The development of innovative internal smart services and suitable business models thus contains the potential to soften the image of production sites as pure cost centers and generate independent revenues.

3.2. Digital Platforms and Data Ecosystems

One of the central value propositions of smart products and services lies in the acquisition, merging, and automated evaluation of different data streams with the aim of being able to offer situational and customer-specific solutions and thus increase the value of the services [34]. In this context, the potential for innovation in developing smart services also increases with the number and variety of available data sources, as the corresponding application context can be mapped more completely by the data. The inclusion of data across domains and departments, therefore, opens up new value-creation potential through services in production. In addition to data from a single production step, data from upstream and downstream process steps, from the physical environment of a machine, or even external data outside the factory (e.g., logistics or incoming orders) can be included. The aim is to break down data silos and use the available data for a wide range of internal smart services.

The technical core for this is provided by software-defined platforms, in which the data collected in various networked physical objects can be stored, merged, and combined with other data [35]. To promote the breaking down of data silos between departments, domains, and production plants, central storage forms that simultaneously enable individual data use for the development of smart services are gaining importance. An exemplary concept is represented by "data lakes", in which raw data is stored and only analyzed when required [36]. In addition to storing, various steps of data pre-processing and data preparation are needed before data can be analyzed. For example, data must be converted into suitable formats, and attention must be paid to data quality management at this point to feed the AI later with valid and meaningful data.

In addition to the technical core, the ecosystem of participating actors is also an integral part of the platform concept, which is responsible for supplying the data. Applying platformization principles on an organizational level, coordinating, and interacting across several organizational entities (e.g., departments), platform-based technical infrastructure can ease the flow of information and data. In order to promote the exchange of data across domains and company divisions, specific governance has to be established, including organizational aspects such as rules, structures, and access rights of platform value creation. As an example, clearly defined processes and steps for the collection and classification of data can be mentioned to ensure the highest possible data quality. The potential gains are to be shared between the actors, as suggested in the collaborative development of a business model in Section 3.2.

3.3. Data Analytics

The development of successful smart services fundamentally depends on the ability to collect relevant data and generate added value from it. In addition to data collection, cleansing, and merging, the key capability for developing and delivering value-adding smart services lies in applying intelligent analytics [37].

In order to develop and deliver value-adding smart services, data needs to be gradually transformed from raw data into insights and knowledge by applying analytical measures [38]. Using appropriate analytical methods, information and recommendations can be obtained from raw data to derive actions or recommendations. This process is often referred to as an information value chain or the data, information, knowledge, and wisdom (DIKW) hierarchy [39]. Raw data, which connote the properties of objects, events, and their environment as simple signs, signals, and symbols, form the basis. Combining and extending the data with meta-data requires meaning and becomes structured information [40]. If the information is put into context by incorporating experience, skills, and knowledge, knowledge is created from which decisions for action can be derived. At the top level of the information hierarchy, wisdom emerges. Here, in addition to a comprehensive understanding of the relationship between knowledge elements, ethical and aesthetic aspects are also taken into account, which can lead to an increase in the attractiveness and effectiveness of smart services [41]. Wisdom in this context can also be understood as the system's judgment to provide actions and decisions not only efficiently but also taking into account collective and individual values. Only in this way can users perceive this information as appropriate and beneficial. In general, however, actions and decisions that lead to the added value of smart services can already be derived from information and knowledge.

In order to prepare data for developing and delivering smart services, different methods of data analysis can be applied. Methods of descriptive data analysis allow historical data to be evaluated and condensed into manageable information [42]. Aggregated reports or cumulative visualizations clarify which events have occurred. In contrast, diagnostic analytics help determine why certain events occurred and thus support people in building knowledge and making better decisions. Predictive data analysis methods represent inductive approaches that lead to predictions about what will happen in the future based on historical data [43]. They play a central role in deriving smart actions and measures based on collected data. Methods of prescriptive data analysis go even one step further and provide recommendations about what should be performed and why [44]. Thus, they play a key role in bringing "smartness" into service offers.

Moreover, different types of analytics also make different contributions to the implementation of smart services (see Figure 2). While descriptive, diagnostic, and predictive procedures support the acting persons by providing information and knowledge, prescriptive analytics enables a level of automation in which decisions or actions can be made or triggered autonomously by information systems [45]. This increasing automation and, in some cases, even autonomization of situation-adaptive service provision represents a central feature of smart services [22]. However, predictive and prescriptive analytic techniques also have higher data quantity and quality requirements that must be met. Companies should, therefore, carefully consider which level of the DIKW hierarchy is possible with the existing data and capabilities of the company.

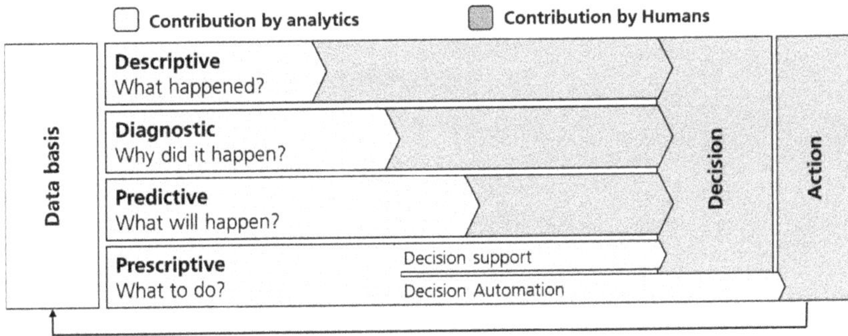

Figure 2. Contribution of analytics for implementing smart services (source: [45]).

The analysis of large volumes of data and the increasing automation of smart services is often made possible by applying technologies and methods of AI, as described in Section 2.3 [46]. Particularly in the case of smart services that use black-box artificial intelligence methods for decision-making, aspects of traceability and transparency often pose a problem. Some machine learning methods (e.g., decision trees or Bayesian classifiers) enable general traceability due to the limited number of paths, rules, and features considered. In contrast, other methods, especially deep learning algorithms (e.g., convolutional neural networks), represent a class of machine learning that prioritize predictive accuracy and, thus, the quality of results of the digital service, sacrificing (at least in part) transparency and traceability [47]. By independently linking mathematically defined entities on a multitude of layers, neural networks can increase the quality of the output, but the exact procedure remains hidden even from data specialists. Here, too, companies should critically weigh which procedures offer the best trade-off between high predictive quality and the need for assurance and acceptance among users.

3.4. Perceived Quality and Acceptance

Internal intelligent services of manufacturing companies are often used to support human work. For example, automated analysis in quality control or AI-supported production control can relieve employees of routine tasks and promote the making of good human decisions (see Figure 2). However, this implies that employees have a high level of acceptance toward the corresponding systems, and smart services are not perceived as a threat [48].

A central prerequisite for the creation of accepted smart services is the knowledge of factors influencing the perceived quality of users and their consideration during development. As a direct determinant of value-in-use and acceptance, a higher perceived quality represents a critical variable for successful smart services [49]. Many companies are currently facing challenges in this respect, as little is known about the quality requirements of smart services [50,51]. However, knowledge regarding perceived quality factors is essential, as the potential benefits are also accompanied by possible risks, such as a lack of protection for sensitive data, a loss of privacy, or the perception of surveillance [52,53]. In addition, increasing automation and the associated loss of personal relationships or the use of complex algorithms may reinforce customer-based uncertainties [25].

In order to gain high-quality perceptions and build acceptance of smart services, the involvement of stakeholders and especially users during all development phases play a central role [54]. Recurring validation cycles in different development stages, from idea generation to the marketing of smart services, can systematically address and ensure a positive perception of quality by users [55]. The so-called quality characteristics of smart services and their influencing factors can provide the basis for a structured assessment with stakeholders. Eight generic quality characteristics are known as smart services for this purpose [56]. The quality characteristics refer to the dimensions of resources and processes and results along the service provision.

With regard to the resources provided, it is apparent that a *function-supporting* role and an *aesthetic appearance* of the physical, digital, and personal artifacts are key drivers of a positive quality assessment. In addition, the capabilities of networked physical objects, digital services (e.g., functionalities), and service staff (e.g., competencies) build a basis for evaluation. Furthermore, the consideration of *security* and *privacy* represents an important quality characteristic in the design of smart services. Security addresses freedom from hazards, risks, and doubts, such as electromagnetic radiation, data access, complex solutions, or interaction with autonomously acting systems. Especially when collecting, storing, and processing personal or sensitive data, ensuring privacy also plays an important role. For example, only data directly related to the value proposition should be collected. The degree of desire for privacy is significantly influenced by the trustworthiness of technologies and acting persons.

In the process dimension, *integration quality* addresses the perception of how well and with how little effort the provided resources can be integrated into the environment and processes of a user during provision. Another characteristic is the *ease of interaction*, which refers to simple, natural, and empathetic interactions between people, technologies, and objects. In the context of smart services, for example, the factors of empathy and naturalness gain relevance in the design of a virtual agent's voice and dialog [57].

Since the value generation of a smart service depends on how well the service components adapt to the situational context of a user, *adaptability* must also be considered as a process quality characteristic [58]. High adaptability is characterized by the degree of flexibility, automated personalization, learning ability, and speed of adaptation. The last quality characteristic of the process dimension addresses the *perception of control*. This describes users' perception of the extent to which they can actively influence service delivery and are not controlled by data collection or automated decisions, or even monitored by sensors [59]. Key influencing factors of a positive perception of control are, for example, the transparency and traceability of service provision, the clear regulation of responsibilities between humans and machines, and the controllability and comfort of using the smart service [55].

In the outcome dimension, the perceived benefit of individual service components, as well as their interaction, is evaluated. The valuable contributions of personal interaction include, for example, the deepening of the customer relationship. In contrast, the contribution of digital services is evaluated along with the improved information situation through data visualization. Since the value proposition of smart services can take very different forms depending on the use case, only the *functional* and *additional benefits* can be distinguished in a generic consideration. The functional benefit addresses the effectiveness, i.e., a customer's effective and holistic task fulfillment or problem solution. Smart services aim to achieve both greater effectiveness through the data-based configuration of service components and to increase efficiency through the targeted use of resources. In addition to being functional, smart services can add value by satisfying emotional, epistemic, or social needs [60]. Emotional value addresses the evaluation of a user's affective state during or after service delivery. For example, through data-driven customization, smart services can satisfy hedonic needs and thus evoke joyful or playful emotions [61].

Furthermore, the use of data to derive evidence-based decisions can contribute to the emotional security of acting individuals. The collection and analysis of data can also lead to novel findings and insights. This can satisfy the epistemic curiosity of users and trigger a change in habitual processes. Another important aspect of added value is social value, which results from the recognition and respect of third parties and positively impacts the self-image and self-confidence of users. For example, the degree of technological innovation or their contribution to a more environmentally friendly solution contributes to social recognition.

The perceived quality has a significant influence on the acceptance of internal smart services in production, which is why the stated quality characteristics should be taken into account during the development of smart services. Moreover, it is highly relevant to emphasize that internal smart services and technologies are not intended to replace employees but to support them. Therefore, user-centered development and design should be of high importance.

3.5. Competence Shift

In particular, the use of the IoT and AI technologies, as well as the increasing service orientation, lead to changing competence requirements in companies [62]. Employees need more diverse competence profiles, which include digital and data-specific skills, such as the effective use of digital tools and different forms of human–machine interfaces [63]. Moreover, employees need a fundamental awareness of the roles, potentials, and limits of AI-driven smart services and knowledge of which data can be collected and converted into added value.

Changes are also occurring in the area of *methodological competencies*, which include, for example, flexible decision-making and problem-solving skills, process, and systems thinking. Innovative smart services with a high degree of complexity require methodological competencies such as the analysis of large volumes of data, the control of complex processes, and the solution of rapidly changing problems. Since internal smart services require a cross-domain and cross-departmental perspective and the integration and coordination of mutually dependent work processes, systems thinking and a distinctive service mindset are also essential areas of competence [64]. Collaborative work on smart services across departmental and domain boundaries requires distinctive *social skills*, influencing effective and efficient collaboration and communication. Since, in many cases, not all service components of a smart service are fully automated and often involve personal interactions, social competencies such as empathy, emotion, and creativity also remain important. Since core technologies, such as AI, have short innovation cycles and are rather untransparent, flexibility, tolerance of ambiguity, and reflectiveness are helpful *organizational skills* [65].

The competencies required at the organizational level are not needed at the individual level by every employee to the same depth and extent. Service literature, therefore, speaks of T-shaped competence profiles, which individual employees should possess [66,67]. Here, a distinction is made between horizontal and vertical competencies. Horizontal competencies describe the skills required to exchange information on a topic with other knowledge areas, specialist departments, or customers and thus be able to connect and engage in dialog. Vertical competencies, on the other hand, describe the in-depth, specialized expertise that is of central relevance to the performance of their activities. As job profiles demand increasing inter-disciplinarity due to the growing thinking and interacting that are involved in smart service ecosystems, interweaving the types of competencies mentioned in the previous section is becoming increasingly important, especially at the horizontal level [34].

This requires, for example, a broader understanding of AI and IoT technologies and their application, service orientation, and overall understanding of processes. Consequently, employees should have technical and service-specific knowledge and be able to translate this into smart service business models. In a dynamic environment with rapidly changing technologies and trends, they are able to move between different disciplines and systems, as well as cross boundaries between disciplines and link them [68].

3.6. Culture and Incentives

Internal smart services can help generate added value from data at different points in the manufacturing process. In principle, they can be applied wherever large amounts of data are generated. Digitization is not a process taking place solely within the organizational boundaries and processes of a company's IT department. In fact, digitization is a transformational process, taking place across all entities of a firm and, therefore, needs the support of many stakeholders (e.g., process owners, affected employees). On the one hand, making it as easy as possible to get started using digital technologies and analytics, for example, using marketable multi-sensor devices connected to low-code platforms, can help employees gain a positive experience and get excited about digital transformation. On the other hand, a successful shift toward more service-oriented value creation in manufacturing companies also depends on the extent to which a corresponding corporate culture, as well as a comprehensive service-oriented mindset, can be achieved among employees [69,70]. This corporate culture is characterized by the following three characteristics.

First of all, the increasingly relevant software elements of physical objects, which are not necessarily developed by a central IT department, require greater agility in production areas themselves [71]. This will also require an open approach to not-yet fully developed solutions, a change in thinking regarding what is possible, and quick short loops to fail and learn quickly for the next releases. In addition, a greater opening of the innovation and work culture is needed, in which internal stakeholders jointly develop new service offerings, and exchange across departmental boundaries and data silos is made possible [17]. A service mentality requires a high degree of user-centricity, where gains and losses of internal customer groups are the starting point for developing data-based services [70]. This also includes, for example, the consideration of emotional and social customer needs, which strengthen the overall experience and increase acceptance of smart services.

In order to maintain the necessary high data quality for the development and provision of smart services, appropriate monetary and non-monetary incentive systems are required, in addition to a corresponding corporate culture and the mindset of the employees. Up to now, many incentive structures in production have been oriented toward cost efficiency, which means that saving costs leads to rewards. However, collecting data and improving its quality often requires time and money, and the benefits of improved data quality may not become apparent until later or elsewhere. Here, holistic and process-oriented incentive systems are needed to overcome such conflicts of interest. Holistic incentive systems to improve data quality and data consistency along the manufacturing process are important areas of digital factory transformation.

4. Conclusions and Discussion

The increasing implementation of digital technologies is accompanied by the need for manufacturing companies to develop smart services in order to create added value from the generated data in the form of quality or productivity benefits and to outweigh necessary investments. This development has been discussed for several years under the term digital servitization with a view to external products and business models of manufacturing companies. However, the digitization of production machines and plants also requires companies to address the central mechanisms and capabilities of internal service-oriented value creation. Against this background, our paper aims to transfer findings from the servitization literature on digital factory transformation to derive impulses.

By focusing on challenges and fields of action in the development of smart services, we adopted a service-specific perspective that broadens the understanding of value creation, where added value is generated not only through industrial factor combination but also through interactive exchange processes and the mutual provision of knowledge and resources by service providers, customers, and cooperation partners [72,73]. The factory of the future is thus transforming itself into a service-oriented factory. In this understanding, the highly flexible series production of quality products forms the basis for developing internal smart services that continuously optimize the manufacturing process and, at the same time, turn the factory into an innovation center for the development and operation of data products and services. Based on the fields of action outlined above, the following recommendations can be derived:

1. For internal servitization to succeed, the smart services developed must be developed, implemented, and operated as professional service products from the outset. It is essential that companies address the question of profitability, i.e., whether smart services can be monetized directly or indirectly, how these effects can be evaluated, and how profits are shared among those involved. In addition, the professional development of smart services also offers the possibility of scaling them beyond the boundaries of the company's production plant. Companies can, for example, offer data, trained algorithms, or fully developed value-added services to other production sites or external companies in outlined XaaS business models that are known for external servitization and thus generate independent revenues. In order to reduce investment and fixed costs against the background of market fluctuations and external uncertainties, companies should also examine the possibility of using XaaS offerings of equipment suppliers themselves for less important and lower-risk production steps.

2. A service-oriented focus on internal value creation in production requires the establishment of integrated and scalable IT infrastructure that supports a high level of data continuity across new and existing systems. In addition, the added value of the individual smart services, as well as the whole service and production system, can be better and more holistically captured by data. The focus should, therefore, be on breaking down and merging internal data silos. This requires digital platforms in the technical sense, but also appropriate data governance and processes in the company that strengthen the sharing of data, and the creation of consistently high data quality is an important component of all departments. In addition, compensation mechanisms for contributions to high data quality should be created and incentive mechanisms for the efficient use of IT infrastructures should be developed to prevent limited resources from not being used efficiently and being threatened by overuse (tragedy of the commons) [74].

3. For shifting toward service-oriented value creation and developing internal smart services, building up suitable technical capabilities for analyzing and processing data into insights and recommendations is necessary. Companies should consider the opportunities and risks of descriptive, diagnostic, predictive, and prescriptive analytics. Early consideration in the light of available data, in-house capabilities, and the need for certainty and transparency can help set realistic goals for service development. Particularly in the production environment, where errors can have critical consequences, companies should take appropriate measures to safeguard the use of prescriptive processes in AI-based smart services.

4. To ensure that the increasing number of smart services within the production site is accepted by the workforce, users of the applications should be regarded as internal customers from the outset and involved during development, testing, and continuous optimization. This also includes focusing on the subjective perception of quality instead of objective quality. In the context of data-intensive, automated, and adaptive services, new quality characteristics must be considered during development, considering smart services' unique character to ensure acceptance.

5. To enable active engagement in service-oriented value creation for many departments in production, additional competencies are needed in the workforce. Following the concept of T-shaped competency profiles, horizontal competencies should be created that promote collaboration across divisional and departmental boundaries, problem-solving thinking, and a basic understanding of data and digital technologies, as well as services. Vertical, in-depth knowledge of the application domains should not be neglected.

6. In addition, the internal service transformation also requires an adjustment of the corporate culture and corresponding incentive systems. This includes, for example, an open approach to errors that is often untypical for manufacturing and the use of data products and services that are not fully developed. Establishing a close and solution-oriented collaboration between departments, a high level of agility in software development processes in various departments, and a constant ability to innovate and adapt are central pillars of an internal service culture. This should be flanked by holistic and process-oriented incentive and motivation systems so that all employees continuously contribute to optimizing data quality and smart service quality along the manufacturing process.

In our paper, we presented selected central findings from the servitization literature and applied them to in-plant applications in manufacturing. Due to the diversity of the topic, our paper does not claim to be exhaustive of the relevant topics and should, therefore, be understood as a first impulse. For example, one important topic is the design of data-based services for the implementation of dynamic manufacturing processes, which are controlled automatically by data and AI applications [75]. We were also unable to consider the process of internal services transformation, i.e., which steps need to be taken in which order and with what intensity. Future research should take up these central questions and empirically investigate the impulses mentioned here in the six fields of action. Finally, it should be noted that the findings and recommendations presented here cannot be applied to all companies in the manufacturing sector to the same extent. Nevertheless, it is hoped that looking at the digital transformation of the factory from a service perspective will provide new insights and support successful implementation for as many companies as possible.

Author Contributions: J.N., M.F., J.K., C.B. and B.B. developed all sections of the paper jointly. All authors have read and agreed to the published version of the manuscript.

Funding: This research received no external funding.

Institutional Review Board Statement: Not applicable.

Informed Consent Statement: Not applicable.

Data Availability Statement: Not applicable.

Conflicts of Interest: The authors declare no conflict of interest.

References

1. Reinhold, J.; Koldewey, C.; Dumitrescu, R.; Rausch, G. Smart Service-Transformation–Den Wandel der Wertschöpfung erfolgreich gestalten. In Proceedings of the 16th Symposium für Vorausschau Und Technologieplanung, Paderborn, Germany, 2–3 December 2021; pp. 337–341.
2. Kagermann, H.; Wahlster, W. Ten Years of Industrie 4.0. *Sci* **2022**, *4*, 26. [CrossRef]
3. Kagermann, H.; Süssenguth, F.; Körner, J.; Liepold, A.; Behrens, J.H. *Resilienz der Fahrzeugindustrie: Zwischen Globalen Strukturen und Lokalen Herausforderungen*; Acatech IMPULS: München, Germany, 2021.
4. Spieske, A.; Birkel, H. Improving supply chain resilience through industry 4.0: A systematic literature review under the impressions of the COVID-19 pandemic. *Comput. Ind. Eng.* **2021**, *158*, 107452. [CrossRef]
5. Schutzbach, M.; Kiemel, S.; Miehe, R. Eco Lean Management—Recent Progress, Experiences and Perspectives. *Procedia CIRP* **2022**, *107*, 350–356. [CrossRef]
6. Urbinati, A.; Rosa, P.; Sassanelli, C.; Chiaroni, D.; Terzi, S. Circular business models in the European manufacturing industry: A multiple case study analysis. *J. Clean. Prod.* **2020**, *274*, 122964. [CrossRef]

7. Thomas, V.J.; Maine, E. Market entry strategies for electric vehicle start-ups in the automotive industry—Lessons from Tesla Motors. *J. Clean. Prod.* **2019**, *235*, 653–663. [CrossRef]
8. Schuh, G.; Anderl, R.; Dumitrescu, R.; Krüger, A.; Hompel, M.T. Using the Industrie 4.0 Maturity Indexin Industry. In *Current Challenges, Case Studies and Trends*; Acatech IMPULS: Munich, Germany, 2020.
9. Binner, H.F. Digitale Revolution—Erfolgreiche Umsetzung über Organisation 4.0. *Z. Für Wirtsch. Fabr.* **2016**, *111*, 152–154. [CrossRef]
10. Kohtamäki, M.; Parida, V.; Patel, P.C.; Gebauer, H. The relationship between digitalization and servitization: The role of servitization in capturing the financial potential of digitalization. *Technol. Forecast. Soc. Chang.* **2020**, *151*, 119804. [CrossRef]
11. Khanra, S.; Dhir, A.; Parida, V.; Kohtamäki, M. Servitization research: A review and bibliometric analysis of past achievements and future promises. *J. Bus. Res.* **2021**, *131*, 151–166. [CrossRef]
12. Rapaccini, M.; Saccani, N.; Kowalkowski, C.; Paiola, M.; Adrodegari, F. Navigating disruptive crises through service-led growth: The impact of COVID-19 on Italian manufacturing firms. *Ind. Mark. Manag.* **2020**, *88*, 225–237. [CrossRef]
13. Bonamigo, A.; Frech, C.G. Industry 4.0 in services: Challenges and opportunities for value co-creation. *J. Serv. Mark.* **2021**, *35*, 412–427. [CrossRef]
14. Ulaga, W.; Kowalkowski, C. Servitization: A State-of-the-Art Overview and Future Directions. In *The Palgrave Handbook of Service Management*; Edvardsson, B., Tronvoll, B., Eds.; Springer International Publishing: Cham, Switzerland, 2022; pp. 169–200, ISBN 978-3-030-91827-9.
15. Kowalkowski, C.; Sklyar, A.; Tronvoll, B.; Sörhammar, D. Digital Servitization: How Data-Driven Services Drive Transformation. In Proceedings of the 55th Hawaii International Conference on System Sciences, Maui, HI, USA, 4–7 January 2022.
16. Veile, J.W.; Schmidt, M.-C.; Voigt, K.-I. Toward a new era of cooperation: How industrial digital platforms transform business models in Industry 4.0. *J. Bus. Res.* **2022**, *143*, 387–405. [CrossRef]
17. Kolagar, M.; Reim, W.; Parida, V.; Sjödin, D. Digital servitization strategies for SME internationalization: The interplay between digital service maturity and ecosystem involvement. *J. Serv. Manag.* **2022**, *33*, 143–162. [CrossRef]
18. Zeithaml, V.A.; Bitner, M.J.; Gremler, D.D. *Services Marketing: Integrating Customer Focus across the Firm*, 7th ed.; McGraw-Hill Education: New York, NY, USA, 2018; ISBN 9780078112102.
19. Sasser, W.E. *Value Profit Chain: Treat Employees Like Customers and Customers LIKE Employees*; Simon and Schuster: New York, NY, USA, 2014; ISBN 978-1-4767-9998-8.
20. Tóth, Z.; Sklyar, A.; Kowalkowski, C.; Sörhammar, D.; Tronvoll, B.; Wirths, O. Tensions in digital servitization through a paradox lens. *Ind. Mark. Manag.* **2022**, *102*, 438–450. [CrossRef]
21. Johnston, R. Internal service—Barriers, flows and assessment. *Int. J. Serv. Ind. Manag.* **2008**, *19*, 210–231. [CrossRef]
22. Korper, A.K.; Purrmann, M.; Heinonen, K.; Kunz, W. Towards a Better Understanding of Smart Services—A Cross-Disciplinary Investigation. In *Towards a Better Understanding of Smart Services—A Cross-Disciplinary Investigation, Proceedings of the Exploring Service Science, Porto, Portugal, 5–7 February 2020*; Novoa, H., Dragoicea, M., Kühl, N., Eds.; Springer International Publishing AG: Cham, Switzerland, 2020; Volume 377, pp. 164–173. ISBN 3030387232.
23. Neuhüttler, J.; Ganz, W.; Liu, J. An integrated approach for measuring and managing quality of smart senior care services. In *Advances in the Human Side of Service Engineering: Proceedings of the AHFE 2016 International Conference on the Human Side of Engineering, Walt Disney World, FL, USA, 27 July–31 July 2016*; Ahram, T.Z., Karwowski, W., Eds.; Springer International Publishing: Cham, Switzerland, 2017; pp. 309–318.
24. Neuhüttler, J.; Fischer, R.; Ganz, W.; Urmetzer, F. Perceived Quality of Artificial Intelligence in Smart Service Systems: A Structured Approach. In *Quality of Information and Communications Technology*; Shepperd, M., Brito e Abreu, F., Da Rodrigues Silva, A., Pérez-Castillo, R., Eds.; Springer International Publishing: Cham, Switzerland, 2020; pp. 3–16, ISBN 978-3-030-58792-5.
25. Neuhüttler, J.; Ganz, W.; Spath, D. An Integrative Quality Framework for Developing Industrial Smart Services. *Serv. Sci.* **2019**, *11*, 157–171. [CrossRef]
26. Satzger, G.; Benz, C.; Böhmann, T.; Roth, A. Servitization and Digitalization as "Siamese Twins": Concepts and Research Priorities. In *The Palgrave Handbook of Service Management*; Edvardsson, B., Tronvoll, B., Eds.; Springer International Publishing: Cham, Switzerland, 2022; pp. 967–989, ISBN 978-3-030-91827-9.
27. Lee, J. *Industrial AI: Applications with Sustainable Performance*; Springer: Singapore, 2020; ISBN 9789811521447.
28. Moll, C.; Lerch, C. KI-basierte Geschäftsmodelle im verarbeitenden Gewerbe—Anwendungspotenziale und Ausgestaltungsmöglichkeiten. In *Künstliche Intelligenz im Dienstleistungsmanagement*; Bruhn, M., Hadwich, K., Eds.; Springer Fachmedien Wiesbaden: Wiesbaden, Germany, 2021; pp. 98–119, ISBN 978-3-658-34323-1.
29. Winter, J. Europa und die Plattformökonomie—Wie datengetriebene Geschäftsmodelle Wertschöpfungsketten verändern. In *Dienstleistungen 4.0*; Bruhn, M., Hadwich, K., Eds.; Springer Fachmedien Wiesbaden: Wiesbaden, Germany, 2017; pp. 71–88, ISBN 978-3-658-17551-1.
30. Bullinger, H.-J.; Neuhüttler, J.; Nägele, R.; Woyke, I. Collaborative Development of Business Models in Smart Service Ecosystems. In Proceedings of the 2017 Portland International Conference on Management of Engineering and Technology (PICMET), Portland, OR, USA, 9–13 July 2017; IEEE: New York, NY, USA, 2017; pp. 1–9. [CrossRef]
31. Hirsch-Kreinsen, H. Industry 4.0: Options for Human-Oriented Work Design. *Sci* **2023**, *5*, 9. [CrossRef]

32. Neuhüttler, J.; Kett, H.; Frings, S.; Falkner, J.; Ganz, W.; Urmetzer, F. Artificial Intelligence as Driver for Business Model Innovation in Smart Service Systems. In *Advances in the Human Side of Service Engineering*; Spohrer, J., Leitner, C., Eds.; Springer International Publishing: Cham, Switzerland, 2020; pp. 212–219, ISBN 978-3-030-51056-5.
33. Evcenko, D.; Kett, H.; Nebauer, S. Risiken managen—Einsatzpotenziale von EaaS-Lösungen in der Produktion. *Z. Für Wirtsch. Fabr.* **2022**, *117*, 872–878. [CrossRef]
34. Bullinger, H.-J.; Ganz, W.; Neuhüttler, J. Smart Services—Chancen und Herausforderungen digitalisierter Dienstleistungssysteme für Unternehmen. In *Dienstleistungen 4.0*; Bruhn, M., Hadwich, K., Eds.; Springer Fachmedien Wiesbaden: Wiesbaden, Germany, 2017; pp. 97–120, ISBN 978-3-658-17549-8.
35. Arbeitskreis Smart Service Welt/acatech. *Smart Service Smart Service Welt: Umsetzungsempfehlungen für das Zukunftsprojekt Internet Basierte Dienste für die Wirtschaft*; Abschlussbericht: Berlin, Germany, 2015.
36. Mathis, C. Data Lakes. *Datenbank Spektrum* **2017**, *17*, 289–293. [CrossRef]
37. Opresnik, D.; Taisch, M. The value of Big Data in servitization. *Int. J. Prod. Econ.* **2015**, *165*, 174–184. [CrossRef]
38. Ardolino, M.; Rapaccini, M.; Saccani, N.; Gaiardelli, P.; Crespi, G.; Ruggeri, C. The role of digital technologies for the service transformation of industrial companies. *Int. J. Prod. Res.* **2018**, *56*, 2116–2132. [CrossRef]
39. Rowley, J. The wisdom hierarchy: Representations of the DIKW hierarchy. *J. Inf. Sci.* **2007**, *33*, 163–180. [CrossRef]
40. Martin, D.; Kühl, N.; von Bischhoffshausen, J.K. System-Wide Learning in Cyber-Physical Service Systems: A Research Agenda. In *Designing for Digital Transformation: Co-Creating Services with Citizens and Industry, Proceedings of the 15th International Conference on Design Science RESEARCH in information Systems and Technology, DESRIST 2020, Kristiansand, Norway, 2–4 December, 2020*; Hofmann, S., Ed.; Springer: Cham, Switzerland, 2020; pp. 457–468, ISBN 978-3-030-64822-0.
41. Spohrer, J.; Bassano, C.; Piciocchi, P.; Siddike, M.A.K. What Makes a System Smart? Wise? In *Advances in the Human Side of Service Engineering*; Ahram, T.Z., Karwowski, W., Eds.; Springer International Publishing: Cham, Switzerland, 2017; pp. 23–34, ISBN 978-3-319-41946-6.
42. Cardoso, J.; Fromm, H.; Nickel, S.; Satzger, G.; Studer, R.; Weinhardt, C. *Fundamentals of Service Systems*; Springer International Publishing: Cham, Switzerland, 2015; ISBN 978-3-319-23194-5.
43. Hunke, F.; Schüritz, R. Smartere Produkte durch analysebasierte Dienstleistungen—Ein methodisches Werkzeug zur strukturierten Entwicklung. *HMD* **2019**, *56*, 514–529. [CrossRef]
44. Delen, D.; Demirkan, H. Data, information and analytics as services. *Decis. Support Syst.* **2013**, *55*, 359–363. [CrossRef]
45. Kühn, A.; Joppen, R.; Reinhart, F.; Röltgen, D.; Enzberg, S.; von Dumitrescu, R. Analytics Canvas—A Framework for the Design and Specification of Data Analytics Projects. *Procedia CIRP* **2018**, *70*, 162–167. [CrossRef]
46. Heuchert, M.; Verhoeven, Y.; Cordes, A.-K.; Becker, J. Smart Service Systems in Manufacturing: An Investigation of Theory and Practice. In Proceedings of the 53rd Hawaii International Conference on System Sciences, Maui, HI, USA, 7–10 January 2020.
47. Rai, A. Explainable AI: From black box to glass box. *J. Acad. Mark. Sci.* **2020**, *48*, 137–141. [CrossRef]
48. Kutz, J.; Neuhüttler, J.; Spilski, J.; Lachmann, T. Implementation of AI Technologies in manufacturing—success factors and challenges. In *The Human Side of Service Engineerin, Proceedings of the 13th International Conference on Applied Human Factors and Ergonomics (AHFE 2022), New York, NY, USA, 24–28 July 2022*; AHFE International: New York, NY, USA, 2022.
49. Medberg, G.; Grönroos, C. Value-in-use and service quality: Do customers see a difference? *J. Serv. Theory Pract.* **2020**, *30*, 507–529. [CrossRef]
50. Klein, M.M.; Biehl, S.S.; Friedli, T. Barriers to smart services for manufacturing companies—An exploratory study in the capital goods industry. *J. Bus. Ind. Mark.* **2018**, *33*, 846–856. [CrossRef]
51. Gonçalves, L.; Patrício, L.; Grenha Teixeira, J.; Wünderlich, N.V. Understanding the customer experience with smart services. *J. Serv. Manag.* **2020**, *31*, 723–744. [CrossRef]
52. Wuenderlich, N.V.; Heinonen, K.; Ostrom, A.L.; Patricio, L.; Sousa, R.; Voss, C.; Lemmink, J.G. "Futurizing" smart service: Implications for service researchers and managers. *J. Serv. Mark.* **2015**, *29*, 442–447. [CrossRef]
53. Mani, Z.; Chouk, I. Consumer Resistance to Innovation in Services: Challenges and Barriers in the Internet of Things Era. *J. Prod. Innov. Manag.* **2018**, *35*, 780–807. [CrossRef]
54. Russo-Spena, T.; Mele, C. "Five Co-s" in innovating: A practice-based view. *J. Serv. Manag.* **2012**, *23*, 527–553. [CrossRef]
55. Neuhüttler, J.; Hermann, S.; Ganz, W.; Spath, D.; Mark, R. Quality Based Testing of AI-based Smart Services: The Example of Stuttgart Airport. In Proceedings of the 2022 Portland International Conference on Management of Engineering and Technology (PICMET), Portland, OR, USA, 7–11 August 2022; pp. 1–10, ISBN 2159-5100.
56. Neuhüttler, J. *Ein Verfahren zum Testen der Wahrgenommenen Qualität in der Entwicklung von Smart Services*; Fraunhofer Verlag: Stuttgart, Germany, 2022.
57. Tuzovic, S.; Paluch, S. Conversational Commerce—A New Era for Service Business Development? In *Service Business Development*; Bruhn, M., Hadwich, K., Eds.; Springer Fachmedien Wiesbaden: Wiesbaden, Germany, 2018; pp. 81–100, ISBN 978-3-658-22425-7.
58. Wixom, B.H.; Schüritz, R. Creating Customer Value Using Analytics. *MIT CISR Res. Brief.* **2017**, *17*, 1–4.
59. Someh, I.; Davern, M.; Breidbach, C.F.; Shanks, G. Ethical Issues in Big Data Analytics: A Stakeholder Perspective. *Commun. Assoc. Inf. Syst.* **2019**, *44*, 718–747. [CrossRef]
60. Huang, M.-H.; Rust, R.T. Artificial Intelligence in Service. *J. Serv. Res.* **2018**, *21*, 155–172. [CrossRef]
61. Wiegard, R.-B.; Breitner, M.H. Smart services in healthcare: A risk-benefit-analysis of pay-as-you-live services from customer perspective in Germany. *Electron Mark.* **2019**, *29*, 107–123. [CrossRef]

62. Ahner, L.; Bayrak, Y.; Cleaver, I.; Ge, M.L.; Neuhüttler, J.; Urmetzer, F. Impacts of Professional Education Measures on the Digital Transformation in Organisations. In *Advances in the Human Side of Service Engineering*; Leitner, C., Ganz, W., Satterfield, D., Bassano, C., Eds.; Springer International Publishing: Cham, Switzerland, 2021; pp. 173–180, ISBN 978-3-030-80839-6.
63. Blumberg, V.S.L.; Kauffeld, S. Kompetenzen und Wege der Kompetenzentwicklung in der Industrie 4.0. *Z. Für Angew. Organ.* **2021**, *52*, 203–225. [CrossRef]
64. Eckert, F.; Heidling, E. *Hybridisierung in der Value Chain und Kompetenzent-Wicklung für Hybride Wertschöpfung in Cloudbasierter Kollaborativer Arbeit*; Frühjahrskongress: Bochum, Germany, 2021.
65. Cimini, C.; Adrodegari, F.; Paschou, T.; Rondini, A.; Pezzotta, G. Digital servitization and competence development: A case-study research. *CIRP J. Manuf. Sci. Technol.* **2021**, *32*, 447–460. [CrossRef]
66. Demirkan, H.; Spohrer, J. T-Shaped Innovators: Identifying the Right Talent to Support Service Innovation. *Res.-Technol. Manag.* **2015**, *58*, 12–15. [CrossRef]
67. Barile, S.; Saviano, M.; Simone, C. Service economy, knowledge, and the need for T-shaped innovators. *World Wide Web* **2015**, *18*, 1177–1197. [CrossRef]
68. Demirkan, H.; Spohrer, J.C. Commentary—Cultivating T-Shaped Professionals in the Era of Digital Transformation. *Serv. Sci.* **2018**, *10*, 98–109. [CrossRef]
69. Tronvoll, B.; Sklyar, A.; Sörhammar, D.; Kowalkowski, C. Transformational shifts through digital servitization. *Ind. Mark. Manag.* **2020**, *89*, 293–305. [CrossRef]
70. Moraes, C.R.; de Cunha, P.R. Enterprise Servitization: Practical Guidelines for Culture Transformation Management. *Sustainability* **2023**, *15*, 705. [CrossRef]
71. Münch, C.; Marx, E.; Benz, L.; Hartmann, E.; Matzner, M. Capabilities of digital servitization: Evidence from the socio-technical systems theory. *Technol. Forecast. Soc. Chang.* **2022**, *176*, 121361. [CrossRef]
72. Grönroos, C.; Ojasalo, K. Service productivity. *J. Bus. Res.* **2004**, *57*, 414–423. [CrossRef]
73. Vargo, S.L.; Webster, F.E.; Bolton, R.N.; Lusch, R.F. *The Service-Dominant Logic of Marketing: Dialog, Debate, and Directions*; CRC: Boca Raton, FL, USA, 2006; ISBN 9780765614902.
74. Wasko, M.M.; Teigland, R. Public goods or virtual commons? Applying theories of public goods, social dilemmas, and collective action to Electronic networks of practice. *J. Inf. Theory Appl.* **2004**, *6*, 25–41.
75. Tombeil, A.-S.; Kremer, D.; Neuhüttler, J.; Dukino, C.; Ganz, W. Potenziale von Künstlicher Intelligenz in der Dienstleistungsarbeit. In *Automatisierung und Personalisierung von Dienstleistungen: Methoden—Potenziale—Einsatzfelder*; Bruhn, M., Hadwich, K., Eds.; Springer Fachmedien: Wiesbaden, Germany, 2018; pp. 135–154, ISBN 978-3-658-30167-5.

Article

Industry 4.0: Options for Human-Oriented Work Design

Hartmut Hirsch-Kreinsen

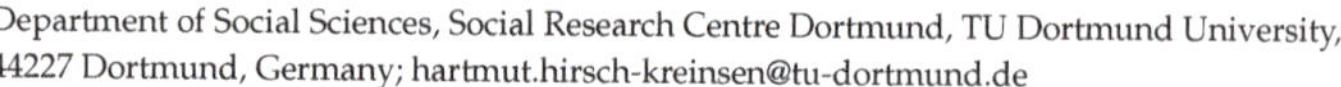

Department of Social Sciences, Social Research Centre Dortmund, TU Dortmund University, 44227 Dortmund, Germany; hartmut.hirsch-kreinsen@tu-dortmund.de

Abstract: This contribution deals with the diffusion of Industry 4.0 technologies and their consequences for work. Additionally, design options for work in Industry 4.0 are discussed. The following are outlined: First, since there are as yet no concrete future prospects for digital work, different development perspectives can be envisioned. Second, the development of Industry 4.0, therefore, has to be regarded as a design project. One theoretical basis for this is the "sociotechnical systems" approach. Third, this approach enables criteria for the design and implementation of human-oriented forms of digitized work to be systematically developed. The empirical basis of this contribution derives from research findings on the implementation of Industry 4.0 technologies and the development of digitized work in German industry. The research results are based on qualitative research methods such as company case studies and expert interviews.

Keywords: Industry 4.0; development of work; sociotechnical systems approach; human-oriented work design

check for
updates

Citation: Hirsch-Kreinsen, H. Industry 4.0: Options for Human-Oriented Work Design. *Sci* **2023**, *5*, 9. https://doi.org/10.3390/sci5010009

Academic Editors: Claus Jacob and Antonio Gil Bravo

Received: 28 November 2022
Revised: 7 February 2023
Accepted: 9 February 2023
Published: 15 February 2023

1. Introduction

The diffusion of Industry 4.0 and digital technologies are changing the world of work in the industrial sector. These changes will become commonplace in the future, but predicting the consequences for jobs and skills is a much more difficult task. In the scientific and public debate, the economic and social consequences of Industry 4.0 are still the subject of intensive discussions, and clear trend statements hardly seem possible.

From a sociological point of view, at least two economic and societal consequences of the now more than ten-year discourse on the vision of Industry 4.0 [1] should be singled out: It has underscored the enormous structural economic importance of the industrial sector for many countries, and with this vision, forward-looking and innovative prospects have opened up for industry. Furthermore, Industry 4.0 puts the long-forgotten question of the future of industrial work back on the political agenda. It can be assumed that Industry 4.0 will by no means offer clear—negative or positive—consequences for jobs and qualifications. However, Industry 4.0 opens up a wide scope of development possibilities for work that can and should be used in its qualification- and human-oriented design.

Following this, the contribution draws on research findings on the development of industrial work as a result of the diffusion of Industry 4.0 systems. Especially, the prerequisites and chances of a human-centered design for industrial work will be outlined. These considerations refer to the approach of the "sociotechnical system", which assumes that new technologies entail both personnel and organizational changes. On this basis, the different and sometimes contradictory perspectives of human, technology and organization are brought together to develop a complementary design approach to industrial work under the conditions of a progressive digitization of industrial processes.

The goal of this article was, therefore, to emphasize design options for industrial work under the conditions of Industry 4.0. This should make clear that the process of change in work is by no means just a technologically caused process. Rather, it is strongly determined by economic and social factors, especially by the decision makers in companies involved in the introduction of digital systems.

2. Materials and Methods

Methodologically, the contribution is based on theoretical considerations and the findings of empirical studies analyzing the diffusion of Industry 4.0 and the development of work in industrial sectors in Germany during the last few years. These are social science studies with a special focus on issues in the sociology of work and labor [2–4].

In detail, the methodical-empirical basis of the contribution includes information that the author and his research group have gained through the digitization discourse at various levels of politics and companies in recent years, secondly, the results of a continuous document and literature research on the digitization of work, and thirdly, research results from relevant, methodologically qualitative empirical research projects that were carried out between 2014 and 2019 at the TU Dortmund University on the change in work in the context of digital technologies.

The examined sample comprised around 23 company case studies. As a rule, expert talks were held with management representatives and works councils, as well as through detailed and sometimes repeated company visits. These were companies in the metal and furniture industry, companies from sub-sectors of the process industry, and logistics companies. The majority of these were medium-sized and small companies. The technical digitization solutions examined are diverse. There were applications from all functional areas of digital technologies in the case-study companies. These ranged from data-based applications, e.g., for process management in real time, to assistance systems for order picking and production planning and various forms of human–robot collaboration and largely autonomous floor conveyor technology to advanced cyber-physical production systems. The digital systems were in different stages of development. Some were still in the pilot phase or were operated as isolated solutions, while others were future-oriented implementation projects that had been approached rather cautiously for technical, economic or work and personnel-related reasons.

3. Divergent Perspectives on Industrial Work

The current debate suggests that digital technologies will change the nature of work in many sectors, especially in manufacturing—from the the shop floor to related areas such as planning, control systems and product development [5,6]. Therefore, leadership and management practices may also change significantly [7,8]. Although studies predicted thorough change in concrete processes of work, they did not agree about how industrial work and employment will change and what this will mean in terms of job structures, worker qualifications and skill requirements [9–11]. This was also shown in our own detailed research results. Companies will follow different approaches to work design and personnel strategies as they introduce Industry 4.0 technologies. No definitive prospects for the development of digital work can be identified, and one has to speak, therefore, of different development scenarios. On the base of our research findings, the development scenarios can be summarized as follows:

One scenario can be characterized as pessimistic about how the future development of industrial work will affect workers. On the one hand, this scenario contends that the demand for many tasks and qualifications will decline and the number of available jobs will be reduced dramatically. In particular, many jobs consisting of routine tasks will be replaced by the new technologies [12,13].

Our empirical findings show also that an increasing diffusion of technology will reduce jobs requiring only medium-level skills, while jobs demanding higher qualifications or low-skilled jobs that cannot be easily automated will benefit. Therefore, this scenario can be interpreted as an increasing "polarization" of high- and low-skilled jobs. In particular, trends towards distinct operational structures were elucidated in our research in the context of intelligent networked logistics systems—automated supply and distribution management systems that leverage digital technologies, such as self-checkout storage systems used by manufacturing companies. A clear polarization of the work is underway: on the one hand, more complex and skilled occupations such as those of managers and supervisors have been established to operate new technology systems. On the other hand, low-skilled tasks and simple operations such as packaging and assembly have been retained, since the cost of automating these tasks has always been higher than the cost of paying cheap labor enough. As our findings show, companies often avoid fully automated systems due to their high technological complexity and high cost, but the tasks they automate are tasks that should be performed by skilled workers. The British sociologists Goos and Manning describe this trend as the emergence of "lousy and lovely jobs" [14].

Another scenario suggests more positive consequences of Industry 4.0 for work: job creation, higher levels of skill requirements, and a general revaluation of jobs, so that a new and more humane turn in work will be take place [15]. This optimistic scenario suggests that efficiency gains, new products, new markets and new employment opportunities will compensate for the short-term negative effects on jobs [16].

In our findings, many company management representatives predicted for Industry 4.0 high productivity gains and higher economic growth rates as well as a positive development in jobs. A majority of experts expect the share of the employed workforce to remain relatively stable and significant over the next few years, with no significant negative employment effects. With respect to skills, the perspective is that Industry 4.0 will bring a growing revaluation of qualifications and skills, the result of an increasing substitution of routinized simple jobs (such as machine monitoring or assembly work) by digital technologies. In line with this trend, nearly all other employee groups will be affected by increasing qualification requirements. The reason for this is that digitization makes a wide variety of information about ongoing processes available to the workers. The resulting complexity and potential applications of new technologies could result in fundamentally new and as yet unknown requirements for all job-related activities.

For example, under these conditions, skilled machine operators can make decisions about workflow sequences based on optimized information and control systems. The new technology provides data and assessment capabilities that allow for a much higher level of transparency in the manufacturing process. The optimistic outlook emphasizes that a general improvement in qualifications in the future is not only possible, but inevitable. In view of this, the pattern of work in manufacturing industries can evolve towards a model that can be characterized by a very limited division of labor, high flexibility, and increasing levels of competence. Therefore, this scenario can be characterized as the "upgrading" of jobs, qualifications and skills.

4. Industry 4.0 as a Design Project

To sum up, while there are no direct consequences of new technologies for work, alternative development perspectives on work in Industry 4.0 should not be overlooked. Of course, the pessimistic outlook presents a possible scenario. Still, our research provides good reasons for optimism, particularly about skills upgrading. The argument is that collaborative work processes, especially characterized by a high degree of autonomy at work, can help skilled workers effectively harness digitized systems to their advantage. However, there are opposing perspectives on how the digitization of work will affect workers of different skill levels and the nature of the work. In other words, there is no linear relationship between new technology and work, but alternative perspectives for the development of work.

This argument can be linked to the common wisdom of labor research, that one cannot speak of "technological determinism", i.e., that technology influences directly the development of work. Rather, the development and design of work are clearly complex and reciprocal relationship shaped not only by technology, but also by multiple economic, social and labor-political factors. Especially, specific company conditions strongly influence the path taken by the technology deployment and work design in each case. The influence and the concrete constellation of these factors determine in what way the new technologies will actually be used and how work will be redesigned [17]. *Therefore, Industry 4.0 has to be regarded as a design project.*

5. Sociotechnical Systems Approach

On our analytical and theoretical considerations, this perspective can be concretized within the "sociotechnical systems" approach. The basic assumption of this approach is that, in any case, both efficient and human-oriented forms of digital work can be realized. Therefore, the "sociotechnical" approach brings together the interactions and interdependences between the technological, human, and organizational dimensions of a work process. Thus, a sociotechnical system can to be regarded as a work unit consisting of interdependent technology, personnel, and organization subsystems [18,19]. The subsystem technology includes the new digital technologies, the human subsystem refers to the employment structures and skill requirements, and the organizational subsystem comprises workplace structures, new management functions, and innovative business models. Furthermore, the sociotechnical system is embedded in strategic and normative framework conditions and societal context factors such as politically established regulations (Figure 1).

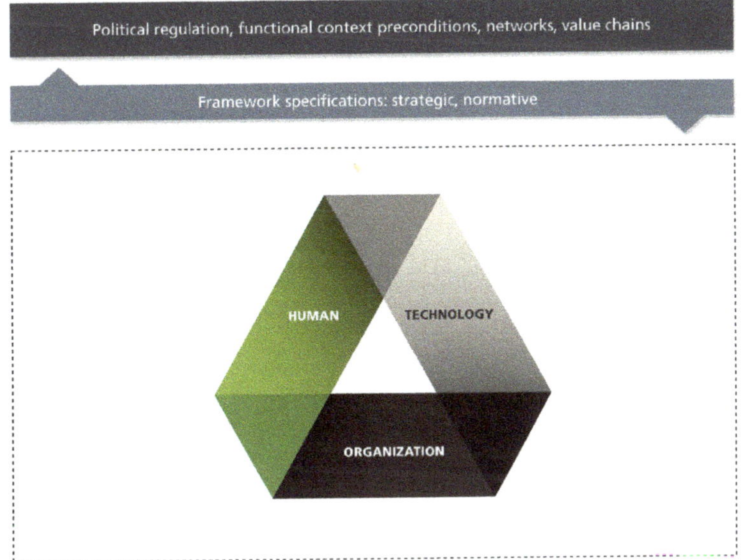

Figure 1. Conceptual representation of the sociotechnical systems approach (own source).

In the sociotechnical approach, it is not a question of *either* technology or work, but rather a *complementary* design of the three subsystems adjusted to one another in a total sociotechnical system. In other words, the specific strengths and weaknesses of the technology and human work should be equally considered to meet the concrete demands of production. Hence the basic principle of the sociotechnical system approach is the *joint optimization* of work, organization, and technology. However, the leading priority should be to exploit as well as possible the potential advantages of a human-oriented work design.

This aspect is also addressed in the catchy formula of the necessary *complementary innovation* (of the organization and personnel) that should be an element of the digital innovation. The argument goes as follows: "Organisations can only fully benefit from technological innovation if it is embedded in a proper work organisation" [20] (p. 138). This criticizes the technology-centered understanding of digital technologies observed among many decision-makers, which relies on the most far-reaching algorithmization of processes without taking into account the organizational and personnel context and necessary adjustments.

Criteria based on these considerations can be systematically deduced for the design and implementation of human-oriented forms of digitized work. The design criteria should not focus on the single subsystems, but rather on the interdependencies between the technology, personnel, and organization, a matter of designing the *interfaces* between the technological, the human, and the organizational subsystems of the entire sociotechnical system. Design options for these interfaces may be highlighted as follows [3,4,21,22].

6. Design Options

6.1. Technology–Human Interface

The technology–human interface is, firstly, a well-known issue of the criteria for ergonomically oriented dialogue design, and secondly, a matter of new criteria for man–machine interfaces. This is because with Industry 4.0 systems, new patterns of *function-distribution* and *interaction between machine and man* are made possible. Their design must be assumed as one of the key issues in the implementation of digital technologies and Industry 4.0. There are currently two alternative solutions to the design of the technology–human interface: First, digital systems can provide strict instructions to workers in order to limit their space for action, and to reduce qualification requirements. This solution can be termed "technology-centered"; second, digital systems can be assistance systems that support workers, allow a variety of work, promote on-the-job learning processes and thereby raise qualification levels.

Of course, from a human-oriented perspective, the second design solution is desirable. In particular, this solution should be based on the following design criteria:

- *Context sensitivity and adaptivity*—These criteria include aspects of the ergonomic adaptation of digital systems to specific load and working conditions, which can be monitoring system loads or automating particularly difficult processes. In addition, it is a matter of optimally delivering situation-specific data and information to ensure uninterrupted workflow and avoid costly and stressful interruptions and delays. Intelligent ability is required to tailor information and support systems to individual, varying levels of competence on the part of workers, to ensure continued learning and enhanced process staff levels. Finally, the implementation of support systems must implicitly support the actual knowledge of employees.
- *Complementarity*—This criterion focuses on two central aspects of human–machine interaction: first, the flexible and situation-specific allocation of functions between humans and machines, and second a sufficiently transparent and controllable system. The relevant design aspects here are: ensuring human–computer interaction through intuitive and fast-learning hardware as well as targeted and situation-specific access to real-time digital information, to enable employees to make decisions and implement digitally supported behavioral preferences in a secure and editable manner.

Generally, an interface design must make possible, above all, a satisfactory functional and economic capability of the total system. This requires a holistic view of the human–machine interaction and the identification of the specific strengths and weaknesses of human work and digital technologies. Importantly, a central prerequisite here is that human work attain and secure control over production processes by gaining and building on the often indispensable practical experience and knowledge supported by smart assistant systems.

This form of interface design leads to broadening the scope of employees' tasks, meeting the demands of challenging and learning-friendly work, and opening up new possibilities for employee engagement in design and decision making. Therefore, the work situation is characterized by an expanding field of tasks and the need for new skills. The interaction between intelligent systems and worker behavior can generally be described as *hybrid*. Contrary to the traditional view of technology as a passive object, in digital technology the role of a behavioral agent is assigned, with the consequence that not only the division of labor but also the decision-making skills in a particular way must be constantly reinvented.

6.2. Human–Organization Interface

The human–organization interface deals with changes in the scope for actions, work-time models, and new demands on skills, qualifications and modes of training. A key issue is how the readily available skills, competencies and experiential knowledge of employees can be used for an optimal utilization of Industry 4.0 systems and a human-oriented work design. The current discourse on Industry 4.0 very often overlooks the organizational design of digitized work that is decisive for the completeness of operational tasks, as well as for the development of scope of action, learning, and qualification opportunities.

From a human-oriented perspective, the human–organization interface can be designed to achieve a sustainable revaluation of activities and skills. There are options for efficient patterns of work organization as well as work situations with particular qualification demands, a high degree of scope for action, the polyvalent deployment of workers, and a multitude of opportunities for "learning on the job", where skills and competencies can be self-acquired. Individual as well as collective learning can take place through job rotation as well as in forms of "learning islands" or "learning factories". Learning-promotive work organization and training measures should take into account the various levels of (existing) experience and skills of the employees. An additional aspect is that the tasks will rarely address only individual workers, but rather teams. That means that "work collectives" should have the scope to act in a self-organizing way and be highly flexible in addressing the problems to be solved in the technological system.

The main criteria for designing work activities at the human and organizational interface can be summarized in the following keywords:

- *Holism*—This criterion means all activities in a dual sense: on the one hand, an activity includes not only operational tasks, but also equipment tasks (organization, planning, and control). On the other hand, this criterion is geared towards a suitable and light mix of tasks that require more and less. For example, this design goal can be realized in the context of new forms of robotics or robot cooperation. Furthermore, the totality of activities is a central requirement for greater freedom of action as well as the ability to self-organize work.
- *Dynamics*—With this criterion, the following issues are addressed. First, the ability to organize work to exchange tasks systematically, in order to create workable learning processes and encourage them. Second, the new social media functions promote interdisciplinary communication and collaboration among employees with different expertise and thus increase creativity in work. Here, it is especially important to be able to "try it out in the workshop" to cope with the rapid development of technology. At the same time, in the context of loosely structured work patterns, it becomes possible to deploy employees with different capacities and production capacities, e.g., in mixed workgroups. Third, loosely structured and dynamic workflows are often a prerequisite for decisions and interventions to effectively deal with emerging unexpected disruptions.

Therefore, such an organizational structure can be described as a "holistic work organization" or metaphorically as a "swarm organization"—a loose network of employees with different qualifications and expertise. The central feature of this organizational model is that there are no defined tasks for each employee. In contrast, the "working collective" operates in a very flexible manner, self-organizing and deciding according to the situation, adjusting its behavior to suit the problems to be solved around the technological system.

6.3. Organization–Technology Interface

At the organization–technology interface, new design options are given due to re-designing the overall work process and even the re-organization of the whole company. This includes changes in the production chain in terms of function and hierarchy, as well as in the structuring and linkage between the direct processes and the indirect planning, engineering, management and support processes. Because of their decentralized and simultaneously networked intelligence, the new digital systems allow a far-reaching departure from the centralized IT systems of previous years.

Therefore, a general shift towards decentralization and de-hierarchization is possible—often in the framework of already relatively "flatly" structured company organizations. Furthermore, the company organization need not be only decentralized, since the digital technologies offer also the option of making the organization (even) more flexible. This suggests a highly individualized production, which is why an organizational structure based on autonomous, self-controlling systems with far-reaching *decentralized control and intelligence* should seriously be taken into account.

This concerns not only the manufacturing sector but also the hierarchical aspect of the company's entire organization, as well as logistics. The features of social media and with them, new forms of communication, also affect indirect areas such as planning, control, and engineering as well as leadership and management functions. In addition, there is a reorganization of management functions, such as in the production and business divisions, due to the change in decision-making power of these divisions and the transfer of responsibilities to subordinates.

Finally, new forms of value-chain structures and *new business models* become possible. In the "networked smart factory", industrial value creation is no longer limited to what happens within traditional organizational boundaries. In contrast, decentralized control is required, and the intelligence is—however—still controllable. As a result of this digitization, new business models are used to address technological and organizational challenges and their interrelationships. Therefore, changes are conceivable in entire value-chains that may significantly transcend previous forms of inter-company division of labor and outsourcing. With that are given the organizational prerequisites for overcoming company barriers to an extended service and customer orientation as well as to change in business models.

7. Conclusions

The findings show that there is no "one best way" or "one single way" in digital work. There are as yet no clear, deterministically derivable consequences for work as Industry 4.0 systems are implemented. Thus, the shape of a framework for the design of digital work becomes recognizable. Following the above-outlined criteria of human, technology and organization and their interdependencies under the requirements of Industry 4.0, forms of work that are characterized by the design of each of the interfaces are conceivable. To sum up, basic criteria for the development of a human-oriented design of work should include: far-reaching monitoring and regulation capabilities, intelligent assistance systems, complete and well-generalized tasks, learnability, high maneuverability, as well as new forms of self-organization with decentralized control. This framework can be applied to social and organizational requirements for high system transparency to human actors, controllability of complex system processes, and thus optimal functioning of the system as a whole. Of course, a successful diffusion and implementation of human-oriented digital work depends on additional conditions on the company level that support this process. Several aspects need to be emphasized here.

First, the acceptance of the new system and its work design capabilities, both on the workforce side as well as management side, must be ensured. The fact that this factor plays an important role has been confirmed many times in the course of our research. To alleviate workforce reticence towards new features of job design—e.g., the need to address concerns about job loss. New sources of stress with the requirements for flexibility are increasing, and the problems arising from data protection as well as the ability to monitor work performance are becoming crucial issues. Anticipated reorganization processes can mask many new and somewhat contradictory demands from workers for flexibility and self-organization. If there is a mismatch between current needs and resources, stressful behavioral dilemmas can arise for employees due to the need to manage immediate needs. Effective approaches to solving these problems can lie in process methods that involve employees and represent their interests in the introduction, design, and implementation of

Industry 4.0. On the management side, there can be frequent protests, especially against far-reaching measures that alter established practices within the hierarchical organization and the company. To overcome these limitations, targeted transfers of knowledge and experience should be introduced accordingly and further developed, presenting exemplary and successful good practice cases and communicating the success potential of human-oriented forms of work.

Second, there are challenges posed by changing management functions and leadership styles. It must be assumed that, in the face of the general challenges of new technology and especially the establishment of human-based forms of work, the traditional hierarchical structures and methods of management will become dysfunctional and outdated. The direction of change needed shows the increasing importance of "soft skills" as well as high ability to communicate and work in a team. Instead of control, now leadership and "motivation from afar", and instead of hierarchical leadership, "coordination" of colleagues and "peer-to-peer" communication and employee participation are now key features of successful management. In general, company leadership, through awareness of the new status quo, must account for trends through digitization and transformation of working forms, as functional and social boundaries between leadership and collaboration members will be eroded, or even reversed. In any case, the distinction between "blue collar" and "white collar" will become increasingly blurred. The aim is to establish new forms of self-organization and control, geared towards corporate goals, of course, but characterized by flexible and problem-oriented forms of management. Admittedly, this fragmentation of past management models and emphasis on bottom-up processes will lead to a certain contradiction: that digital transformation is successful and sustainable companies emerge simultaneously through top-down functional processes. However, because there are still many open questions, this issue needs more in-depth research in the future.

At the societal level, certain factors play a role in coping with change and further developments in regulatory forms of labor and social policies that at least indirectly influence the introduction of human-oriented digital forms of work. This includes, for example, the regulation of flexibility, working hours, co-determination and further training. These areas often require a new alignment of labor policy interests. This is the only way to avoid obstacles and reservations about work transformation arising from unresolved conflicts and objections. The importance of numerous further training measures as well as the development of skills for spreading human-centered forms of work in the context of Industry 4.0 cannot be overestimated.

A central goal of such measures should be to overcome multiple "digital divides". First, there is a need to balance skill differences between technology-intensive and non-technology-intensive firms. Second, differences in skills and performance between different employee groups (qualifications, age, etc.) must be balanced. Low-skilled jobs in particular must be taken into account so that these employees are not cut off from the general development of their qualifications. Overall, however, "capacity development" should be understood to mean the central means of education and social policy necessary to implement competence-oriented and people-oriented forms of work across a wide range of social levels.

All in all, a socially responsible development perspective on "Industry 4.0" is also the best way to ensure that future industrial work is suitable for an aging workforce. Furthermore, this design perspective can increase the attractiveness of industrial work and thus counteract the urgent demographic shortage of skilled workers in many industrialized countries. To overcome these pressing social challenges, key players in business, science and politics are investigating the social and organizational conditions necessary to realize the potential of people-oriented "Industry 4.0" designs and integrate them into comprehensive perspectives of sociotechnical integration. The EU's new Industry 5.0 concept [23] systematically addresses these requirements. In particular, it puts worker well-being at the center of the production process and uses new technologies to create wealth beyond work and growth.

Funding: This research received no external funding.

Institutional Review Board Statement: Not applicable.

Informed Consent Statement: Not applicable.

Data Availability Statement: The empirical data referred to in this paper are available on request from the corresponding author, but are not public due to privacy restrictions.

Conflicts of Interest: The author declares no conflict of interest.

References

1. Kagermann, H.; Helbig, J.; Hellinger, A.; Wahlster, W. *Recommendations for Implementing the Strategic Initiative INDUSTRIE 4.0: Securing the Future of German Manufacturing Industry*; Final Report of the Industrie 4.0 Working Group; Forschungsunion: Berlin/Heidelberg, Germany, 2013.
2. Hirsch-Kreinsen, H. Digitization of industrial work: Development paths and prospects. *J. Labour Mark. Res.* **2016**, *49*, 1–14. [CrossRef]
3. Hirsch-Kreinsen, H. *Digitale Transformation von Arbeit. Entwicklungstrends und Gestaltungsansätze*; Kohlhammer: Stuttgart, Germany, 2020.
4. Hirsch-Kreinsen, H.; Ittermann, P. Digitalization of Work Processes: A Framework for Human-Oriented Work Design. In *The Palgrave Handbook of Workplace Innovation*; McMurray, A., Muenjohn, N., Weerakoon, C., Eds.; Palgrave: Cham, Switzerland, 2021; pp. 273–293.
5. Alcácer, V.; Cruz-Machado, V. Scanning the Industry 4.0: A Literature Review on Technologies for Manufacturing Systems. *Eng. Sci. Technol. Int. J.* **2019**, *22*, 899–919. [CrossRef]
6. Meyer, U.; Schaupp, S.; Seibt, D. (Eds.) *Digitalization in Industry*; Palgrave Macmillan: Cham, Switzerland, 2020.
7. Oeij, P.R.A.; Dhondt, S. Theoretical approaches supporting workplace innovation. In *Workplace Innovation: Theory, Research and Practice*; Oeij, P.R.A., Rus, D., Pot, F.D., Eds.; Springer: Cham, Switzerland, 2017; pp. 63–78.
8. Howaldt, J.; Kopp, R.; Schultze, J. Zurück in die Zukunft? Ein kritischer Blick auf die Diskussion zur Industrie 4.0. In *Digitalisierung Industrieller Arbeit*, 2nd ed.; Hirsch-Kreinsen, H., Ittermann, P., Niehaus, J., Eds.; Nomos: Baden-Baden, Germany, 2018; pp. 347–363.
9. Balliester, T.; Elsheikhi, A. *The Future of Work: A Literature Review*; Research Department Working Paper No. 29; International Labour Office: Geneva, Switzerland, 2018.
10. Larsson, A. A journey of a thousand miles. An introduction to the digitalization of labor. In *The Digital Transformation of Labor*; Larsson, A., Teigland, R., Eds.; Routledge: Oxford, UK, 2020; pp. 1–11.
11. Rainnie, A.; Dean, M. Industry 4.0 and the future of quality work in the global digital economy. *Labour Ind. J. Soc. Econ. Relat. Work* **2020**, *30*, 16–33. [CrossRef]
12. Frey, C.B.; Osborne, M.A. The future of employment: How susceptible are jobs to computerisation? *Technol. Forecast. Soc. Change* **2017**, *114*, 254–280. [CrossRef]
13. Dengler, K.; Matthes, B. *Auch Komplexere Tätigkeiten können Zunehmend Automatisiert Werden. Wenige Berufsbilder Halten mit der Digitalisierung Schritt*; IAB-Kurzbericht 13/2021; Institut für Arbeitsmarkt-und Berufsforschung (IAB): Nürnberg, Germany, 2021.
14. Goos, M.; Manning, A. Lousy and lovely jobs: The rising polarization of work in Britain. *Rev. Econ. Stat.* **2007**, *89*, 118–133. [CrossRef]
15. Zuboff, S. Creating Value in the Age of Distributed Capitalism, McKinsey Quarterly. 2010. Available online: http://glennas.files.wordpress.com/2010/12/creating-value-in-the-age-of-distributed-capitalism-shoshana-zuboff-september-2010.pdf (accessed on 5 November 2019).
16. Pettersen, L. Why artificial intelligence will not outsmart complex knowledge work. *Work. Employ. Soc.* **2019**, *33*, 1058–1067. [CrossRef]
17. Evangelista, R.; Guerrieri, P.; Meliciani, T. The economic impact of digital technologies in Europe. *Econ. Innov. New Technol.* **2014**, *23*, 802–824. [CrossRef]
18. Trist, E.; Bamforth, K. Some social and psychological consequences of the long wall method of coal-getting. *Hum. Relat.* **1951**, *4*, 3–38. [CrossRef]
19. Rice, A. *The Enterprise and Its Environment*; Tavistock: London, UK, 1963.
20. Brynjolfsson, E.; McAfee, A. *The Second Machine Age: Work, Progress, and Prosperity in a Time of Brilliant Technologies*; Norton: New York, NY, USA, 2014.
21. Dregger, J.; Niehaus, J.; Ittermann, P.; Hirsch-Kreinsen, H.; ten Hompel, M. The Digitization of Manufacturing and its Societal Challenges. A Framework for the Future of Industrial Labor. In Proceedings of the 2016 IEEE International Symposium on Ethics in Engineering, Science and Technology (ETHICS), Vancouver, BC, Canada, 13–14 May 2016; pp. 1–3.

22. Kadir, B.A.; Broberg, O. Human-centered design of work systems in the transition to industry 4.0. *Appl. Ergon.* **2020**, *92*, 103334. [CrossRef] [PubMed]
23. European Commission; Directorate-General for Research and Innovation; Breque, M.; De Nul, L.; Petridis, A. *Industry 5.0: Towards a Sustainable, Human-Centric and Resilient European Industry*; Publications Office: Luxembourg, 2021.

Article

Industry 4.0 as a Challenge for the Skills and Competencies of the Labor Force: A Bibliometric Review and a Survey

Abdelkarim Alhloul [1] and Eva Kiss [2,*]

[1] István Széchenyi Doctoral School, Faculty of Economics, Sopron University, Erzsébet út 9., 9400 Sopron, Hungary

[2] Geographical Institute, Research Centre for Astronomy and Earth Sciences, RCAES: MTA Centre for Excellence, Budaörsi út 45, 1112 Budapest, Hungary

* Correspondence: kisse@helka.iif.hu

Abstract: The latest technological development called Industry 4.0, like the previous industrial revolutions, has also brought a new challenge for people as a labor force because new technologies require new skills and competencies. By 2030 the existing generation in the labor market will have a skill gap threatening human replacement by machines. Based on bibliometric analysis and systematic literature review the main aims of this study are, on the one hand, to reveal the most related articles concerning skills, competencies, and Industry 4.0, and on the other hand, to identify the newset of skills and competencies which are essential for the future labor force. Determining the model of new skills and competencies in connection with Industry 4.0 technologies is the main novelty of the study. A survey carried out among the workers of mostly multinational organisations in Hungary has also been used to explore the level of awareness about those skills and Industry 4.0 related technologies, and this can be considered the significance of the empirical research.

Keywords: Industry 4.0; skills; competencies; bibliometric analysis; survey; Hungary

Citation: Alhloul, A.; Kiss, E. Industry 4.0 as a Challenge for the Skills and Competencies of the Labor Force: A Bibliometric Review and a Survey. *Sci* **2022**, *4*, 34. https://doi.org/10.3390/sci4030034

Academic Editor: Johannes Winter

Received: 23 June 2022
Accepted: 31 August 2022
Published: 5 September 2022

Publisher's Note: MDPI stays neutral with regard to jurisdictional claims in published maps and institutional affiliations.

1. Introduction

The economic structure has changed over time because of technological development. This development started with the dawn of the first industrial revolution (1760–1840). After that, the world witnessed two more revolutions at the end of the 19th century and between the 1970s and 1990s [1,2]. The Fourth Industrial Revolution (4IR), which started at the turn of the millennium, is also called Industry 4.0 (I4.0), and it has been accelerated in the last decade. The term Industry 4.0 was originally mentioned as Industrie 4.0 at the Hannover Fair in 2011 and indicated a programme for the digitisation and strategic development of the German industry [3,4]. Since then, it has been widespread, although it has several definitions. For example, "Industry 4.0 is nothing more than a digital transformation" or "The next phase in the digitalisation of the processing industry". According to a different view, "Industry 4.0 is a vision sponsored by the German government for a more advanced processing industry". In a narrower sense and most often, as can be seen from the above definitions to some extent, it is related to industry and includes the new technologies that will result in a radical transformation of industrial production [5]. According to Reischauer [6], Industry 4.0 represents a major technological revolution, which takes place primarily in industry, factories, and production. This is why Industry 4.0 and thus the Fourth Industrial Revolution are often referred to as "smart factory", "intelligent industry" or "advanced manufacturing".

The use of Industry 4.0 emerging technologies to fulfill the requirements of production has caused a rapid change in the labor market which has been defined as a digital influence on the labor market. Industry 4.0 has affected many jobs, replacing humans with machines, as we can see e.g., during the check-in process at the airport and many other routine jobs.

160

Previous studies have confirmed that only highly skilled and qualified human resources will be able to control Industry 4.0 technologies [7–11].

Industry 4.0 emerging technologies require more than just performing a task or resolving a problem in each field, which is exactly the definition of the skill. Rather, the capability to meet complex demands, including interpersonal attributes to be self-driven for lifelong learning in each field as the competency definition states [12,13] and to be able to understand what the required skills and job profiles are, as well as having an understanding of the emerging technologies of Industry 4.0 is important. Therefore, Industry 4.0 has ten main technologies which are the driving forces of this revolution as follows:

1. Industrial Internet of Things (IIoT) is a communication technology which makes the connectivity between the things possible. "Things can be anything like an object or a person." [14].
2. Cloud Computing (CC) is an alternative technology which enable sharing the storage of each data using on the internet for the companies which are outsourcing IT services as well as individuals [15].
3. Big data is a huge amount of data generated in a homogenised way as objects on the network. This data can be structured, semi-structured and unstructured. The value of big data is that it is organised with accessibility [16].
4. Simulation is an essential element of Industry 4.0, as it is a powerful tool to draw and evaluate many scenarios, not only in the manufacturing systems. It is also a powerful tool in the field of knowledge sharing and training [17,18].
5. Augmented reality is a system able to process information by combining real and virtual objects in a real environment in an interactive way combining 3D in real-time [4,18,19].
6. Additive manufacturing can be described as a rapid prototyping, solid freeform manufacturing, layer manufacturing, digital manufacturing or 3D printing [20,21].
7. Horizontal and vertical systems integration: I4.0 systems integration has two approaches, [22] which are enabling real-time sharing [18].
8. Autonomous robots refer to Artificial Intelligence (AI) [3,23].
9. Cybersecurity (CS) may serve as a new term for a high level of information security, and through the word "cyber" it spreads to apply also to industrial environments and IoT [4,18].
10. Cyber-physical systems (CPS) can be viewed as an innovative technology that permits control by integrating physical and computational environments of interconnected systems [18,24].

In the operation of these, Industry 4.0 technology operators play an important role. The concept Operator 4.0 became popular among studies referring to the qualified persons for those technologies. Operator 4.0 is also known as a smart operator, and it defines this as "a smart and skilled operator who performs not only 'cooperative work' with robots, but also 'work aided' by machines and if needs employing of human cyber-physical systems, advanced human-machine interaction technologies, and adaptive automation towards human-automation symbiosis work systems" [11,25,26]. In order to achieve the concept of Operator 4.0, which represents the future of workplaces, a set of skills is needed to integrate the workforce into I4.0. This integration can be called human-cyber-physical systems (H-CPS). Those systems are created to enhance the human-machine relationship [27].

Operator 4.0 knowledge transfer methodologies are aimed to create an environment to reach the concept (CPS) to improve the abilities of the workforce by allocation of tasks to machines and operators overseeing the ructions to the machine, which can be programmed into a machine, as an aid to handle uncertain events [11,28]. That can sum up the abilities of the human and machine in optimised manufacturing. To infer the cognitive states and emotions associated with the decision-making and operator behavior, the Operator 4.0 concept requires precise chronological time-harmonisation of the operator actions, sensory data and psychophysiological signals [29]. Moreover, the study aims to upskill and train the existing labour to be able to use the Industry 4.0 technologies in an innovative way. In the future, the number and composition of employees will also transform [30,31].

The most visible consequences of the use of new technologies can most likely be expected in industrial employment. On the one hand, the increase in automation, digitalisation and robotisation will reduce the demand for living labour in industrial production; thus, a smaller number of people will work in the manufacturing industry. On the other hand, thanks to new technologies, the quality of the workforce is also changing. Among the few industrial employees, there will be fewer low-skilled, physically employed and more qualified, intellectually skilled employees. According to a survey conducted in 37 countries, as the use of industrial robots increases, the proportion of people doing routine work among the employees, who are usually less educated, decreases [32]. Some 80 million low-skilled workers in the EU could lose their jobs as a result of automation and robotisation, while in the US, it is estimated that 47% of jobs could disappear [33,34]. Not only are old jobs and occupations transformed or eliminated, but new ones (e.g., data scientists) appear. Some of the new jobs will have different requirements on the workforce than the current ones. That is why new or different knowledge, abilities and skills will be needed more than before, and this will also place a heavy burden on education at all levels. It is likely that there will be high demand in the labour market for those who have adequate competencies in software development and information technology, as well as in info-communications, because the use of software, connectivity and analysis will increase [35]. In addition, many other skills and capabilities (e.g., flexibility, creativity, problem solving, decision-making, etc.) are needed to meet the labour market challenges of the coming decades. This also shakes the world of work to its foundations and may lead to serious problems [36]. Consequently, it is very important to identify the new skills and competencies, which can be relevant in the future. That is the research gap that this study intends to fill by replying to the following study questions:

RQ1: What are the top-cited articles concerning Industry 4.0 jobs related to needed skills and competencies?

RQ2: What are the trends of required skills and competencies in Industry 4.0 jobs among the different sectors of the economy according to the top-cited articles in relation to the topic?

RQ3: What is the level of awareness about Industry 4.0 emerging technologies among the employees of mostly multinational companies in Hungary?

The replies to these questions contribute to set up a new model for Industry 4.0 skills and competencies, and this can be considered the main novelty of the study. The empirical research is significant because it makes an attempt from a practical viewpoint to reveal the level of current awareness of skills and competencies related to Industry 4.0 technologies.

The study has five major parts. After the "Introduction", the "Materials and Methods" are presented with particular regard to the major steps of the research process. Section 3 demonstrates the results of the bibliometric analysis, which describes the database of the study and the results of the survey concerning the level of skills and the awareness level concerning Industry 4.0 technologies and needed competencies. Section 4 is the discussion of the results, and, finally, the conclusions follow.

2. Materials and Methods

2.1. The Process of the Research

Reaching the study aim requires going through study goals. Thus, reaching the best profile fit for the human workforce to meet the requirement of Industry 4.0 needs to explore the most related scientific studies on the given topic. Therefore, a hybrid method of a bibliometric analysis on the Scopus database and systematic literature review (SLR) was applied on the most cited articles. After reaching the results of the conducted search, a survey was carried out mostly among employees of multinational organisations in Hungary to reveal the awareness level concerning Industry 4.0 technologies and the new required skills. The major phases of the research work were the following:

The first step was to formulate the study questions, which allowed us to screen the data sets and include and/or exclude the desired documents.

The second step was data collection. Study data were collected from the Scopus database using the following query in the advanced search: TITLE-ABS-KEY (("human factor") OR ("operator") OR ("smart operator") OR ("workforce") OR (operator 4.0)) AND (("Industry 4.0") OR ("4th industrial revolution") OR ("smart factories") AND ("training") OR ("education")) AND (("skill*") OR ("Competenc*")) AND (EXCLUDE(PUBYEAR,2022)) AND (LIMIT-TO(LANGUAGE, "English") OR LIMIT-TO (LANGUAGE, "German")) to create illustrative maps. Then, other searches were also conducted to reveal data more related to the topic to avoid excluding important research from the discussion and conclusion. Those search queries were made concerning the job advertisements of Industry 4.0-related technologies. Also, a search query took place in the normal search instead of the advanced to compare the results with the study search. The last search was made to ensure the information novelty. Only one study was found in a peer reviewed journal which adopted a similar concept and method in terms of Industry 4.0 skills and competencies as well as using a bibliometric review. That research, however, used a different software together with different inclusion and exclusion approaches.

The third step was to apply the above-mentioned search query in the Scopus database and download the data sets for further analysis.

As a fourth step, exclusion and inclusion criteria were defined. Exclusion criteria were: (1) not English and/or German; (2) not related to Industry 4.0 related skills and competencies; (3) articles related to chemistry, biology, hydrology, medical and psychology aspects; (4) no full text available. Inclusion criteria were the following:

1. Peer reviewed manuscript in an impact factor journal or conference proceeding.
2. Related keywords have occurred at least three times in the title, abstract and keywords.
3. The document has been cited at least three times.

It is also necessary to note that the information for the documents that meet the requirements were the year of publication, language, journal, title, author, affiliation, keywords, document type, abstract and counts of citation which were exported into (CSV) format for the Scopus data set (This data set is compatible with VOS-viewer software.).

The fifth step was reporting the results using descriptive analysis. The software VOS viewer and Excel was used for bibliometric analysis.

As a sixth step, a small survey was carried out in 2022 among expats working in national and multinational organisations in Hungary to reveal the level of awareness concerning Industry 4.0 requirements for the new labor market using Google forms and Microsoft excel. Testing the awareness level in practice is also a new kind of approach in this topic.

The time span of the search was 2015–2021. The search was conducted from the middle of January with continuous updates until the beginning of April 2022.

2.2. Justification for the Methodology Used

VOS viewer (version 1.6.18) was used to analyse the co-authorship, co-occurrence, citation, bibliographic coupling, co-citation and themes. The research questions have been set up to make a bibliometric investigation of the needed skills and competencies in the Industry 4.0 paradigm. Using these methods and software such as VOS viewer helps to explore the relationships through visualising and mapping that can help in reaching the answers to the study questions in a logical way [13,19]. VOS viewer and equivalent software can supply a clustering mapping that can be a powerful tool for reaching the most important studies by knowing the citation strength, which explains the document's importance. Why have the top cited articles been used for the analysis? Studies have proved that concrete answers are more likely to be found in the top cited articles [37,38]. The stronger the citation position is the more valuable information the document holds in regard to the chosen topic.

The study objective is to find the most related set of skills and competencies that must exist in the workforce of the existing generation to cope with Industry 4.0. For those reasons, many keywords concerning the topic have been reviewed to reach the most related keywords. The explanation for choosing them is that the concept of Industry 4.0 is known in a decent number of studies as the Fourth Industrial Revolution. The other variable of the study is the human factor, which is known in most of the studies as the Operator or Operator 4.0. Sustainability was chosen as a keyword combined with the rest of the keywords because the studies which are concerned with the replacement of humans by machines (human-centered studies) have the keywords of training and education. The last variable is the skills and competencies combined. The reason for choosing, for example (Competenc*) is to relate all the studies that have competence or competencies all at once. Those areas of research will help in revealing the most related skills and competencies that are needed for Operator 4.0 in the era of Industry 4.0 [39]. They will form the skills and competencies model of the study.

3. Results

3.1. Bibliometric Analysis Results

In this section, first, the results will refer to the first RQ1. Running the search on the Scopus database, 588 documents were found, covering the years of 2015–2021 in all the fields, except for those mentioned in the exclusion criteria. Then, their number was decreased to 266 using the condition of exclusion (three citations at least per document). Dates of the search were in January 2022.

After that, using VOS viewer software, the citation analysis of the documents was conducted to determine the top cited articles in the given topic and to create a map depicting how much they are connected by the citation links of the documents and authors. Then, the articles were examined to help to create an image of what are the most suitable skills that can be built through training to reach the efficiency of the competencies in the workplace to cooperate with Industry 4.0. The final result of the top 20 cited articles in concern of Industry 4.0 skills and competencies is given in Table 1.

Table 1. Top twenty cited documents in relation to skills and/or competencies, training, and Industry 4.0 emerging technologies.

Ref. Number	Document Title	Number of Citations	Links	Journal Name	Journal Impact Factor	Journal Cite Score
[14]	Scanning the Industry 4.0: A Literature Review on Technologies for Manufacturing Systems	378	0	Engineering Science and Technology, an International Journal	4.336	9
[9]	Holistic Approach for Human Resource Management in Industry 4.0	297	21	Procedia CIRP	0.6	3.3
[10]	Smart operators in industry 4.0: A human-centered approach to enhance operators' capabilities and competencies within the new smart factory context	249	29	Computers & Industrial Engineering	5.431	7.9

Table 1. *Cont.*

Ref. Number	Document Title	Number of Citations	Links	Journal Name	Journal Impact Factor	Journal Cite Score
[16]	Big data analytics as an operational excellence approach to enhance sustainable supply chain performance	126	1	Resources, Conservation and Recycling	10.204	14.7
[19]	Supporting Remote Maintenance in Industry 4.0 through Augmented Reality	119	5	Procedia Manufacturing	1.794	1.39
[24]	Placing the operator at the center of Industry 4.0 design: Modelling and assessing human activities within cyber-physical systems	107	5	Computers & Industrial Engineering	5.431	7.9
[11]	Enabling Technologies for Operator 4.0: A Survey	85	6	Applied Sciences	2.679	3
[30]	Industry 4.0 and the human factor—A systems framework and analysis methodology for successful development	71	4	International Journal of Production Economics	7.885	12.2
[40]	Influences of the Industry 4.0 Revolution on the Human Capital Development and Consumer Behavior: A Systematic Review	71	10	Sustainability	3.251	3.9
[31]	Empowering and engaging industrial workers with Operator 4.0 solutions	71	3	Computers & Industrial Engineering	5.431	7.9
[39]	A training system for Industry 4.0 operators in complex assemblies based on virtual reality and process mining	69	0	Robotics and Computer-Integrated Manufacturing	5.666	12.5
[37]	Text mining of industry 4.0 job advertisements	68	4	International Journal of Information Management	14.098	18.1
[41]	Ageing workforce management in manufacturing systems: state of the art and future research agenda	62	3	International Journal of Production Research	8.568	10.8
[42]	Rethinking Human-Machine Learning in Industry 4.0: How Does the Paradigm Shift Treat the Role of Human Learning?	57	4	8th CIRP Sponsored Conference on Learning Factories (CLF 2018)	N/A	N/A
[43]	Estimating Industry 4.0 impact on job profiles and skills using text mining	55	2	Computers in Industry	7.635	12

Table 1. *Cont.*

Ref. Number	Document Title	Number of Citations	Links	Journal Name	Journal Impact Factor	Journal Cite Score
[44]	Augmented reality-assisted robot programming system for industrial applications	54	0	Robotics and Computer-Integrated Manufacturing	5.666	12.5
[45]	A framework for operative and social sustainability functionalities in Human-Centric Cyber-Physical Production Systems	53	2	Computers & Industrial Engineering	5.431	7.9
[46]	Visual computing technologies to support the Operator 4.0	49	0	Computers & Industrial Engineering	5.431	7.9
[47]	Social Factory Architecture: Social Networking Services and Production Scenarios Through the Social Internet of Things, Services and People for the Social Operator 4.0	48	1	IFIP International Conference on Advances in Production Management Systems	N/A	N/A
[48]	Dynamic task classification and assignment for the management of human-robot collaborative teams in work cells	44	0	The International Journal of Advanced Manufacturing Technology	3.226	N/A

Source: based on Scopus database edited by authors.

According to the Scopus database, the most cited article had 378 citations, while the least cited one had only 44. However, the article published in the journal International Journal of Information Management with the highest impact factor had only 68 citations. At the same time, the second most cited article had the lowest IF. Thus, it seems that there is no close correlation between the number of citations and the value of impact factor. Figure 1 of the database shows the connectivity among the documents using the document citations as the unit of analysis because the software has excluded some documents and shows only the connected ones. Bigger dots or circles represent more cited documents (Figure 1).

The following results refer to RQ2: What are the required trends of the skills and competencies concerning Industry 4.0 that can fit different professions among the different economic sectors?

VOS viewer was used for mapping the data extracted from Scopus and the top twenty cited articles to help to create the most adaptable skills and competency attributes model. A competency model is a descriptive tool that identifies what are the required competencies to perform a job effectively [1,8,9,37,43,49–54]. Therefore, these four sets of competencies are the most used and accepted in the related studies of the study database (Table 2).

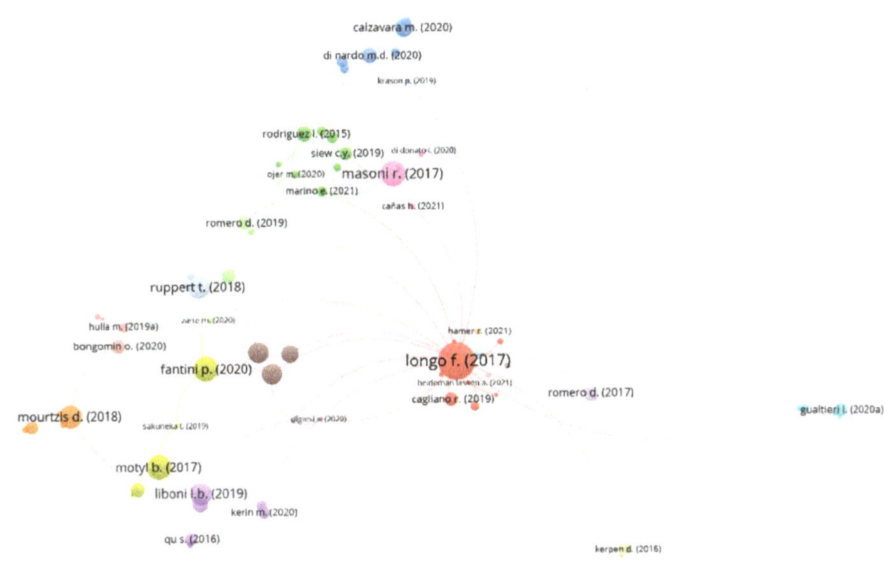

Figure 1. The most connected documents of the database by citations. Source: based on Scopus database edited by authors.

Table 2. A model for Industry 4.0 competences.

Competency Category	Competencies	Related Studies
Personal	(Flexibility, ambiguity tolerance, motivation to learn, ability to work under pressure, sustainable mindset)	[1,8,52,54,55]
Social/Inter-personal	(Intercultural skills, language skills, communication skills, networking skills, teamwork, ability to transfer knowledge, leadership skills)	[1,8,51,54,55]
Technical	(Technical skills, media skills, coding skills).	[1,8,54–57]
Methodological	(Creativity, research skills, problem-solving, conflict solving, decision making).	[1,8,54–57]

Source: based on Scopus database edited by authors.

The competencies mentioned in Table 2 are required by many companies for the new jobs, which are related to Industry 4.0, for example, supply chain analyst, supply chain engineer, CPS and IoT for a robotised production engineer. More results on the new jobs which have been created by Industry 4.0 can be found in only four studies [37,58–60]. The bibliometric analysis has revealed the top twenty countries in this field of the research using citation as the basis of comparison. It is presumable that the majority of these countries are advanced in the application of Industry 4.0 technologies and/or in their research (Table 3).

Table 3. Top twenty contributed countries in the given field.

Country	Documents	Citations	Total Link Strength
Italy	78	1311	133
United States	77	604	46
Germany	49	399	53
India	37	224	12
United Kingdom	35	381	9
Spain	29	348	37
Malaysia	24	78	8
Australia	23	217	13
Austria	22	112	10
Sweden	22	277	36
Portugal	21	451	16
Poland	20	54	12
Russian Federation	19	64	2
South Africa	19	85	11
Brazil	17	182	43
Turkey	15	83	6
China	14	230	15
Canada	13	161	17
France	13	178	16

Source: based on Scopus database edited by authors.

Most of the publications were published in Italy, the US and Germany, and the number of citations was also the highest in these countries. The total link strength means the connection between one document and another by a different author/s in Industry 4.0 topic. The stronger it is, the more citations it has from more than three authors in more than three documents.

A special spatial pattern of countries can be created by the database using the citation links between the documents as the unit of the analysis as well as the authorship analysis concerning Industry 4.0 skills and competencies (Figure 2).

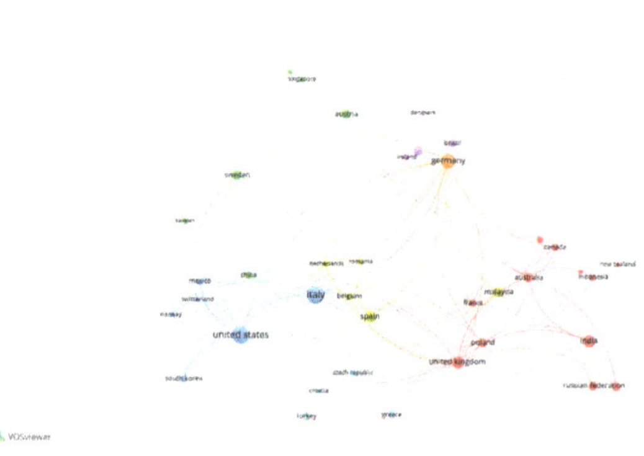

Figure 2. The most connected countries of the database by citations. Source: based on Scopus database edited by authors.

The most important keywords which were mentioned in each document at least 3 times and the connectivity among them also have a special pattern (Figures 3 and 4).

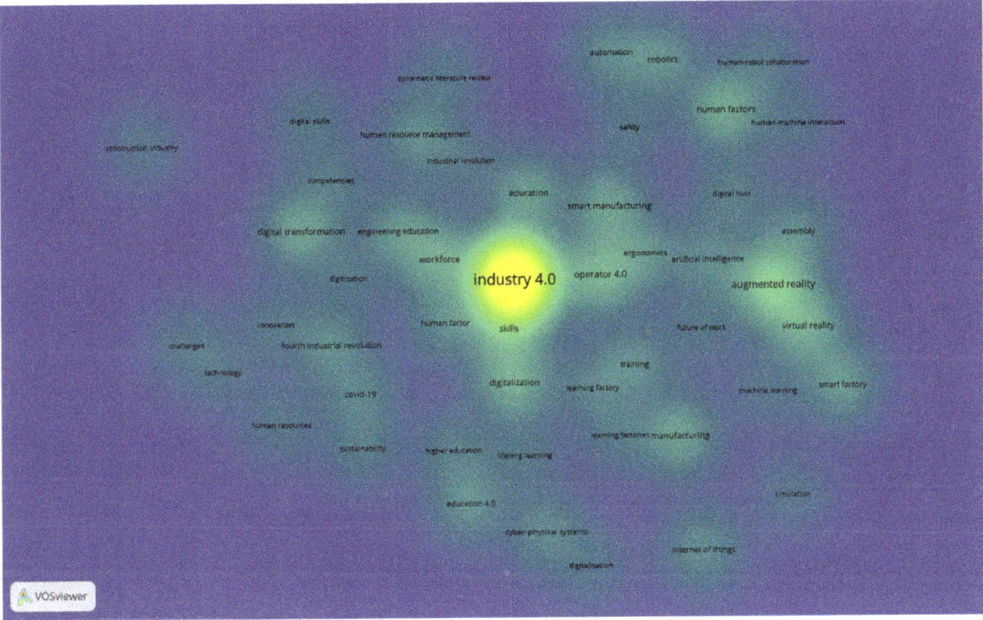

Figure 3. Database by keywords. Source: based on Scopus database edited by authors.

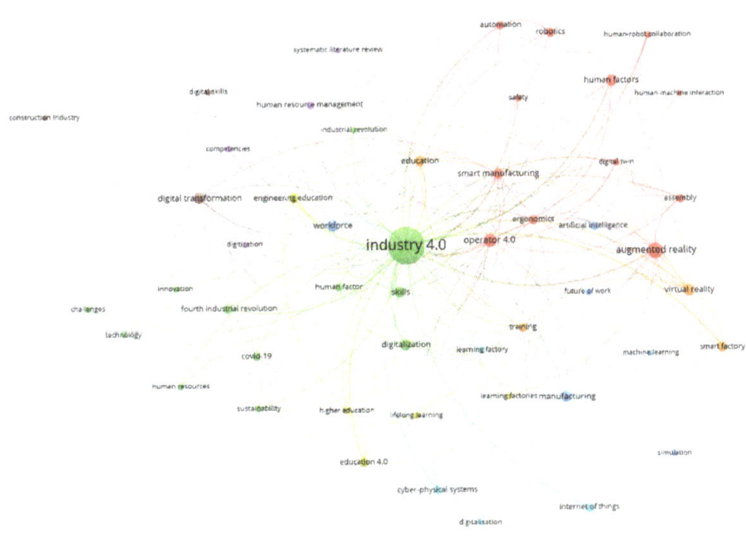

Figure 4. Database by keywords and the connectivity among them. Source: based on Scopus database edited by authors.

All keywords are connected to a big dot, which is Industry 4.0. This is probably because of the popularity of the word itself. Augmented reality is the other, most common keyword, which takes the second place in the intensity of connections and the third is the Operator 4.0, and the reason for this is that most of the studies were related to skills and competencies.

3.2. Industry 4.0 Awareness and Its Impacts on the Labor Force Based on a Survey

To answer RQ3 (What is the level of awareness about Industry 4.0 emerging technologies?), a survey was conducted in March 2022 among expatriates working primarily at multinational organisations in Hungary. The survey was shared using Google forms, which is a popular method as it makes it possible to create and share the questionnaire as the study requires.

The questionnaire consists of four major parts: respondents' personal data, Industry 4.0 concept, impacts of Industry 4.0 and skills and competencies. The survey had open and close questions and Likert scale questions to reveal the awareness level and the skills and competencies level among the sample. Also, the sample was asked about the relationship between technology and the COVID-19 crisis.

Only twenty expatriates took part in the survey. Their number or this sample is too small to generalise any of the results, but they are sufficient to indicate the level of awareness concerning I4.0. The participants of the study were selected because they were occupying positions related to I4.0 technologies in different sectors of the economy, and the reason that the employees of multinational companies were asked is that those companies attract talent from all countries. They are also known for their innovation, research and development, which makes them the best place for such technologies and have already started using these new technologies.

The major characteristics of the respondents are the followings:

- Gender: 25% female, 75% male.
- Age: 75%, 25–34 years old; 20%, 35–44 years old; 5%, 45 years and older.
- Education level: 50%, postgraduate; 35%, graduate; 15%, non-graduate
- The respondents worked in different positions in different fields of the economy. They were the following: marketing, computers, discrete elements methods, English literature, crisis management, senior submission and information specialist management, industrial control systems, transportation mechanics, mechanical engineer, architect, English studies, philologist, medicine, electrical engineering, structural engineering, communication, mathematics, environmental engineering (composting). These fields can give some ideas about their knowledge regarding Industry 4.0.
- Work experience: 50% of them had more than 3 years of experience in the given field.
- Twenty percent of the sample had not heard about the Fourth Industrial Revolution before.

Responses to the different questions are the following:

- To measure if COVID-19 crisis has accelerated the dependency on related Industry 4.0 technologies: 70% strongly agreed with the statement that "COVID-19 pandemic has increased the level of dependency on IT-related systems among the people".
- For the question "Whose responsibility is it to educate the people in order to meet the new requirements?" 40% replied and strongly agreed that it is the government's responsibility, while 45% agreed it is a lifelong learning and it is the people's own responsibility.
- When the sample was asked about robots replacing humans in the labour market and whether it is in the initial stages to say so, respondents estimated positively with the statement, "Robots are replacing humans in the routine jobs (for example: self-check-in at the airport, self-checkout at the supermarket), with 50% strongly agreeing, while 45% agreed on replacing humans in complicated jobs".
- The next question considered which set of the four skills is more important. Respondents estimated that Technical (technical skills, media skills, coding skills) and Methodological (creativity, research skills, problem solving, conflict solving, decision making) are the most important.

- As most of the new jobs related to Industry 4.0 require and/or prefer coding and programing skills, the study sample was asked about the ability of programing. Fifty percent responded that they cannot use programing languages, but the other 50% indicated the knowledge of more than one programing language.
- The responses for the question "How do you imagine your work 10 years later in terms of these technologies?" show that the majority of respondents imagined working from a home office and/or in hybrid form. However, someone wrote for the open-end question that:

"I work in a multinational company in technology business as a Service Desk Analyst, some parts of the system are already automated, I can imagine that my work will be less and less important". This reply also calls attention to the fact that in the future, not only will new skills and competences be required, but several jobs may also disappear. At the same time, new jobs, although in smaller numbers, will also emerge [34]. Some employees may also not be able to work because they cannot meet the requirements or because there will not be enough jobs as machines take over more work.

4. Discussion

This study has made an attempt to determine what kind of new skills and competencies will be required by Industry 4.0. Based on the bibliometric analysis and the questionnaire survey, it has become obvious from theoretical and practical viewpoints that the labour force has to be trained in order to be able to use the new technologies. For that reason, previous studies have focused on putting humans at the center of Industry 4.0 [9].

There is no doubt that having humans at its center is the key to the success of Industry 4.0. Thus, Operator 4.0 has a minimum requirement of the skills that those studies discussed [8,9,11,13,24,25,49,60–62], and they all agreed on a similar model described clearly in [9], which divided the skills into four main categories. They are the following:

- Personal (flexibility, ambiguity tolerance, motivation to learn, ability to work under pressure, sustainable mindset),
- Social/Interpersonal (intercultural skills, language skills, communication skills, networking skills, teamwork, ability to transfer knowledge, leadership skills),
- Technical (technical skills, media skills, coding skills),
- Methodological (creativity, research skills, problem-solving, conflict solving, decision making).

Different studies have discussed more than the four categories of skills considering scenario-based learning (SBL), Education 4.0 and vocational training [1,52,63]. In connection with these, the main question is: Which of them is believed to be the most suitable way of training the new workforce to meet the requirements of the labour market? Another study besides the ones which used the text mining techniques [37,43] compared most of the models resulting in "Five dimensions of worker readiness competencies model" [8] discussing most of the studies which have proposed other models of competencies to meet the requirements of Industry 4.0 [1,8,9,37,43,49–54], and all those studies agreed on the model used in this study. At the same time, other studies' models have focused on the skills needed to enhance the machine-human relationship [46]. The need for new behaviours in the machine-human relationship is important, and at the same time, the trust in the machine, the system and their connectivity can be challenging for the communication infrastructure in the era of cybersecurity.

Finally, we also have to mention that many studies highlighted how important it is to have the skill of decision-making as it appears in most of the studies as a soft skill, while other studies find it more related to AI systems. The question remains on what is the most important skill to have: programming or decision-making. As this study can conclude that both are indeed needed, decision-making can be more accurate and effective with the use of machine learning (ML) as one of the AI applications, as well as the use of the ML, which needs the ability to work with the cloud systems and big data that both require programming languages. This study has not mentioned anything related to programming so far; the needed programming languages for use in I4.0 applications, based on a study made on the LinkedIn database, were C, C++, assembly and JavaScript [37,64]. The results of the survey also highlighted the importance of using programming and coding skills in the age of Industry 4.0.

Parallel to the spread of Industry 4.0 technologies, a marked transformation will occur in all areas of life. New technologies first appeared in the manufacturing industry and continue to spread throughout the economy and society as a whole. The use of new machines and IT tools will require many new skills and competencies. This will most likely be a challenge for the workforce. Those who will have these new skills and competencies, which the study also revealed, will be in a more advantageous position in the labour market. There may be more to these in the future, as Industry 4.0 is constantly developing and making demands on the workforce. However, it is not only the workers and subordinates who have to constantly adapt to the new expectations through the development of their various skills and knowledge, but also the managers of the enterprises. In the age of the Fourth Industrial Revolution, a particularly large responsibility falls on managers, who are responsible not only for the training of the workers, but also for the development of their own expertise and skills. A great variety of knowledge, skills, abilities and competencies are necessary for them in order to be successful and for their business to function effectively. It is likely that, thanks to new technologies, certain skills (e.g., digital skills, communication skills, quick adaptation, system-level thinking, problem solving) will become more valuable, the absence or modest level of which may have unfavourable consequences for the development and future of the enterprise.

The empirical research also confirmed that the new skills and competencies will not be needed to the same extent in different sectors of the economy. Those interviewed considered technical and methodological skills to be the most important. The COVID-19 epidemic probably also contributed to the former, because the use of ICT increased, which required the development of technical skills. In the following years, in line with new technologies and the transformation of education, not only the number of the workforce, but also its quality (training, skills, competences, knowledge, etc.), will change. The labour market, the operation of enterprises and the management of human resources, as well as the economy as a whole, are being transformed.

The result of the study was narrow because of the inclusion criteria of the research, which resulted in excluding many of the most important documents on the topic of Industry 4.0. These articles were mostly in German and discussed the technologies rather than the skills and competencies. Examining [9] in relation to the most crucial citations for the definitions of Industry 4.0 as well as [14], which led this study to elaborate the definitions of Industry 4.0, another finding is that the citation score could be related to the name of the author and the connections between the authors rather than the in-depth information of the document. A further result is that most of the highly cited documents are not necessarily in high impact factor journals. It is probably because non-IF journals are used in a larger circle than the journals with impact factor.

The theory of the study that claims the COVID-19 crisis accelerated and increased the dependency on IT-related systems and Industry 4.0 emerging technologies is supported by the fact this crisis has opened our eyes to the ability of those technologies as crisis response and contingency plans as those studies have discussed before [65]. However, the remaining important issues and questions are related to the education of the coming generation: Is Education 4.0 in developed countries enough? Will the Operator 4.0 be able to control Industry 4.0 technologies according to the risk assessment of volatility, uncertainty, complexity and ambiguity? Answering the questions and applying Industry 4.0 necessitates the development of education and training. In the era of the Fourth Industrial Revolution, a significant transformation of the structure of education will be necessary at all levels (e.g., new subjects must be introduced, new methods must be used, the role of the teacher will change). This is a huge challenge for the current education system everywhere because it is necessary to provide trainings that provides marketable and competitive knowledge and professions, while it is impossible to know exactly what occupations there will be and what knowledge, skills and competencies will be required in the next years and decades.

5. Conclusions

Since the First Industrial Revolution, the labour force has had to adjust to the requirements of the labor market. In each industrial revolution, a new set of skills and competencies had to be developed. Since the Fourth Industrial Revolution has already begun, it is important to explore what the new expectations of the labour market are and what new human resource capabilities are necessary for the workforce to meet them. Based on bibliometric analysis and systematic literature review, this study determined the most related articles concerning skills, competencies, and Industry 4.0, and identified the new set of skills and competencies which are important for the future labour force. It has also evaluated several skill and competency models referring to the top-cited articles in the topic and more models referring to more recent articles published in 2020 that did not have enough time to reach a high citation score.

According to the models which have focused on interpersonal and technological skills, the most important skills and competencies are interpersonal skills, as many studies have confirmed that these are necessary in the workforce on all levels [23,31,38]. Interpersonal skills are important because they are the crucial area where human can surpass the machine. This study also focuses on the innovation competencies. These kinds of competencies can enhance the ability of the human to use the machine relationship to create and invent using the AI and ML at the maximum application, which will make a place for the human workforce in the workplace [62].

Fewer studies have focused on the technical and domain skills, which are more important in regard to the programming language, which is the way to communicate with the computer. It is presumable that in the upcoming ten years, these competencies will be categorized as communication skills and considered as a language, not a technical skill. Moreover, interpersonal skills and programming competencies will be necessary in all job profiles in the future, and the technical and domain skills to be developed based on the job profile. The question here is: What are the most important interpersonal skills? It can be found in the model of [52,66,67]. The experiences of the survey also confirmed that technical and methodological skills are the most important, the significance of which will probably continue to increase in parallel with the spread of Industry 4.0. For the latter, however, significant infrastructural, especially info-communication technology, developments are also necessary. In the future, what we cannot ignore is that the real application of this industry on all levels would need the glue of communication, which cannot be provided without the fifth generation of communication. Although humans will play the key role in the success of this transition, ensuring the material availability that helps in producing, for example, the chipsets of the computers theyalso have to face several other challenges (e.g., shortage of energy and water, climate change, different epidemics, economic crisis, digitalised world), the effects of which affect all fields of life and people's skills and competencies to varying degrees, encouraging them to continuously develop and adapt. In order for humans to cope with and adjust to them, it is extremely important to see clearly what kind of new skills and competencies will be important in the future, and this study has made an attempt at this.

The results of this research, on the one hand, draw attention to the new skills and competencies, which should be emphasized more at different levels of education. On the other hand, they can contribute to the development of human resources of enterprises and the elaboration of new training programs of educational institutions. Based on the lessons learned from the research, the study recommends that more attention be paid to the development of analytical skills in education and vocational training, which may be important in the adaptation of Industry 4.0, as well as to the teaching of subjects related to information and communication technologies.

There are several options for continuing the research. One of these could be to use another database (e.g., WOS) and make a comparison to confirm and refine the current results. Another possibility to continue the research is to widen the scope of the participants in the survey and to examine how much the various sectors and enterprises of the economy are prepared for the new challenges posed by Industry 4.0 technologies to the skills and competencies of the workforce. These studies can also contribute to defining the necessary structural changes in the economy, labour market and education in the age of the Fourth Industrial Revolution.

Author Contributions: Conceptualization: A.A. and E.K.; making bibliometric analysis and survey: A.A.; writing original draft: A.A.; rewriting, reviewing and editing: E.K.; supervising and project administration: E.K. editing figures and tables: A.A. All authors have read and agreed to the published version of the manuscript.

Funding: This research was supported by the National Research, Development and Innovation Office (project number K125 091).

Institutional Review Board Statement: Not applicable.

Informed Consent Statement: Not applicable.

Data Availability Statement: The questionnaire and replies are available for request.

Acknowledgments: The authors of the study would like to thank the three anonymous reviewers and editors for their valuable comments and suggestions, which contributed to the final form of the study.

Conflicts of Interest: The authors declare no conflict of interest.

References

1. Erol, S.; Jäger, A.; Hold, P.; Ott, K.; Sihn, W. Tangible Industry 4.0: A Scenario-Based Approach to Learning for the Future of Production. *Procedia CIRP* **2016**, *54*, 13–18. [CrossRef]
2. Winter, J. The evolutionary and disruptive potential of Industrie 4.0. *Hung. Geogr. Bull.* **2020**, *69*, 83–98. [CrossRef]
3. Kagermann, H.; Lukas, W.-D.; Wahlster, W. Industrie 4.0: Mit dem Internet der Dinge auf dem Weg zur 4. industriellen Revolution. *VDI Nachr.* **2011**, *13*, 2–3.
4. Kagermann, H.; Helbig, J.; Hellinger, A.; Wahlster, W. *Recommendations for Implementing the Strategic Initiative INDUSTRIE 4.0: Securing the Future of German Manufacturing Industry*; Final Report of the Industrie 4.0 Working Group; Forschungsunion: Berlin, Germany, 2013.
5. Nagy, J. Az Ipar 4.0 fogalma és kritikus kérdései-vállalati interjúk alapján. *Vezetéstud* **2019**, *50*, 14–27. [CrossRef]
6. Reischauer, G. Industry 4.0 as policy-driven discourse to institutionalize innovation systems in manufacturing. *Technol. Fore. Soc. Chang.* **2018**, *132*, 26–33. [CrossRef]
7. Samad, T.; Annaswamy, A.M. *The Impact of Control Technology*; IEEE Control Systems Society: Gainesville, FL, USA, 2011.
8. Blayone, T.J.B.; VanOostveen, R. Prepared for work in Industry 4.0? Modelling the target activity system and five dimensions of worker readiness. *Int. J. Comput. Integr. Manuf.* **2021**, *34*, 1–19. [CrossRef]
9. Hecklau, F.; Galeitzke, M.; Flachs, S.; Kohl, H. Holistic Approach for Human Resource Management in Industry 4.0. *Procedia CIRP* **2016**, *54*, 1–6. [CrossRef]
10. Longo, F.; Nicoletti, L.; Padovano, A. Smart operators in industry 4.0: A human-centered approach to enhance operators' capabilities and competencies within the new smart factory context. *Comput. Ind. Eng.* **2017**, *113*, 144–159. [CrossRef]
11. Ruppert, T.; Jaskó, S.; Holczinger, T.; Abonyi, J. Enabling technologies for operator 4.0: A survey. *Appl. Sci.* **2018**, *8*, 1650. [CrossRef]
12. Ananiadou, K.; Claro, M. *21st Century Skills and Competences for New Millennium Learners in OECD Countries*; OECD Education Working Papers; Organisation for Economic Co-Operation and Development: Paris, France, 2009.
13. Kipper, L.; Iepsen, S.; Forno, A.J.D.; Frozza, R.; Furstenau, L.; Agnes, J.; Cossul, D. Scientific mapping to identify competencies required by industry 4.0. *Technol. Soc.* **2021**, *64*, 101454. [CrossRef]
14. Alcácer, V.; Cruz-Machado, V. Scanning the Industry 4.0: A Literature Review on Technologies for Manufacturing Systems. *Eng. Sci. Technol. Int. J.* **2019**, *22*, 899–919. [CrossRef]
15. Mell, P.; Grance, T. *The NIST Definition of Cloud Computing*; NIST Special Publication; National Institute of Standards and Technology: Gaithersburg, MD, USA, 2011; Volume 800.
16. Bag, S.; Wood, L.C.; Xu, L.; Dhamija, P.; Kayikci, Y. Big data analytics as an operational excellence approach to enhance sustainable supply chain performance. *Resour. Conserv. Recycl.* **2019**, *153*, 104559. [CrossRef]
17. Stachová, K.; Stacho, Z.; Cagáňová, D.; Starecek, A. Use of digital technologies for intensifying knowledge sharing. *Appl. Sci.* **2020**, *10*, 4281. [CrossRef]

18. Boston Consulting Group. Industry 4.0: The Future of Productivity and Growth in Manufacturing Industries. 2015. Available online: https://www.bcg.com/publications/2015/engineered_products_project_business_industry_4_future_productivity_growth_manufacturing_industries.aspx (accessed on 19 May 2019).
19. Masoni, R.; Ferriseb, F.; Bordegonib, M.; Gattulloc, M.; Uvac, A.E.; Fiorentinoc, M.; Carrabbad, E.; Di Donatoe, M. Supporting Remote Maintenance in Industry 4.0 through Augmented Reality. *Procedia Manuf.* **2017**, *11*, 1296–1302. [CrossRef]
20. Chang, J.; He, J.; Mao, M.; Zhou, W.; Lei, Q.; Li, X.; Li, D.; Chua, C.-K.; Zhao, X. Advanced material strategies for next-generation additive manufacturing. *Materials* **2018**, *11*, 166. [CrossRef]
21. Kagermann, H.; Anderl, R.; Gausemeier, J.; Schuh, G.; Wahlster, W. *Industrie 4.0 in a Global Context: Strategies for Cooperating with International Partners*; Herbert Utz Verlag: Munich, Germany, 2016.
22. Bai, C.; Dallasega, P.; Orzes, G.; Sarkis, J. Industry 4.0 technologies assessment: A sustainability perspective. *Int. J. Prod. Econ.* **2020**, *229*, 107776. [CrossRef]
23. Rodriguez, L.; Quint, F.; Gorecky, D.; Romero, D.; Siller, H.R. Developing a Mixed Reality Assistance System Based on Projection Mapping Technology for Manual Operations at Assembly Workstations. *Procedia Comput. Sci.* **2015**, *75*, 327–333. [CrossRef]
24. Fantini, P.; Pinzone, M.; Taisch, M. Placing the operator at the centre of Industry 4.0 design: Modelling and assessing human activities within cyber-physical systems. *Comput. Ind. Eng.* **2020**, *139*, 105508. [CrossRef]
25. Kadir, B.A.; Broberg, O. Human-centered design of work systems in the transition to industry 4.0. *Appl. Ergon.* **2020**, *92*, 103334. [CrossRef]
26. Wang, K.J.; Rizqi, D.A.; Nguyen, H.P. Skill transfer support model based on deep learning. *J. Intell. Manuf.* **2021**, *32*, 1129–1146. [CrossRef]
27. Romero, D.; Bernus, P.; Noran, O.; Stahre, J.; Fast-Berglund, Å. The Operator 4.0: Human Cyber-Physical Systems & Adaptive Automation Towards Human-Automation Symbiosis Work Systems. In Proceedings of the IFIP International Conference on Advances in Production Management Systems, Iguassu Falls, Brazil, 3–7 September 2016; pp. 677–686.
28. Bousdekis, A.; Apostolou, D.; Mentzas, G. A human cyber physical system framework for operator 4.0—Artificial intelligence symbiosis Author links open overlay panelaba. *Manuf. Lett.* **2020**, *25*, 10–15. [CrossRef]
29. Nasoz, F.; Alvarez, K.; Lisetti, C.L.; Finkelstein, N. Emotion recognition from physiological signals using wireless sensors for presence technologies. *Cogn. Technol.* **2004**, *6*, 4–14. [CrossRef]
30. Neumann, W.P.; Winkelhaus, S.; Grosse, E.H.; Glock, C.H. Industry 4.0 and the human factor—A systems framework and analysis methodology for successful development. *Int. J. Prod. Econ.* **2021**, *233*, 107992. [CrossRef]
31. Kaasinen, E.; Schmalfuß, F.; Özturk, C.; Aromaa, S.; Boubekeur, M.; Heilala, J.; Heikkilä, P.; Kuula, T.; Liinasuo, M.; Mach, S.; et al. Empowering and engaging industrial workers with Operator 4.0 solutions. *Comput. Ind. Eng.* **2020**, *139*, 105678. [CrossRef]
32. de Vries, G.J.; Gentile, E.; Miroudot, S.; Wacker, M.K. The rise of robots and the fall of routine jobs. *Labour Econ.* **2020**, *66*, 101885. [CrossRef]
33. Arntz, M.; Gregory, T.; Zierahn, U. *The Risk of Automation for Jobs in OECD Countries, OECD Social*; Employment and Migration Working Papers OECD; OECD: Paris, France, 2016; Volume 189.
34. Ford, M. *The Rise of the Robots: Technology and the Threat of Jobless Future*; Basic Books: New York, NY, USA, 2015.
35. Fonseca, L.M. Industry 4.0 and the digital society: Concepts, dimensions and envisioned benefits. In Proceedings of the 12th International Conference on Business Excellence, Bucharest, Romania, 22–23 March 2018; Volume 12, pp. 386–397.
36. Csoba, J. Flexibilitás a munkacrőpiacon. *Munkaügyi Szle.* **2018**, *61*, 7–20.
37. Pejic-Bach, M.; Bertoncel, T.; Meško, M.; Krstić, Ž. Text mining of industry 4.0 job advertisements. *Int. J. Inf. Manag.* **2020**, *50*, 416–431. [CrossRef]
38. Jaturanonda, C.; Nanthavanij, S. Analytic-based decision analysis tool for employee-job assignments based on competency and job preference. *Indust. J. Indust. Eng.* **2011**, *18*, 58–70.
39. Roldán, J.J.; Crespo, E.; Martín-Barrio, A.; Peña-Tapia, E.; Barrientos, A. A training system for Industry 4.0 operators in complex assemblies based on virtual reality and process mining. *Robot. Comput. Integr. Manuf.* **2019**, *59*, 305–316. [CrossRef]
40. Sima, V.; Gheorghe, I.G.; Subić, J.; Nancu, D. Influences of the Industry 4.0 Revolution on the Human Capital Development and Consumer Behavior: A Systematic Review. *Sustainability* **2020**, *12*, 4035. [CrossRef]
41. Calzavara, M.; Battini, D.; Bogataj, D.; Sgarbossa, F.; Zennaro, I. Ageing workforce management in manufacturing systems: State of the art and future research agenda. *Int. J. Prod. Res.* **2020**, *58*, 729–747. [CrossRef]
42. Ansari, F.; Erol, S.; Sihn, W. Rethinking Human-Machine Learning in Industry 4.0: How Does the Paradigm Shift Treat the Role of Human Learning? *Procedia Manuf.* **2018**, *23*, 117–122. [CrossRef]
43. Fareri, S.; Fantoni, G.; Chiarello, F.; Coli, E.; Binda, A. Estimating Industry 4.0 impact on job profiles and skills using text mining. *Comput. Ind.* **2020**, *118*, 103222. [CrossRef]
44. Ong, S.K.; Yew, A.W.W.; Thanigaivel, N.K.; Nee, A.Y.C. Augmented reality-assisted robot programming system for industrial applications. *Robot. Comput. Integr. Manuf.* **2020**, *61*, 101820. [CrossRef]
45. Pinzone, M.; Albè, F.; Orlandelli, D.; Barletta, I.; Berlin, C.; Johansson, B.; Taisch, M. A framework for operative and social sustainability functionalities in Human-Centric Cyber-Physical Production Systems. *Comput. Ind. Eng.* **2020**, *139*, 105132. [CrossRef]
46. Segura, Á.; Dieza, H.V.; Barandiaran, I.; Arbelaiz, A.; Álvarez, H.; Simões, B.; Posada, J.; García-Alonso, A.; Ugarte, R. Visual computing technologies to support the Operator 4.0. *Comput. Ind. Eng.* **2020**, *139*, 604–614. [CrossRef]

47. Romero, D.; Wuest, T.; Stahre, J.; Gorecky, D. Social factory architecture: Social networking services and production scenarios through the social internet of things, services and people for the social operator 4.0. *IFIP Adv. Inf. Commun. Technol.* **2017**, *513*, 265–273. [CrossRef]

48. Bruno, G.; Antonelli, D. Dynamic task classification and assignment for the management of human-robot collaborative teams in workcells. *Int. J. Adv. Manuf. Technol.* **2018**, *98*, 2415–2427. [CrossRef]

49. Jerman, A.; Pejić Bach, M.; Aleksić, A. Transformation towards smart factory system: Examining new job profiles and competencies. *Syst. Res. Behav. Sci.* **2020**, *37*, 388–402. [CrossRef]

50. Adolph, S.; Tisch, M.; Metternich, J. Challenges and Approaches to Competency Development for Future Production. *J. Int. Sci. Publ.* **2014**, *12*, 1001–1010. Available online: https://www.scientific-publications.net/en/article/1000581/ (accessed on 25 October 2021).

51. Razali, H.; Ismail, A.A. Challenge and Issues in Human Capital Development Towards Industry Revolution 4.0. In Proceedings of the International Conference on the Future of Education, Penang, Malaysia, 10–12 July 2018; pp. 716–727.

52. Dworschak, B.; Zaiser, H. Competences for cyber-physical systems in manufacturing-First findings and scenarios. *Procedia CIRP* **2014**, *25*, 345–350. [CrossRef]

53. Abbasi, M.; Hosnavi, R.; Tabrizi, B. Application of Fuzzy DEMATEL in Risks Evaluation of Knowledge-Based Networks. *J. Optim.* **2013**, *2013*, 913467. [CrossRef]

54. Mourtzis, D. Development of skills and competences in manufacturing towards education 4.0: A teaching factory approach. In *Proceedings of the 3rd International Conference on the Industry 4.0 Model for Advanced Manufacturing*; Springer: New York, NY, USA, 2018; pp. 194–210. [CrossRef]

55. Wang, B.; Ha-Brookshire, J.E. Exploration of Digital Competency Requirements within the Fashion Supply Chain with an Anticipation of Industry 4.0. *Int. J. Fash. Des. Technol. Educ.* **2018**, *11*, 333–342. [CrossRef]

56. Gökalp, E.; Şener, U.; Eren, P.E. Development of an assessment model for industry 4.0: Industry 4.0-MM. *Commun. Comput. Inf. Sci.* **2017**, *770*, 128–142. [CrossRef]

57. Canetta, L.; Barni, A.; Montini, E. Development of a Digitalization Maturity Model for the Manufacturing Sector. In Proceedings of the IEEE International Conference on Engineering, Technology and Innovation (ICE/ITMC), Stuttgart, Germany, 17–20 June 2018; pp. 1–7. [CrossRef]

58. Papoutsoglou, M.; Rigas, E.S.; Kapitsaki, G.M.; Angelis, L.; Wachs, J. Online labour market analytics for the green economy: The case of electric vehicles. *Technol. Forecast. Soc. Chang.* **2022**, *177*, 121517. [CrossRef]

59. Wadan, R.; Bensberg, F.; Teuteberg, F.; Buscher, G. Understanding the Changing Role of the Management Accountant in the Age of Industry 4.0 in Germany. In Proceedings of the 52nd Hawaii International Conference on System Sciences, Maui, HI, USA, 8–11 January 2019; pp. 5817–5826. [CrossRef]

60. Caputo, F.; Garcia-Perez, A.; Cillo, V.; Giacosa, E. A knowledge-based view of people and technology: Directions for a value co-creation-based learning organization. *J. Knowl. Manag.* **2019**, *23*, 1314–1334. [CrossRef]

61. Papetti, A.; Gregori, F.; Pandolfi, M.; Peruzzini, M.; Germani, M. A method to improve workers' well-being toward human-centered connected factories. *J. Comput. Des. Eng.* **2020**, *7*, 630–643. [CrossRef]

62. Whysall, Z.; Owtram, M.; Brittain, S. The new talent management challenges of Industry 4.0. *J. Manag. Dev.* **2019**, *38*, 118–129. [CrossRef]

63. Subekti, S.; Ana, A.; Barliana, M.S.; Khoerunnisa, I. Problem solving solving improvement through the teaching factory model. *J. Phys. Conf. Ser.* **2019**, *1402*, 022044. [CrossRef]

64. Landherr, M.; Schneider, U.; Bauernhansl, T. The Application Center Industrie 4.0—Industry-driven Manufacturing, Research and Development. *Procedia CIRP* **2016**, *57*, 26–31. [CrossRef]

65. Nickinson, A.T.O.; Carey, F.; Tan, K.; Ali, T.; Al-Jundi, W. Has the COVID-19 Pandemic Opened Our Eyes to the Potential of Digital Teaching? A Survey of UK Vascular Surgery and Interventional Radiology Trainees. *Eur. J. Vasc. Endovasc. Surg.* **2020**, *60*, 952–953. [CrossRef] [PubMed]

66. Ismail, F.; Kadir, A.A.; Khan, M.A.; Yih, Y.P. The Challenges and Role Played among Workers of Department Human Resources Management towards Industry 4.0 in SMEs. *KnE Soc. Sci.* **2019**, 90–107. [CrossRef]

67. Fitsilis, P.; Tsoutsa, P.; Gerogiannis, V. Industry 4.0: Required Personnel Competences. *Intern. Sci. J. Ind. 4.0* **2018**, *3*, 130–133.

Review

Development and Practice of Industrie 4.0 in China—Practical Experience of a German Industrial Software Company in China

Kuan-Lun Lee *, Andrea Roesinger and Uwe Hommel

FORCAM GmbH, 88214 Ravensburg, Germany; roesinger.andrea@web.de (A.R.);
uwe.hommel@max-it.solutions (U.H.)
* Correspondence: leekuanlun@outlook.com

Abstract: Industrie 4.0 has stirred turbulences in China since its birth in 2011. The struggles of the Chinese manufacturing enterprises towards realizing and adapting Industrie 4.0 in their production processes have given us many new perceptions. These insights and findings can in turn serve as inputs for academics and policy makers to structure or fine tune the development of the next generation of Industrie 4.0. The authors of this paper summarize the knowledge and understandings from their personal engagement assisting the Chinese manufacturing enterprises with digitalization in their production processes. A real-life example shows how a typical Chinese mid-size manufacturing enterprise ended up with new business models when they started out the digitalization journey with a simple goal to increase efficiency. We conclude that the Chinese market will continue to be relevant for the future development of Industrie 4.0.

Keywords: Industrie 4.0; intelligent manufacturing; China Manufacturing 2025; Industrial Internet; Cloud Manufacturing; digitalization; small-medium enterprises; new business models

check for
updates

Citation: Lee, K.-L.; Roesinger, A.; Hommel, U. Development and Practice of Industrie 4.0 in China—Practical Experience of a German Industrial Software Company in China. *Sci* **2022**, *4*, 28. https://doi.org/10.3390/sci4030028

Academic Editor: Johannes Winter

Received: 28 May 2022
Accepted: 4 July 2022
Published: 13 July 2022

Publisher's Note: MDPI stays neutral with regard to jurisdictional claims in published maps and institutional affiliations.

1. Introduction

The birth of Industrie 4.0 has incurred a lot of thoughts and discussions in both Germany as well as in China. From the western point of view, trying to grasp the grand old Chinese culture alone will take some effort. Since "openness" and "integration" are the major philosophies of Industrie 4.0, how does one integrate these with China, which is one of the most important manufacturing bodies in the world? From the Chinese point of view, the concepts and philosophies of Industrie 4.0 have inspired them in their search for better living, but the question of how to deploy these concepts to suit their mentality and environment remains a field that requires further investigations from academics and the market.

This paper reviews the development of Industrie 4.0 in China. We look at Industrie 4.0 in China from a macroeconomic point of view, as well as looking into concrete digitalization projects to discuss the problems that enterprises encounter. We also look at the history up to the present situation and discuss the possible future development of Industrie 4.0 in China. Through a better understanding of the market and how it would further develop, each enterprise or individual could have a better judgement of its/their strategy of engagement with China.

We make our observations and conclusions on the Chinese market based on engagements with large state-owned-enterprises (SOEs) and small-and-medium enterprises (SMEs) predominantly in automotive discrete manufacturing (OEM and supplier). We dive into deeper insights to the SMEs as they define to a higher degree the phenomenon of the Chinese economy. We feel that more attention should be directed to the SMEs than the huge conglomerates.

The authors intentionally used the term "Industrie 4.0" in its original German spelling throughout the paper. The intention was to stress the affinity of the Chinese to the German technologies and philosophies in its authentic form. As Prof. Henning Kagermann, co-Father of Industrie 4.0, once remarked after his visit to China, "The Chinese spell Industrie 4.0 with 'ie'."

Methodology

This paper presents research results from a market perspective, the paper therefore does not represent research in a narrow sense. This practice-oriented paper provides important and helpful insights for applied research on smart manufacturing.

The results come from real-life customer projects, customer surveys and personal experiences of the authors as experienced IT/OT managers and entrepreneurs with long years of industry experiences in both Germany and China. The insights to the market need for digitalization in China are concluded from over 150 cases of customer engagements at the pre-sale phases between 2018 and 2022.

We have carried out dozens of digitalization projects in the manufacturing industry in China. We conducted surveys, collected feedbacks and carried out discussions with the customers during and after every project. The observations, conclusions and recommendations are also made from this specific industry (industrial software) point of view but most of them are generally applicable to any other industries.

For a topic like Industrie 4.0, the problems and feedback from the market and politics very much build the foundation to shape its development. Therefore, the case studies presented in this paper can be valuable to both researchers and practitioners of the field. The conclusions derived from these projects can also be generally applicable to many other similar fields of studies and regional applied research.

2. Overview of Industrie 4.0 in China

Industrie 4.0 represents the interconnection of people, intelligent objects and machines, the use of service-oriented architectures and the composition of services and data from different sources into new business processes. Industrie 4.0 is the basis for the future of industrial value creation. The focus is on data-based value creation, digitally enhanced business models and forms of organization, but also new solutions in areas such as energy, finance, health, and mobility. Economically, it initially involved a shift from traditional automation with predetermined outcomes to learning and self-adapting machines and environments that react in real time to changes in customer demand as well as to unexpected disruptions. This is accompanied by a move from mass production to customization, i.e., the competitively priced production of individual, tailor-made products [1,2].

In 2013, with the publication of the revolutionary white paper *"Recommendations for implementing the strategic initiative INDUSTRIE 4.0. Securing the future of German manufacturing industry. Final Report of the Industrie 4.0 Working Group"* [3], Prof. Henning Kagermann and Prof. Wolfgang Wahlster introduced the concept of "Industrie 4.0"to the world. The beauty of Industrie 4.0 to the Chinese was that it initiated them to begin pondering the question "What is the future of manufacturing?" and gave the answer "Digitalization and new business models create values" at the same time [3,4]. Since then, Industrie 4.0 has become a hype word in China. There has been a rush to learn, implement and practice Industrie 4.0 technologies and methodologies in the industry. Figure 1 shows the growth of the market for intelligent manufacturing in the past few years.

Lately, compared with the rush and the attempts for a radical change in manufacturing a few years back, the major players of Industrie 4.0 in China are becoming calm and thoughtful about their endeavor. After numerous explorations from different angles, it has become obvious that this road would not be an easy one. Patience, experience, accumulation of technology foundations and knowledges, as well as sufficient monetary investments are critical factors for the success.

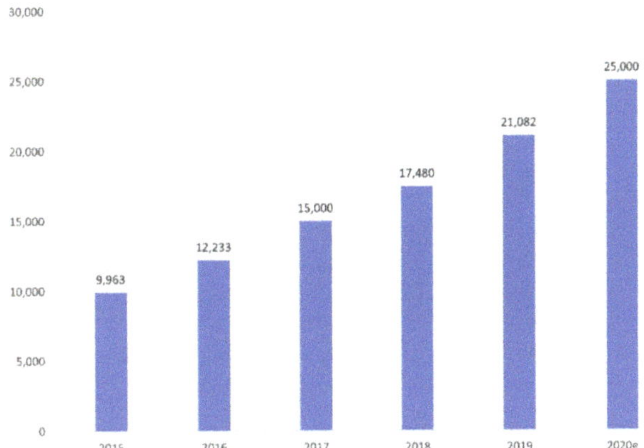

Figure 1. Output scale of China's intelligent manufacturing industry; unit: 100 million RMB. (Source: Forward Industry Research Institution).

2.1. Variants of Chinese Versions of Industrie 4.0

A paradigm like Industrie 4.0 is bound to be influenced by the practical deployment of relevant projects in the market. Problems and issues encountered in the market provide feedbacks to help reshape the development of Industrie 4.0. In China, where the economy, technological progress and cultural environment are very much different than in Germany, implementing Industrie 4.0 in the Chinese market demands constant modification to the model and adapting the local needs, what is commonly called "localization". Parallel to deploying Industrie 4.0 in the Chinese industrial environment, several variants of Industrie 4.0 have emerged to suit different market demands at different stages.

Figure 2 shows the timeline of the emergence of the variants of Industrie 4.0 in China. These variants of Industrie 4.0 dominated the major philosophy of smart manufacturing in the Chinese market at the different stages.

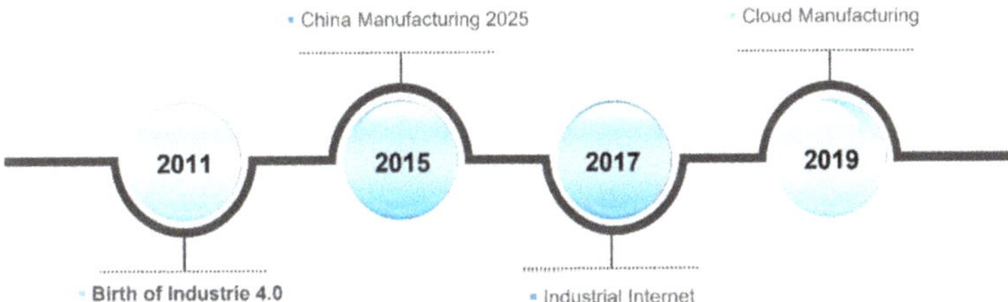

Figure 2. Variants of Industrie 4.0 and the respective time of market emergence in China.

2.1.1. China Manufacturing 2025

"China Manufacturing 2025" was a strategic initiative brought out by Prime Minister Li Keqiang. An official document was issued by the State Council in May 2015 meant to strengthen the manufacturing capability of the Chinese economy.

The China Manufacturing 2025 Plan is also known as the "China Version of The Industrie 4.0 Plan". The concept of "China Manufacturing 2025" was first proposed by the Chinese Academy of Engineering. Under the overall planning of the State Council, the Ministry of Industry and Information Technology (MIIT) took the lead in working with more than 20 ministries and commissions such as the National Development and Reform Commission, the Ministry of Science and Technology and more than 50 academicians to formally compile the "China Manufacturing 2025 Plan" [5].

The core goal of "China Manufacturing 2025" was to transform China from a manufacturing supplier to a manufacturing driver. The year 2025 was chosen after careful analysis of China's status of manufacturing capabilities and the overall market environment at that time (2015). It was set as a goal to transform, in 10 years, its capability from simply providing low-cost labor for manufacturing outsourcing, to the form where demands of the new era will challenge the supply end and thus bring added value to the industry by driving innovations in the supplier technologies.

2.1.2. Industrial Internet

The concept of "Industrial Internet" actually originated in America. It was first proposed by General Electric (GE) in 2012 through the release of the white paper *"Industrial Internet: Pushing the boundaries of minds and machines"* [6], which elaborated on the connotation and future vision of the Industrial Internet from the aspects of technical architecture, development opportunities, potential benefits, application conditions, etc. In essence, industrial interconnectivity is the integration and innovation of industrial and IT capabilities. Subsequently, five industry giants in the United States joined forces to form the Industrial Internet Consortium (IIC), which vigorously promoted the concept of Industrial Internet and set off a wave of Industrial Internet movement in the world.

In 2015, MIIT issued the documents *"Implementation of the State Council on actively promoting the Internet + Guidance action plans (year 2015–2018)"* [7], clearly stating the guidelines for the internet to be widely integrated into the entire process of production and manufacturing, the entire value chain and the whole life cycle of products, cultivating and developing an open innovative R&D model, and accelerating the development and application of industrial big data.

In February 2016, China established the Alliance of Industrial Internet (AII). Jointly initiated by the China Academy of Information and Communications Technology and related enterprises such as manufacturing, communications industry, and the Internet, the alliance actively carries out work in the research of major issues of the Industrial Internet, the development of standards, technical test verification, and industry promotion. In the same year, AII released version 1.0 of the Industrial Internet Architecture.

Since then, hundreds of industrial internet platforms have been founded and have entered the market, providing solutions and services to the manufacturing industry.

2.1.3. Cloud Manufacturing

In 2009, the research team led by Prof. Li Bohu, an academician of the Chinese Academy of Engineering and Professor Zhang Lin, then deputy dean of the School of Automation Science and Electrical Engineering of Beihang University, realized that the deep integration of advanced information technologies such as cloud computing and the Internet of Things with the manufacturing industry will bring profound changes to the manufacturing industry and they took the lead in proposing the concept of "Cloud Manufacturing" [8]. They pointed out that "cloud manufacturing was a kind of network and service platform, organizing online manufacturing resources according to user needs", to provide users with a variety of on-demand manufacturing services.

Cloud manufacturing provides a new model and means for China to move from a manufacturing outsourcer to a manufacturing driver. The initiative was jointly supported by more than 300 researchers and developers from 28 organizations of universities, research institutes and enterprises across the country, and achieved a number of pioneering research results [9,10].

Prof. Li Bohu, Prof. Zhang Lin and their team members have publicized and introduced the concept and research results of cloud manufacturing by organizing international conferences or forums, publishing academic papers, carrying out academic exchanges and other forms, which have made important contributions to the recognition and attention of international counterparts for cloud manufacturing research, and gradually forming a new research direction in the world. Cloud manufacturing is one of the very few academic directions initiated by China that attracted the attentions of international academies.

3. The Need of Digitalization

In the eyes of most people, a factory has the means to receive orders, manufacture and ship produced goods. In this process, the business owners and the employees made a certain profit out of it. A need for digitalization was not very straight forward.

However, in the past few years, especially after the 2008 financial crisis, a higher level of perception to the means of a factory became more and more widespread. In the entire industrial chain, the raw materials were processed and turned into finished products to complete the value-adding. A factory was beginning to be perceived as the means of value discovery, value creation and value transmission.

In the past over 30 years, the economic environment in China has favored the growth of a manufacturing market [11]. Most manufacturing businesses founded have encountered relatively low founding hurdles and most of them lacked core competitive technologies. Thus, they would compete in the market with product cost performance (mainly price). Businesses with better management gradually outperform their competitors and those with poor management faced the threat of being discarded by the market competition. With the continuous rise in labor, rent and raw material prices, as well as intensive homogeneous competition, the pressure of efficiency in factory operation was increasing. The COVID pandemic has also made many traditional manufacturing practitioners understand the importance of digital operation. At the same time, international trade war, the financial crisis, fluctuations in exchange rates or other unpredictable factors threatened the future of manufacturing businesses.

Huge conglomerates have the expertise and capabilities to take measures to mitigate the risks of uncertainties. However, a large majority of the Chinese businesses came in existence simply because the market has been generous. These companies grew in an environment without much risk, and they have not been able to build up the necessary risk mitigation capabilities [12]. They are facing challenges today that were never encountered in the past 30 years. This is an obvious driving factor for a growing need to seek for solutions to ensure their survival in the market.

3.1. China's SMEs in Urgent Need of Digitalization

Digital transformation of SMEs in China is a very important aspect of the overall digital transformation in China's economy. The big enterprises alone will not be able to bring forward the entire economy to a digitalized one. Therefore, it is crucial to look into and understand the problems and difficulties of the SMEs in their digitalization transformation endeavor.

One point to note is the difference in perception of the different cultures and markets. This very often leads to differences in usage of vocabularies and semantics for mutual understanding. For example, when we mention "Small and Medium Enterprises" in Germany, we envision companies which the ownership, management and liability are traditionally in one hand. The German SMEs can be strong partners for large companies worldwide, and they bear responsibility for the skilled workers of tomorrow and stand for innovative strength, sustainable growth and job creation [13]. In an associated recommendation of the Commission of the European Communities from 6 May 2003, SMEs have up to 249 employees and make a yearly turnover of less than 50 million Euros [14]. Many German SMEs are even world leading in the technologies of their field, the so-called "hidden champions". However, the perception of SMEs in China is a completely different concept. Very often, a company with several thousand employees will still be regarded as an SME. The majority of SMEs in China have not overcome the phase of struggling for survival. Such conceptional differences increase the difficulties for communication and critical messages are very often misinterpreted.

According to the *"Analysis Report on the Digital Transformation of Small and Medium-sized Enterprises (2020)"* [15] published by the Chinese Electronics Standardization Institute, 89% of China's SMEs are exploring digital transformation, but only 3% of SMEs are in the stage of more extensive applications of digital transformation. This number, according to the research done by ZEW (Leibniz-Zentrum für Europäische Wirtschaftsforschung) and Infas (Institut für angewandte Sozialwissenschaft), sits at 21% in Germany [16]. These figures highlight the expectations and dilemmas of current SMEs in China regarding digital transformation.

Through the digital transformation of manufacturing enterprises, the Chinese manufacturing industry aims to achieve cost reduction and an increase in efficiency, energy saving, improve product quality, improve product added value, shorten the time for products to go to market, address the customization needs of customers, and seeking profitability from providing services, etc., and ultimately enhance the core competitiveness and profitability of enterprises. The demands fall mainly in the following three aspects.

3.1.1. Adapting to Rapidly Changing Market Conditions

China's manufacturing industry has roughly gone through three periods. In the 1980s, there was generally a shortage of goods, and thus the manufacturing industry only focused on "producing" to meet market demand; from 1990 to 2010, commodities became abundant, customers began to have a higher demand of goods, and enterprises began to pay attention to marketing, branding, quality and service. To achieve this, it needed the support of information technology. Since 2010, with a general surplus in production, there began the need for upgrading in consumptions. Customers began to be more focused on customization, user experience, and posed high demand on the supplier's rapid response to products and services. Thus, enterprises needed to constantly innovate to meet the needs of consumers, and they had to quickly adapt to market changes, improve work efficiency as well as improve the customer experience.

3.1.2. Improve Product Quality and Production Management Efficiency

Although China's manufacturing industry is large in scale and complete in offerings, it has been relatively weak in terms of innovativeness for a certain period of time, and there are still great inefficiencies in the management of the production processes. With the drastic changes in today's manufacturing environment where the costs of labor and raw materials are rising rapidly, the Chinese manufacturing industry must focus on improving product quality and production management efficiency in order to regain a competitive advantage in the global market.

In addition, with the increasing popularity of the Internet, the rapid development of computing and storage capabilities, the wide application of the Internet of Things and sensor technology, and the continuous evolution of industrial software, the technical foundation has been laid for the acquisition, storage, transmission, display, analysis and optimization of data. At this point, digital transformation supported by technologies such as mobile internet, cloud computing, big data, and artificial intelligence is an important way to increase the competitiveness of the manufacturing industry.

3.1.3. Reduce Cost of Labor and Shortage of Skilled Workers

In the past years when there was a surplus of workforce in the labor market, it was a common practice for enterprises to rely on adding human workforce for scaling up their businesses. Therefore, many enterprises, especially in the manufacturing industries, developed on the foundation of labor-intensive business models.

However, the "one child policy" that was implemented in the 1980s has taken effect and made a drastic impact on the overall population and thus on the supply of the labor workforce. The Chinese economy turned from a market with a labor surplus to a market with a labor shortage within a decade. The business models that were working very well just a few years ago suddenly broke down, and the enterprises are facing challenges that they had never experienced before.

On the one hand, the problems of blue-collar recruitment difficulties and the shortage of high-skilled workers in the production workshop are frequent; on the other hand, there is a high demand for skilled talent to accomplish challenging tasks like digital transformation of enterprises. In many locations it becomes harder and harder to hire qualified workers.

Software developers in China with a relatively demanding skill level are already very expensive. Many of the SMEs cannot afford this cost. Even if they are determined to cultivate their own talents, it will take a long time until their problems are solved.

3.2. The Difficulties for SMEs in Digital Transformation

Industrie 4.0 has clearly shown the path for the development of the manufacturing industry. The core and basic part are the digitalization that brings changes to the way enterprises do business, as well as the intelligence that came along with it, built on the foundation of digitalization.

However, there are many concepts surrounding the idea of digitalization. For example, big data, cloud manufacturing, internet, AI. There are even more complicated concepts like "Integration of IT & OT", "Smart Manufacturing Total Solution", "Internet + Manufacturing", "Industrial Big Data", "Industrial APP" etc. The multitude of hype words enveloping the digitalization paradigm has been very confusing, and the industries have not been able to standardize their vocabulary. This creates even more obstacles for SMEs, who are trying to find ways into the digitalization world.

Unlike big enterprises, which have extensive resources of professionals for production, IT, for internet platform and for services, the SMEs generally face the problems of lack of financing and expertise. When the SMEs turn to look for assistance from third parties, for example, by outsourcing or engaging external consultants, they find that the market has not developed such experience and expertise sufficiently.

For most of the SMEs in manufacturing industries in China, there are many semi-mechanized, outdated technologies, processes and equipment, and traditional management methodologies and mentalities that are still dominating a large proportion of the businesses. Due to the shortage of funds and talents, many of their immediate pain-points are about the survival, so much of the efforts invested in digital transformation have to be short-termed and produce immediate results. There is little tolerance for ambiguity of the returns they can expect.

On top of that, due to the nature of the manufacturing industry itself, where standardization in the business processes has been difficult, a lack of standardized methodology in digital transformation poses a dilemma to the SMEs. On the one hand, they are educated by the big companies through pilot and lighthouse projects which demand huge investments that they can never afford; on the other hand, the success stories they see from the market are difficult to duplicate, and learning of countless failures they hear in the market make them hesitant to go forward in their attempts at digital transformation.

3.2.1. Sales—The Major Stumbling Block of the Chinese SMEs

For many of the vendors, their Industrie 4.0 solutions originate from the manufacturing end. Many solution providers believe that the owner of the businesses would need efficiency in the production processes. This is of course true. However, it is also important to understand that in reality, the boss is not the only deciding factor for the success of a digitalization program.

Making changes in an enterprise is normally painful, because there are always existing mechanisms or interested parties that are affected by the change. Thus, very often, we came across projects where the boss was very enthusiastic, but the employees would not have the matching passion to support the projects. In fact, they felt that the introduction of the solutions brought more troubles to their work.

Therefore, some Industrie 4.0 solution providers in China came into the market with very innovative offerings. They bundled shop floor digitalization solutions together with B2B sales platform solutions. The logic was to address the greatest pain of the business owner. The business owner could have access to the sales channels if they deployed the shop floor solutions as well, which were fully integrated into the sales platform.

3.2.2. Lack of Expertise in IT Strategy

Unlike in the mature western countries, China has a short history of cultivating experts well trained in IT strategy. This leads to the phenomenon that most of the CIOs (regardless of company size) tend to look for vendors who can provide total solutions to their digitalization needs. In other words, the capabilities in integration are demanded, and there is a tendency to shift the responsibility of integration to the vendor.

Figure 3 shows the talent gap for intelligent manufacturing in China. The gap will persist through 2025. Because of the lack of expertise in IT strategy, it is very common that the enterprises, especially the SMEs, carry out digitalization projects without proper consideration for long and short-term goals. Therefore, they very often run into difficulties in the middle of the projects.

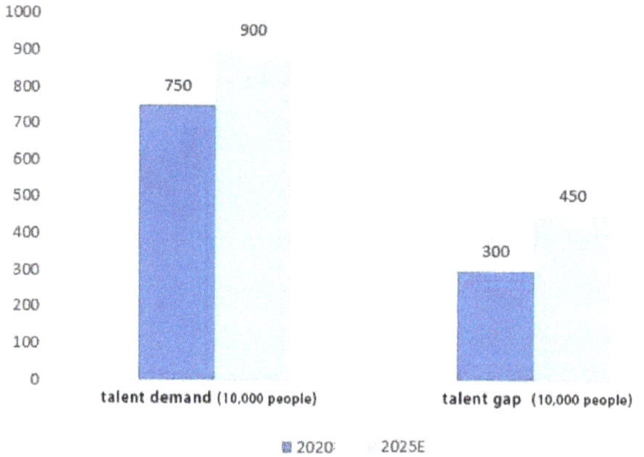

Figure 3. Talent demand and talent gap in intelligent manufacturing field in China; unit: 10,000 persons (Source: China Ministry of Human Resource and Social Security).

3.3. Pursuing for New Business Models—Small Lot Manufacturing and Mass Customization

The introduction of Industrie 4.0 gives the Chinese economy the hope of reformation in its economy. With digital transformation, many new business models become possible. The emergence of new business models gives key players (especially the investors) hope, and thus patience for the future.

One of the biggest dreams of manufacturing enterprises is to connect directly to the end consumer market. If one can connect directly to the end consumer, one can maximize the efficiency of customized production, thus increasing value and profitability. In terms of a business model, the biggest wish of manufacturing companies is to use factories to directly connect to end-consumers as this is the only way to maximize the efficiency of customized production.

The Chinese adopted the first experience with C2B (Customer to Business) from the USA in 2006. As the concept that was very successfully driven by Groupon in USA spread to China, Groupon-duplicate companies like Meituan emerged and flourished in the Chinese market as well. At that time, the C2B model was more or less restricted to the consumer market. Most players did not have the confidence to migrate the business model to the manufacturing end.

The concept of C2M went a step further than C2B, bringing the end consumer directly to the manufacturer. Here, there was a huge emphasis in "dis-intermediary", i.e., bypassing the need of agents, distributors, resellers, thus reducing the cost of the value chain.

Many manufacturing enterprises were ambitious to achieve a breakthrough with the C2M model. They firmly believed that future consumers would pay less attention to product branding. Instead, the trend was that consumers would look for "uniqueness" of customized products and were willing to sacrifice the branding values. Therefore, their weaknesses in marketing and attracting sales traffic would no longer be an obstacle to their success in business.

So, for many years, there was a rush to establish their own portals for end consumers, wechat shops, webshops etc. There was a movement where the manufacturers insisted on building their own sales channels and setting up their own banners.

However, realities are often crueler than dreams.

Except for big brands like Haier, Midea, Huawei or such, who had a huge manufacturing capacity and had established their own sales channels, the majority of the manufacturing enterprises had no or very limited influence at the sales end. They had no access to the end consumers, and the effort in trying to do so became too much of a burden. The ideal C2M did not arrive as promised.

The lessons learned were painful. The field of attracting traffic and sales was a totally different business than that of manufacturing. Even big brands like Haier with killer platforms like COSMOPlat would still rely on third party sales channels for their sales processes. Professional businesses should be handled by professional people. Bypassing the sales value chain was fated to fail from the very beginning.

In contrast, the internet giants that owned a large wave of consumer traffic began to penetrate the manufacturing end at a very high speed.

In the years between 2016–2018, internet giants like Netease, Alibaba, and JD.com launched their own C2M platforms. These platforms introduced to the consumer market the concept of ODM (Original Design Manufacturer). These internet giants took advantage of their consumer-end traffic to get into a preliminary exploration of the C2M business model.

The returns were not very promising. After engaging in uncountable price competitions among the internet giants, they finally returned to the more profitable platform model. The internet platforms began to probe the OBM (Original Brand Manufacturing) model, where the manufacturers were allowed to promote their own brands on these platforms. Alibaba went further to deploy simple technologies like bar codes, RFID, cameras etc., in the factories, attempting to integrate the factories' production data with the sales orders in order to realize the model of "Manufacture to Order".

These were courageous attempts to probe the new business models with the SMEs. Values were created for both the customers as well as the platforms. The SMEs who were thirsty for sales traffic welcomed such platforms as they brought additional values targeted at their immediate pain points—sales. And furthermore, these internet companies were eager to learn and reacted very fast to the market. Starting from the SMEs, successful test pilots could be iteratively developed into more mature solutions and finally penetrate the bigger enterprises. This could be a different path to achieve a country-wide penetration of Industrie 4.0 solutions and practices.

A more detailed description of this part of C2M history can be found in [17].

4. Case Study: Experimenting Industrie 4.0 of a Chinese Manufacturing Enterprise

The following is an overview of a typical Chinese SME in its journey throughout a digital transformation. The example given here comes from a very typical medium Chinese manufacturing enterprise, with a kind of "mass customization" production model.

4.1. Background of GuoMao Reducer

Founded in 1993, GuoMao has grown into a mid-size company with 2200 employees, making a revenue of almost RMB 3 billion (about Euro 450 million) in 2021. Since its public listing on the stock market in 2019, GuoMao has achieved strong growth, even amidst the COVID pandemic. It is now a market leader in the China's market of general reducer machine industry.

GuoMao is headquartered in Changzhou City in the Jiangsu Province. Its main products are industrial reducers with many product models. GuoMao produces more than a dozen series of varieties of transmission machineries, tens of thousands of varieties such as standard, customized, large and medium-sized non-standard, new products, high-precision transmission, etc. At the same time, GuoMao is developing its business overseas such as Southeast Asia, Europe and the United States.

4.2. Challenges and Motivation for Digital Transformation

GuoMao has the ambition to become a "world class transmission producer". After careful analysis and considerations, it is decided that digitalization would be the means to achieve this goal. GuoMao believes that the long-term goal of digital transformation is to capture growth and drive value, and the adoption and implementation of all digital technologies should also revolve around this goal.

The immediate pain point that GuoMao wanted to address was the bottleneck to achieve large scale personalized customization of the customer orders. Due to the nature of the business, GuoMao intended to seek solutions to customize on demand plus a certain degree of flexible manufacturing. The ultimate goal would be to be able to deliver on a large scale, while ensuring that the quality and cost of each product unit were well controlled.

For GuoMao, gaining a competitive advantage in the market required improvement in the following capabilities: individual customization, agile planning, precise logistics, production transparency, flexible manufacturing, full traceability and energy savings.

The targets that GuoMao set for the digitalization programs were as follows:

- Making a change in the existing production management model.
- Achieve transparent and traceable production processes.
- Improve productivity and operational efficiency in the production process.
- Achieve reliability and efficiency in warehouse logistics, in combination with lean logistics.

4.3. The Struggle for Digital Transformation

The road through the digitalization transformation was not without challenges. First of all, there were few or no success stories similar to the GuoMao case which the task force could learn from. CIO Kong Donghua saw the greatest challenges coming from the lack of expertise and experience in the IT team to realize the desired targets, whereas the operations departments mentioned that it was generally difficult to understand the large number of professional jargons that appeared in the context of digitalization.

4.3.1. Designing the Overall Roadmap and Looking for Solutions

After defining the overall targets with the management, the task force set out to define the overall roadmap for the digitalization program. There were 3 possible lines of thoughts to meet a variety of "smart manufacturing" system architecture requirements:

1. The traditional architecture: ERP on the top floor + MES on the shop floor;
2. Advance architecture: middleware + professional system architecture with "connectivity" as the focus;
3. Ultimate architecture: business applications running on an IOT enabled scenario.

Unable to determine what was the most suitable way for itself, the digitalization task force decided to probe each possibility and learn on its way. GuoMao started seeking for solutions in the market. From his personal previous experience, CIO Kong Donghua knew that interfacing and integration among systems would be the most challenging part of the technology implementation. Therefore, special attention was given to the completeness and feasibilities of integration of the solutions.

4.3.2. Top Floor—Implementation of a Pricing System

The digitalization task force defined the implementation program into two phases and defined the tactical targets for each phase.

The pain points to be addressed in the first phase of implementation were:

1. Huge amount of manual work for the calculation of product pricing; GuoMao had a total of 1.3 billion individually customized products. Considering freight, insurances and such, there were billions of pricing calculations that it was no longer possible to build price lists in the traditional way and manage them in the traditional way of a database.
2. Historical product pricing and the management of current pricing policies were not conformed. Multiple versions of the pricing were left unmanaged in the field.
3. Lack of product pricing life cycle management, this resulted in difficulties to analyze the pricing fluctuation.

In the second phase of the projects, the pain points to be addressed were split into business and IT scenarios.

Pain points in the business scenario:

1. The overall production plan and cycle time were not fast enough to respond to the market demand.
2. Lack of end-to-end tracking of the materials (from procurement to warehousing and production and then to shipping).
3. Material allocation and logistics were managed manually, error prone and high cost.

Pain points in the IT landscape:

1. Efficiency and cost in operation of the data center.
2. The capabilities of designing architectures for both cloud native and non-cloud native systems.
3. Ensuring the end-to-end definition for cloud native systems, as well as the delivery and operational capabilities.
4. Leading the partners to reduce cost while ensuring quality during collaborative delivery.

The digitalization task force started to get in contact and learned from the different cloud infrastructure vendors in 2019. In the beginning, it was not easy for them to clearly differentiate which of the providers would be the best choice to solve their problems. However, during the long engagement of communication and market research, they were convinced by the consultants of the Amazon Cloud Technology team. Other than technical expertise and the positive and responsible attitude which were also visible with other vendors, the ultimate deciding factor was the feeling of the willingness of the Amazon consultants to understand and learn from GuoMao's business. Furthermore, the Amazon consultants were also willing to impart their expertise to enable a traditional machining production enterprise like GuoMao. This was very important as the task force was aware that the road ahead would not be smooth, and a vendor who positioned itself as a partner could help to overcome unforeseen difficulties and ensure the smooth implementation of the project. After careful analysis, the digitalization task force decided on AWS to be their cloud backbone architecture. AWS provided the foundation platform to enable cloud-native applications and infrastructures from scratch.

With the help of the implementation partner, GuoMao deployed the flexible and extensible architecture for multiple quotation. They implemented the cloud-native price management solutions with the corresponding APIs. For the infrastructure, the project team used the Amazon VPC, EKS, DynamoDB, S3 and cloud-native safety and security measures, among other technologies. They integrated the CI/CD lines and deployed the monitoring and logging functionalities of the infrastructure. For the people empowerment, the project team introduced the trainings of the management tools and coached the workers in the usage of the applications. In addition, further trainings had been given to employees for agile product design, architecture, development, testing, process documentations, as well as trainings for cloud-native applications, infrastructure, coding, testing and system security.

The project team sought for an application with the minimal delivery value (so called MVP) as the entry point for the digital transformation. The project was divided into phases, each phase gradually carried out, making full use of the advantages of microservices and cloud services. At the same time, the consultants worked with GuoMao employees hand-in-hand to implement the project, meanwhile empowering the GuoMao employees in the implementation capabilities. The IT staff were trained in several fields, such as basic operations of cloud technology, business driven methodologies, project implementation methods and tools usage. During this time, communications to the management was crucial, this was the way to close the gaps with between the management and the employees for such traditional, large-scale manufacturers.

The project team encountered various problems of technical difficulties. For example, the problems of integration among the systems, the synchronization of the master data, and so on. The capabilities of the consultants were very crucial, not only must they be able to resolve the technical problems, they must also have the capabilities to empower the customer to resolve the problems on their own in the long run.

4.3.3. Shop Floor—Implementation of a MES System

The pain points on the shop floor were very much different than those on the top floor. In the shop floor, GuoMao was confronted with an outdated existing production management model. There was generally a lack of transparency in the production process and the production efficiencies could not be properly measured. At the same time, it took a lot of manpower to collect the data necessary for financial accounting. The managers of the shop floor could not perceive and understand the problems and actual situation of the production line in time, this resulted in poor management performances and a huge loss to the company's resources. For a traditional discrete manufacturing enterprise like GuoMao, where there is high labor intensiveness and low level of automation, data acquisition and reporting on the production front were done mostly manually, the accuracy and integrity of the data in the reports were very much questionable.

Searching for a solution for the shop floor was a lot more challenging than that for the top floor. The countless number of solutions offered in the market made it difficult to distinguish what the best solution could be. The road to digitalization on the shop floor was not as smooth as that on the top floor, GuoMao had to suffer setbacks and learned from its mistakes before it could find the most suitable solution for itself.

Stage 1: Realizing MES Functionalities with ERP

In the first stage, the digitalization task force chose to extend the shop floor digitalization from the existing ERP system. The task force established a team internally and assessed the shop floor situation, defining the necessary MES functionalities. Then a project was defined, budget allocated, and the MES project implementation started.

The project implementation took several months, and finally failed. There were fundamentally two reasons for the failure.

Firstly, because of the nature of the business (mass customization of the products), the system had to process a huge amount of data on the shop floor. Attempts were made to extend the shop floor functionalities from the ERP but it ended up with the performance of the entire ERP system deteriorating. The application extended from the ERP systems were generally not meant to process data of such volumes. The team could not resolve the performance issue with any ERP technology which was very frustrating for them.

Secondly, the fundamental conflict of the ERP logic to the production logic was more or less the showstopper. This was a general issue that occurred in many cases and so far, it was not obvious that such conflicts could be reconciled. GuoMao had to terminate the project and look for other solutions.

Stage 2: Implement MES Functionality Using Customized Development from Scratch

Since realizing MES functionalities through ERP extensions did not work, the digitalization task force turned to the idea of developing the functionalities by customized development. As the requirements had already been determined in the first stage, they just needed to convert the requirements into detailed functional specifications and engage programmers to realize the functionalities.

The attempt failed again after one year.

The task force analyzed the reasons for the failure and drew the following conclusions:

1. For a project in the field of MES, a platform was indispensable. The MES solution would not be scalable without a development platform. A simple issue like program version control could easily grow into a nightmare caused by the diversification of the business processes.
2. Functionalities were developed from scratch. The long development cycle resulted in the sponsors losing faith in its success.
3. Lack of professional bug handling mechanisms posed too many risks, especially when the MES systems were to be deployed at the production sites.
4. Lack of transparency in the production processes resulted in difficulties to track the products.

Overall, instability in the system that was developed, poor scalability, long development cycle time, high development costs resulted in the failure, and a loss to all stakeholders involved.

Stage 3: Engaging Professional Shop Floor Solutions

Summarizing the lessons learned from the previous failures, the digitalization team realized that a professional MES solution would be needed to solve their problems on the shop floor. Thus, they began looking in the market for such a solution.

After communicating with several MES vendors in the market, GuoMao decided to select FORCAM as their partner for the shop floor digitalization. The product FORCAM FORCE® IIoT (FORCAM GmbH, Ravensburg, Germany) was deployed in different phases of several projects.

The major reasons for a decision in favor of the FORCAM solution were:

1. Clear and simple user interface. User friendliness of the solution was an important factor as it had direct impact on the willingness of the workers to use the system in their daily work.
2. Productized solution with modular design. This made it convenient for users with various needs for individual selection.
3. The system parameters were configurable and flexible enough to address varying customers' needs, furthermore, quick implementation was desired.
4. Powerful reporting functions, the team was looking for best practices that were verified by other customers. GuoMao could learn from the various dimensions of reporting how other users used the data to manage their production processes.
5. Customizable reporting functions to cater for diversified management reporting needs.
6. B/S architecture for the solution. This largely reduced the operational and maintenance costs of the IT team.
7. The overall architecture of the IIoT platform was suitable for the expansion of enterprise needs, and it was convenient to hook up new application software and realize information exchange between application software through API.

GuoMao began implementing FORCAM FORCE® IIoT in March 2019. A total of 5 production lines (gear and tooth axis lines of the pre-heating workshop, grinding lines of the post heating workshop) were chosen for the pilot project. The pilot was completed and went live after three months of project implementation.

4.3.4. Results and Achievements

The first phase of the project was the implementation of a pricing system. Traditionally, the quotation was done manually, the workload was huge, the result was error-prone and difficult to scale. Without proper product pricing life cycle management, it was not possible to analyze the pricing strategies. The new system enabled the automation of price calculation processes, ensured the consistency of the product prices globally, and saved an estimated 16 h/person/month of the workload. GuoMao was able to increase the frequency of product releases and upgrades from quarterly to at least once a week. GuoMao planned to further integrate the quotation system with other systems such as WMS, APS and the MES.

After the completion of the second phase of the project, the following problems were solved:

1. The cycle time of production planning and capacity was shortened from T+7 to T+3.
2. The number of order-taking personnel in one of the final assembly lines was reduced from 5 to 0.
3. The flow of materials was fully traceable from end to end.

After the pricing system went live, GuoMao has even constructed a new service offering to the market, providing a price management service to its suppliers and end customers. This new business model became GuoMao's target for the next phase of digital transformation. This is a very strong supporting argument for the original decision for the digital transformation.

The implementation of the MES solution on the shop floor achieved the following benefits:

1. Realization of an IIoT platform integrated with open APIs.

This subverted the conventional production management model. Deploying a MES system that meets the characteristics of GuoMao's business and products, plus an open IOT architecture platform that meets the ultimate needs of "smart manufacturing", helped to achieve real-time transparency and availability on the shop floor.

2. Realize transparencies in equipment availability and visualization in production processes.

The actual operation status and product order information of each machine became clear at a glance; the production data was collected in a real-time manner and was accurate. The managers on the shop floor could keep abreast of the actual situation on the production lines, which greatly improved the efficiency of management and provided a reliable basis for accurate and correct decision-making. The cycle time for the transmission of production instructions, on-site confirmation and production preview was reduced from originally 4–8 h to the current 10 min.

3. Various production KPIs were refined and optimized.

Based on the "true availability" data collected, managers could better calculate downtime and maintenance costs. Control and calculation of production indicators such as OEE, availability, performance, quality, production cycle time) on the shop floor were optimized. Not only was the production efficiency increased by over 30%, the digitalization of the shop floor production process also generated an impetus to the surrounding systems. These indirect gains in efficiency were invaluable to GuoMao because they allowed GuoMao to gain an overall competitive advantage when producing similar products.

4. Continuous CIP reduced the manufacturing costs for GuoMao

Using digital monitoring of the machines in production, the digitalization task force continuously deployed lean management to make improvements in a timely manner when abnormalities were discovered. At the same time, the introduction of continuous improvement mechanisms, improvement of bottlenecks in production management, shortening of time to delivery, ultimately reduced the total manufacturing costs.

4.4. Lessons Learned

The GuoMao case study shows an example of a fast-growing Chinese manufacturing enterprise that has a very dynamic culture. Though GuoMao is a very typical mid-size manufacturing enterprise in China, the digitalization journey it went through was extraordinary. Even without a clear roadmap ahead, it took the courage to take the first steps to test out different possibilities. It was not afraid to fail and kept readjusting its strategy by learning from the mistakes. These were characteristics that were seldom found in manufacturing enterprises of its size.

The digitalization transformation was just at the beginning. There were still many plans ahead. Nevertheless, it was already worth the effort to summarize the lessons learned so far.

4.4.1. Short-Term ROIs Are Very Important

It is important to focus on the problems one wants to solve with digitalization, and also to analyze in advance one's own capabilities. SMEs normally have relatively low tolerance for making mistakes, so short-term ROIs are very important. It is crucial for the sponsors to see what the returns are in a timely manner. Though there has been a rush for "Dark Factories" where full automation throughout the production processes was anticipated, GuoMao chose the path to move forward in phases, weighing the pros and cons of each phase, and analyzing the investment-benefits ratio. Thus, even after the first phases of digitalization, one can still see a lot of manual processes in GuoMao where workers still enter some data onto dashboards manually. This is the result of careful consideration of the benefits and investment needed to turn it digitalized.

4.4.2. Digitalization Does Not Equal Informatization

There are many IT providers who are vigorously promoting the concept of "digitalization". However, some of them are simply driven by selling IT solutions that turn offline business processes into online processes.

GuoMao learned during the exploration that informatization is just one of the means used in the digitalization process. It mainly uses software systems to achieve process optimization and efficiency improvement within the organization, which is still one of the technical paths of management improvement. For example, GuoMao implemented ERP, but only used it to solve the problem of financial accounting, invoicing and inventory; it went to the MES to solve the problem of on-site production management.

In the past, management in the factories were done with huge amount of manual workload, it was error-prone, and the processes were not clear. It was difficult to analyze the KPIs from the results as the processes were not transparent. Furthermore, the inhouse IT team was comparatively weak in digitalization know-how, but all these underwent a great change after the projects.

4.4.3. Digitalization Does Not Equal Data

A common misconception in the Chinese SMEs was that through deployment of IOT technologies, enterprise hard- and software systems, digitalization can be achieved by means of various large-screen dashboards and explicit BI visualizations.

GuoMao learned from the digitalization process that such visualizations are just a different way that business data is presented. Analogous to the manufacturing processes, the data collected from the business processes is like the "raw materials". If GuoMao could not change the way the company's business are operated, then this change of data presentation will remain at the "raw material" stage. What GuoMao as an enterprise needed was to process these "raw material" into finished products which it can then consume. Only then was value created, and only then transformation occurred.

Digitalization is the "digitization of business", and the starting point of transformation is the integration of business and IT, which cannot be separated from business application scenarios. GuoMao learned that digital transformation only occurred when new business opportunities are created under the support of new technologies, leading to access to incremental markets. There should be significant changes in corporate strategies, business processes, organizational capabilities and profit models.

4.4.4. People Are the Most Important Success Factor

Generally, in the Chinese market there exists IT solutions that are still built on outdated architectures from decades ago. When GuoMao began looking for solutions in the market, they were worried about this and paid special attention to the technological background of the vendors. During the projects, they realized that it makes a big difference to the final result whether one learns from the projects, so a good teacher taking them alongside and teaching them hand-in-hand became a very critical requirement. All project development is also the process of developing the people, and the cultivation of digital talents will yield very high benefits, even if a project might not have fully achieved its targets in the short run.

Furthermore, the digitalization processes introduced a lot of flexibilities in the business operations [18]. For example, the prices of each product were encapsulated in the product definitions in the past. After the digitalization, GuoMao is now able to change and amend the prices anytime. This gives them a lot of freedom and they even introduced agile methodologies to manage the processes. The people are thus empowered, and their capabilities grow together with the business.

The purpose of intelligent manufacturing has to be people-centric, i.e., to solve the problems for people. Whether for managers or for front-line workers, it is a person's common desire to pursue a better life, and intelligent manufacturing is a means to assist people to achieve this goal.

4.4.5. Management Methodology of Software Products and Integration Projects Is Generally Applicable to Enterprise Management

By introducing the agile development methodology and actively learning from the professional software providers how to build good products, GuoMao found that the management philosophy of software products and integration projects could be applied to any area of the enterprise management. This was a windfall benefit for GuoMao in this digitalization transformation journey.

4.4.6. Creation of New Business Model as a Highlight for Industrie 4.0

A major highlight of digitalization was the creation of a pricing management system which GuoMao is currently incubating into a new business model. This pricing management is very useful in dealing with very complex quotation scenarios.

Guomao has always invested efforts in the product quality and cost control, but the quotation process has always been done manually, relying on spreadsheets and staff experience. When GuoMao developed more and more new products, the customers began to doubt the inconsistencies of the price quotations. An example was the criticism that two reducers with very different functionalities had the same price simply because they looked similar, or because they had the same cost of production.

With digitalization, GuoMao was able to integrate the rules, conditions and logic of the pricing mechanism into a rule engine. Adding visualization with a web interface to the rule engine, GuoMao was able to offer its customers and suppliers a pricing management system. This not only created a new profit center, but also strengthened the image and thus market leadership of GuoMao.

5. Discussions on Future Development of Industrie 4.0 in China

Being one of the largest manufacturing economies in the world, China will inevitably move forward into the smart manufacturing world. We have seen from the above examples that enterprises on both end of the industrial chain are striving towards the next level of Industrie 4.0. We summarize here some of our observations of the Chinese smart manufacturing movements, especially in the discrete manufacturing market.

Observation 1. *China local customized solutions necessary to address the market requirements.*

An interesting observation with the industrial software market is that the market is split in two directions. In the market where specific professional technologies are dominant, this market is dominated by technologies from the western countries where the few global leaders take up the absolute majority of the market. This we see in the field of ERP and in specialized industrial software like Product Lifecycle Management (PLM), Computer-aided Design/Engineering (CAD/CAE), etc. On the other hand, the market for general-purpose industrial software like Warehouse Management System (WMS), Manufacturing Execution System (MES), etc. is full of local competitions. The foreign software and platform providers face very often the challenge of the requirement of "China only" solutions.

The market challenges are two-fold. In the specialized industrial software field, the market leaders will have to expect to face intensive competitions from big major players in China. It is necessary on the national strategic consideration to have a certain level of local providers on the specific industry. For the general-purpose software industry, the main challenge will only come from market competitions.

In both cases, to remain competitive, local solutions to a certain degree is indispensable. When there is a surplus of product offerings in the market, the tolerance of the end-user to adapt to systems not localized for Chinese usage is becoming less and less.

Almost all the digitalization projects we encountered demanded localized solutions to the standard products. In the GuoMao case study, additional customized development was necessary in both top floor and shop floor digitalization projects so that the requirement gaps could be covered up. Therefore, flexibility to support localization is an indispensable requirement for all Industrie 4.0 solutions aiming to penetrate the Chinese market.

Observation 2. *Machine Connectivity/Digital Twin not yet the top priority. Machine Connectivity is perceived as non-issue in the SMEs.*

The concept of Digital Twin and the related "Factory of the Future" seemed not to be on the top of the mind of most discrete manufacturing leaders. The lack of standardization at the machine connectivity level posed a lot of difficulties in the digitalization of the shop floor. Most of the digitalization projects started with very ambitious top-level designs, and finally ended up in only partial realization of the digital processes. This resulted in the majority of the manufacturing industries positioning the shop floor digitalization with a "try and see" attitude. More strategic focuses were positioned at resolving the bottlenecks at the sales order front. So far, the machine data to be collected was mostly focused on operations. The scope for that was limited and the machine connectivity were mostly implemented by local suppliers in a customized manner. No doubt, this will definitely change in the near future. We believe that the key driver for a change would be a successful standardization process in the machine connectivity issue.

Observation 3. *The market development is technology driven instead of management methodology driven. Lean Management is not yet in focus.*

Unlike most of the western developed countries where the experience of management in conjunction with software technologies to improve efficiencies have been very much developed, China is still generally undergoing the transition phase when upgrading in hardware technologies which alone will make a great change in the production efficiencies. This resulted in the phenomenon that Lean methodology and OEE KPI-driven improvements are not a top priority in the majority of the market. In the GuoMao case study, the original intention was to deploy digitalization technologies and expecting obvious value from the deployment. Throughout the projects, GuoMao learnt that the technologies were simply tools which have to be supplemented by management methodologies. GuoMao set off to learn and figured out Lean methodologies with inhouse managers after realizing that data acquisition alone could not bring the value they were expecting. As is generally the case in the Chinese market, the values of management methodologies are palpable only after technologies are successfully deployed. We can also see that when the market leaders in Lean Management barely generated revenue worth mentioning. As the market matures over time, we believe that the tipping point for that will sooner or later arrive. As from past experience from other industries, once it does, it probably will take no time to burst into a huge market demand.

Observation 4. *The demand for small-lot manufacturing technologies is very high in China SMEs.*

The Chinese history of economic reform mentioned above resulted in a huge market of Chinese SME manufacturing industry offering comparatively lower value of production goods. Instead of production in mass quantity, these manufacturers normally produce a larger variety of goods in much smaller quantity. The inefficiency is the major reason for low profit margins. With the advancement in internet and IOT technologies, and with the introduction of Industrie 4.0, the concepts like "make-to-order" or "flexible manufacturing" became more and more feasible to the SMEs. This has released a huge market demand for technologies that would help SMEs with digitalization catered to small-lot manufacturing.

6. Conclusions

The 3rd Industrial Revolution was the biggest beneficiary of globalization. The first ten years of Industrie 4.0 also benefitted from globalization as its influence got widespread throughout the world [19]. Globalization originated from the need to optimize the allocation of global resources and the formation of industrial value chains under the impetus of the international division of labor [20]. Currently, globalization is facing more and more challenges in the international economic development. Geopolitical tensions increase supply chain and technological decoupling risks. Foreign companies face more and more strategic challenges in China.

Despite these challenges, we believe that China will continue to open up and integrate to the international community. The Chinese have painstakingly learned of the consequences that followed when the country shut its door to the world. Rejecting communication and integration to the international community 300 years ago did hinder industry competition, protected its market and brought tentative stability and prosperity to its country, but it also became a curse to the country in which the damage persisted until even today [21]. Furthermore, the opening up of its market has brought more benefits than ever before. Therefore, integration with the international community is the only logical and feasible path for its future development.

To reach its ultimate goal, Industrie 4.0 has still a long way to go, and it is not a disruptive transformation that can be completed overnight. China, with its 1.4 billion population, sound infrastructure and a rich market of industrial application scenarios, offers a very good opportunity to test, validate, and stimulate further development of Industrie 4.0. New technologies, new products, new methodologies, and new business models are emerging endlessly in a very dynamic way, and it is monetarily rewarding. The talents for design, R&D, manufacturing, sales and other fields are sufficient and the cost is comparatively low.

China will continue with a high speed of change in the market demand, the thirst for innovation is never ending. China has passed the time when they were simply learning from their counterparts in the west. They are now offering a lot of learning opportunities for how the development could proceed, like in the area of artificial intelligence, autonomous driving, circular economy etc.

The Chinese market has similar demands in Industrie 4.0 like Germany on the strategic level. However, at the execution and operation levels, the problems encountered are very different. Filling up these gaps is on one hand challenging to the technology providers, but on the other hand also strengthens their capabilities. We can expect that Industrie 4.0 technologies that successfully survive the challenges in the Chinese market will provide better and more robust solution offerings to the world.

The technology innovations of Chinese companies are very much driven by monetary powers. The impatience of capital for returns result in the flow of investments into businesses which are innovative in business models instead of technologies. We see the internet businesses in China has developed in a light-speed manner, with all unimaginable ways of doing businesses. But there is generally a lack of market leaders who are leading in technologies, with very few exceptions.

The opportunities for collaborations between Germany and China thus arise here. Germans are the leaders for building engines and foundational technologies. Combining this capability with Chinese user applications would result in a kind of best breed solutions—applications with fancy user interfaces and yet powerful and stable engines. From the point of R&D, it is very appropriate to adopt non-competitive R&D in the initial stage of scientific exchanges and co-operation between the western countries and China. It is only when the two countries conduct joint R&D based on their respective strengths that they will find the possibility of entering into specialized R&D together [22].

From the projects and experience we encounter in the field, we conclude that the China market offers a lot of value for the development of Industrie 4.0 and China will continue to be relevant to the development of Industrie 4.0 for the next ten years.

7. Future Directions

We see from the market that localization of Industrie 4.0 solutions is in strong demand in China. The existence of the variants of Industrie 4.0 indicates that even the development model of Industrie 4.0 itself requires localization. This is due to the fact that after decades of development, China has so far acquired the capabilities in almost every aspect of technology development. According to the United Nations, China is the only country in the world which possesses all industrial categories listed in the United Nations, i.e., one can produce anything imaginable in China. This resulted in vendors of a specific technology having to deliver more efficiently and at much lower price. Those who managed to innovate to meet such demands are rewarded with tremendous success from the huge market. Therefore, in the 21st century, any foreign companies which wish to be successful in China need a new strategy suitable in the new environment. Those companies who failed to adopt a localized strategy inevitably are unsuccessful in the Chinese market.

It would be beneficial to have an understanding of the major parameters (for example, pricing models, industry know-how availability, user interfaces, ability to scale etc.) that have to be localized in order for an Industrie 4.0 technology developed in Germany to be successful in the China market.

On the other hand, research of Industrie 4.0 requires the collaboration between theorists and practitioners, where the fundamentals of the theories are verified in the field by the projects. Countless numbers of digitalization projects are being carried out every day in China. Many innovations occur every day as new knowledge is gained in the execution of the projects. Some of the projects are successful, but the majority of them suffered setbacks and ended in failures. Failed digitalization projects are valuable in the sense that they provide market feedback to pitfalls and difficulties that can occur in the field. An interesting research direction could be derived from generalizing the projects and identify the parameters that lead to the success and failures of the projects under different scenarios. In addition, it would be also interesting to determine how to generalize these findings into structured knowledge so that future Industrie 4.0 projects could learn from the failures.

To make the vision of Industrie 4.0 visible it is recommendable to establish a standard "Factory of the Future" model to explain and demonstrate very clearly the different aspects of the Industrie 4.0 initiative. From a Chinese market point of view, it is important to stress the fact that Industrie 4.0 was initiated in Germany to make sure that manufacturing in Germany (Europe) stays competitive to other markets (and especially Asia).

Since a shortage of skilled labor and the increasing cost of labor is a fact in China it appears that China has also to adopt Industrie 4.0, though probably in a different variant. For that, the "Digital Twin" and AAS (Asset Administration Shell) should gain more visibility and importance in China. On top, material provisioning must be automated, inventory be managed end to end, the supply chain tightly integrated.

Author Contributions: Conceptualization, K.-L.L.; methodology, K.-L.L., A.R. and U.H.; validation, K.-L.L., A.R. and U.H.; formal analysis, K.-L.L., A.R. and U.H.; writing—original draft preparation, K.-L.L.; writing—review and editing, K.-L.L., A.R. and U.H.; project administration, K.-L.L. All authors have read and agreed to the published version of the manuscript.

Funding: This paper received no external funding.

Institutional Review Board Statement: Not applicable.

Informed Consent Statement: Informed consent was obtained from all subjects involved in the study.

Data Availability Statement: The datasets from Figures 1 and 3 are available at https://www.qianzhan.com/ (accessed on 29 April 2022).

Acknowledgments: Special thanks to Johannes Winter for the guidance and supervision that made this paper possible. Thanks to Kong Donghua for the input to the GuoMao case. Thanks to Candy Jiang for the contribution to many market insights and suggestions to this paper. Thanks to Gu Haiwen for the intriguing opinions and ideas of international R&D collaboration.

Conflicts of Interest: The authors declare no conflict of interest.

Abbreviations

AAS	Asset Administration Shell
AII	Alliance of Industrial Internet
API	Application Programming Interface
APS	Advanced Planning System
AWS	Amazon Web Services
B/S	Browser/Server (architecture)
B2B	Business to Business
BI	Business Intelligence
C2B	Consumer to Business
C2M	Consumer to Manufacturer
CAD/CAE	Computer-aided Design/Engineering
CIO	Chief Information Officer
CIP	Continuous Improvement Process
COVID	Corona Virus Disease
ERP	Enterprise Resource Planning
IIC	Industrial Internet Consortium
IIoT	Industrial Internet of Things
IOT/IoT	Internet of Things
IT/OT	Information Technology/Operation Technology
KPI	Key Performance Indicators
MES	Manufacturing Execution System
MIIT	Ministry of Industry and Information Technology
MVP	Minimal Viable Product
OBM	Original Brand Manufacturing
ODM	Original Design Manufacturer
OEE	Overall Equipment Efficiency
OEM	Original Equipment Manufacturer
PLM	Product Lifecycle Management
RFID	Radio Frequency Identification
ROI	Return on Investment
SOEs	State owned Enterprises
SMEs	Small and/or Medium Enterprises
WMS	Warehouse Management System

References

1. Kagermann, H.; Anderl, R.; Gausemeier, J.; Schuh, G.; Wahlster, W. (Eds.) *Industrie 4.0 in a Global Context. Strategies for Cooperating with International Partners*; Herbert Utz Verlag: Munich, Germany, 2016.
2. Kagermann, H.; Winter, J. The second wave of digitalization. Germany's chance. In *Germany and the World 2030. What Will Change, How We Must Act*; Mair, S., Messner, D., Meyer, L., Eds.; Econ Verlag: Berlin, Germany, 2018.
3. Forschungsunion Wirtschaft—Wissenschaft and Acatech. Recommendations for implementing the strategic initiative INDUSTRIE 4.0. Securing the future of German manufacturing industry. Final Report of the Industrie 4.0 Working Group. 2013. Available online: https://en.acatech.de/publication/recommendations-for-implementing-the-strategic-initiative-industrie-4-0-final-report-of-the-industrie-4-0-working-group/download-pdf?lang=en (accessed on 1 July 2022).
4. Kagermann, H.; Wahlster, W. 10 years of Industrie 4.0. *Science* **2022**, *4*, 26. [CrossRef]
5. State Council of the People's Republic of China: China Manufacturing 2025 Plan. GuoFa (2015) No. 28. Available online: http://www.gov.cn/zhengce/content/2015-05/19/content_9784.htm (accessed on 1 July 2022).
6. Evans, P.; Marco, A. *Industrial Internet: Pushing the Boundaries of Minds and Machines*; Technical Report; General Electric: Boston, MA, USA, 2012.
7. Ministry of Industry and Information Technology (MIIT) of the People's Republic of China. *Implementation of the State Council on Actively Promoting the Internet + Guidance Action Plans (Year 2015–2018)*; Ministry of Industry and Information Technology (MIIT) of the People's Republic of China: Beijing, China, 2015.
8. Li, B.; Zhang, L.; Xudong, C. Introduction to Cloud Manufacturing. *ZTE Commun.* **2010**, *8*, 6–9.
9. Li, B.; Chai, X.; Hou, B.; Zhang, L.; Zhou, J.; Liu, Y. New Generation Artificial Intelligence-Driven Intelligent Manufacturing (NGAIIM). In Proceedings of the 2018 IEEE SmartWorld, Ubiquitous Intelligence & Computing, Advanced & Trusted Computing, Scalable Computing & Communications, Cloud & Big Data Computing, Internet of People and Smart City Innovation (SmartWorld/SCALCOM/UIC/ATC/CBDCom/IOP/SCI), Beijing, China, 8–12 October 2018; pp. 1864–1869.
10. Li, B.; Chai, X.; Zhang, L.; Hou, B.; Liu, Y. Accelerate the Development of Intelligent Manufacturing Technologies, Industries, and Application under the Guidance of a New-Generation of Artificial Intelligence Technology. *Chin. J. Eng. Sci.* **2018**, *20*, 73–78. [CrossRef]
11. Ash, R.; Howe, C.; Kueh, Y.Y. *China's Economic Reform: A Study with Documents*, 1st ed.; Routledge: Oxfordshire, UK, 2002.
12. Dassisti, M.; Panetto, H.; Lezoche, M.; Merla, P.; Semeraro, C.; Giovannini, A.; Chimienti, M. Industry 4.0 paradigm: The viewpoint of the small and medium enterprises. In Proceedings of the 7th International Conference on Information Society and Technology ICIST, Kopaonik, Serbia, 12–15 May 2017.
13. Industrie-Contact AG: Internationaler Tag des Mittelstandes am 27. Juni. Available online: https://industrie-contact.de/blog/internationaler-tag-des-mittelstandes-am-27-juni/ (accessed on 28 April 2022).
14. European Commission. Commission Recommendation of 6 May 2003 Concerning the Definition of Micro, Small and Medium-sized Enterprises (Text with EEA Relevance) (Notified under Document Number C(2003) 1422); 20.5.2003. Available online: http://data.europa.eu/eli/reco/2003/361/oj (accessed on 28 April 2022).
15. Chinese Electronics Standardization Institute. *Analysis Report on the Digital Transformation of Small and Medium-sized Enterprises*; Chinese Electronics Standardization Institute: Beijing, China, 2020.
16. Institute for Advanced Social Sciences (INFAS) and Centre for European Economic Research (ZEW): Innovation Survey. 2022. Available online: https://www.infas.de/projekte/infas-projekt/die-innovationserhebung/ (accessed on 28 April 2022).
17. Sheng, M. *Creating High Valuation: Creating a Value-based Internet Business Model*; China Machine Press: Beijing, China, 2020.
18. Küting, K.; Snabe, J.H.; Rösinger, A.; Wirth, J. (Eds.) *Geschäftsprozessbasiertes Rechnungswesen—Unternehmenstransparenz für den Mittelstand mit SAP Business ByDesign*, 2nd ed.; Schäffer-Poeschel: Stuttgart, Germany, 2011.
19. Jerath, K.S. Globalization and the Third and Fourth Industrial Revolution. In *Science, Technology and Modernity*; Springer: Cham, Switzerland, 2021. [CrossRef]
20. Lim, G.C. Globalization, spatial allocation of resources and spatial impacts: A conceptual framework. In *Globalization and Urban Development. Advances in Spatial Science*; Richardson, H.W., Bae, C.H.C., Eds.; Springer: Berlin/Heidelberg, Germany, 2005. [CrossRef]
21. Pomeranz, K. *The Great Divergence: China, Europe, and the Making of the Modern World Economy*; Princeton University Press: Princeton, NJ, USA, 2000.
22. Kuo-Feng, H.; Chwo-Ming, J.Y. The effect of competitive and non-competitive R&D collaboration on firm innovation. *J. Technol. Transf.* **2011**, *36*, 383–403.

Article

COVID-19 as a Jump Start for Industry 4.0? Motivations and Core Areas of Pandemic-Related Investments in Digital Technologies at German Firms

Florian Butollo [1,*], Jana Flemming [2], Christine Gerber [2], Martin Krzywdzinski [2], David Wandjo [2], Nina Delicat [2] and Lorena Herzog [2]

[1] Weizenbaum Institute for the Networked Society, 10623 Berlin, Germany
[2] Berlin Social Science Center, 10785 Berlin, Germany; christine.gerber@wzb.eu (C.G.); lorena.herzog@wzb.eu (L.H.)
* Correspondence: florian.butollo@wzb.eu

Abstract: Academic studies prior to the pandemic rather emphasized that the progression towards Industry 4.0 happened in an incremental manner. However, the extraordinary circumstances of the pandemic have led to considerable investments that were widely interpreted as a (generalized) digitalization push. However, little is known about the character of such investments and their effects. The goal of this contribution is to provide an empirically based overview of recent investment in digital technologies in six economic sectors of the German economy: mechanical engineering, chemicals, automotives, logistics, healthcare, and financial services. Based on 36 case studies and a survey at 540 companies, we investigate the following questions: 1. How much did the COVID-19 pandemic reduce existing obstacles for investments in digitalization measures? 2. Is there a universal digitalization push due to the COVID-19 pandemic that differs from the trajectory before the pandemic? The results show that the pandemic affected investment in an unequal manner. It was driven by the immediate need to sustain business operations through the virtualization of communication among employees and with external partners. However, there was less dynamism in shop-floor-related digitalization, as it was less related to epidemiological concerns and is more long-term in nature.

Keywords: Industry 4.0; digitalization; remote work; COVID-19; investment

1. Introduction

In numerous commentaries, COVID-19 has been described as a tipping point with regard to the digital transformation of the economy and the workplace [1]. Indeed, there is little doubt that COVID-19 acted as a catalyst, as enterprises were forced to instantly move many activities online, thereby overcoming obstacles and reservations that had previously limited substantial steps towards virtual collaboration. As millions of white-collar workers were forced to work from home, it became evident that the technological foundations for remote collaboration were ready to use and that doing so would entail numerous opportunities to facilitate cooperation in the post-pandemic economy. However, remote work was not the only area in which such instantaneous innovation of business practices emerged. Restaurants, shops, public institutions, and industrial companies all had to quickly adapt to delivering goods and services especially under the condition of politically imposed contact restrictions, resorting to a number of digital tools, such as online shops and virtual customer interaction, to do so. The pandemic certainly ushered in a wave of improvisation and experimentation that helped digitalization to flourish.

"Digitalization", however, has become a catch-all term that remains analytically shallow. According to a definition by the German business development bank KfW, digitalization can be understood as "projects for the introduction of digital technologies for the implementation or improvement of processes, products and services of a company and in

Citation: Butollo, F.; Flemming, J.; Gerber, C.; Krzywdzinski, M.; Wandjo, D.; Delicat, N.; Herzog, L. COVID-19 as a Jump Start for Industry 4.0? Motivations and Core Areas of Pandemic-Related Investments in Digital Technologies at German Firms. *Sci* **2023**, *5*, 28. https://doi.org/10.3390/sci5030028

Academic Editor: Johannes Winter

Received: 31 March 2023
Revised: 7 June 2023
Accepted: 19 June 2023
Published: 7 July 2023

contact with customers and suppliers" [2]. This definition is narrow in the sense that it only is concerned with the economic aspects that are touched by the application of digital technologies. Even so, it is a very broad definition referring to all kinds of application fields of digital—i.e., binary data processing—technologies. The term leaves open whether we deal with the introduction of industrial robots, the interaction of agents through the internet, remote communication through cloud infrastructures, the connection of devices through internet-of-things technology, or other aspects from the myriad of applications that are somehow connected to the term "digital".

A more differentiated understanding of the recent changes in economic sectors, however, is meaningful for assessing their likely outcomes. It can be assumed that not every field of application in which digital technologies matter is affected equally. Can the pandemic even be characterized, as a recent survey concluded, as a push that is largely limited to aspects of communication, i.e., virtual interactions within and between companies ([3])? Even if we assume that the pandemic affected a broader set of digitalization themes, are there any new trends and technological fields that experienced particular investment? In particular: How do these changes relate to the fundamental assumption of a new era of digitalized manufacturing associated with the term "Industry 4.0"? These questions point to the need to go beyond linear assumptions of a comprehensive acceleration of digitalization through the pandemic. They concern the *quality* and the *direction* of the digital transformation of enterprises.

The goal of this contribution is to provide an empirically based anatomy of recent developments related to investment in digital technologies in industries, with a particular focus on so-called "Industry 4.0" technologies, by which we understand applications that take advantage of new technological possibilities through the internet of things and/or artificial intelligence. This article summarizes empirical data from six sectors of the German economy: mechanical engineering, chemicals, automotives, logistics, healthcare, and financial services. In particular, we want to answer the following questions:

1. How much did the COVID-19 pandemic reduce existing obstacles to investments in digitalization measures?
2. Is there a universal digitalization push due to the COVID-19 pandemic that differs from the digitalization trajectory before the pandemic?

The answers to these questions are of great relevance both at the level of theory and for practitioners. Theory-wise, they help to grasp the socio-technical realities of digitalization in the present period, i.e., the combination of technical, economic, and social factors that condition the ability of corporate actors to implement digital applications. The resulting empirically grounded assessment of current digitalization trends at companies also helps to assess the state of implementation and the scope of the so-called "Industry 4.0", which we interpret as a narrative and a metaphor for diverse approaches related to new technological approaches based on the IoT and AI. Concrete assessments of the main trajectories of transformation in each economic sector that go beyond such narratives are important for practitioners in management, trade unions, and politics as they identify constraints, potentials, and possible fields of action.

This contribution is structured as follows: In Section 2, we discuss the state of digitalization in the German economy by introducing a pragmatist perspective that emphasizes the economic, institutional, and social conditions for progress towards "Industry 4.0". In the third section, we provide an assessment of the pandemic as a dual crisis, i.e., a crisis in public health and the rippling effects of an economic crisis that was soon to be followed by recovery and growth. This perspective informs the empirical analysis in the sections thereafter, in which we discuss possible impacts on digitalization strategies that are related to the economic, institutional, and social fallout of the pandemic: We briefly summarize (Section 4) the qualitatively and quantitatively oriented research design and the chosen methods before (Section 5) discussing the mechanisms by which the pandemic affected the digitalization of enterprises based on an inductive analysis of material on the drivers and obstacles of digitalization from 36 company cases. In the penultimate section (Section 6), we identify and discuss the main areas of investment in digital technologies during the pandemic based on the qualitative and quantitative data from our investigation. In the discussion (Section 7), we interpret the findings as a layered progression of digitalization with investment in remote work and infrastructures at its core, complemented by the automation, digitalization, and virtualization of cognitive work, but with little activity at the level of the automation of physical processes and shop-floor-related applications. These results, as emphasized in the conclusion (Section 8), point to the need to contextualize the focus on manufacturing technologies inherent in the discourse on Industry 4.0 with a stronger consideration of the transformation of cognitive work and the associated organizational effects across different functions in enterprises.

2. Techno-Centric Narratives and Incremental Socio-Technical Change

The rapid development of a broad range of digital applications based on comprehensive methods for recording and processing data has created manifold possibilities concerning the automation, interconnection, and virtualization of processes in enterprises [4–6]. Beyond raising productivity, new possibilities for capturing economic value from data have resulted in the emergence of new business models through the supply of digital services and the rationalization of distribution through platforms [7,8].

In Germany, a coalition of industry associations, research institutions, and government agencies were vocal in framing certain aspects of this transformation as "Industry 4.0" and popularizing it as a stylized narrative, depicting a distinct stage of industrial development. It singled out the IoT and AI as base technologies and highlighted a bundle of digital applications that would engender leaps in productivity and enable companies to reconcile the conflicting imperatives of flexibility and productivity [9,10]. This narrative has a distinctively German flavor, emphasizing the legacy of "diversified quality production" [11] and identifying manufacturing as the main area of the digital transformation.

According to several accounts, the new industrial revolution did not live up to the initial expectations in the period preceding the pandemic. The patterns of change have been much more gradual than the narrative of Industry 4.0 would have it [12–14]. The diffusion of the new possibilities usually proceeds not in a revolutionary way, but rather on a pragmatic and incremental trajectory, through trial and error, as enterprises experiment with single applications from a 'bundle of new technologies' that mostly constitute specific modifications of existing production models, rather than a new paradigm [15]. A leading representative of the machine builders' association (VDMA) even summarized the state of affairs in this key sector of the German economy by speaking of "ten lost years" since the initial proclamation of Industry 4.0. He emphasized that productivity growth has remained flat and said that there had been little progress in the bulk of enterprises, except for the flagship projects by technological leaders [16]. If the progress of digitalization across all sectors in the economy is concerned, the state of affairs in Germany seems to be even more critical: For an index on the levels of digitalization in Europe published by the European Commission, Germany is listed in the 18th position, with a substantial gap between it and the leading economies of Ireland, Finland, Belgium, and the Netherlands, and even behind economies such as those of the Czech Republic, Spain, and Slovenia (DESI 2020). In the period preceding the pandemic, several studies reported a slowdown in investment in digitalization at enterprises due to the then-looming recession (KfW 2021).

Gradualism is a key characteristic of any major period of socio-technical change. In fact, industrial revolutions should be thought of not as a big bang with immediate impact, but as a long-term period of accelerated innovation that involves experimentalism and micro-innovations that amount to a predominance of incremental modes of change [17,18]. However, the gap between high expectations about an imminent industrial revolution and the slow and uneven implementation of new digital technologies at enterprises puts the very essence of techno-optimist predictions into question. It warrants explanations that concern not only the state of technological innovation, but also the social processes concerning the implementation of technologies.

Existing knowledge about the barriers to digitalization processes in industrial companies is being generated from a vast pool of case studies from diverse industries and regions. Despite the high significance of the issue for practitioners and its prominence in political discourse, there have been few meta-analyses that have caused the findings to converge towards a more general understanding. In Table 1, we compare the frameworks of two available meta-studies on the subject [19,20].

Table 1. Meta-studies on categories for barriers to digitalization in industries.

Lammers et al. (2019) [20]	Horváth/Szabó 2019 [19]
Financial	Shortage of financial resources
Knowledge and skills	Human resources and work circumstances
Regulatory	Standardization problems
Technological	Technological integration
Environmental	Coordination across organizational units
Organizational	Organizational resistance
Cultural	Concerns about cybersecurity and data ownership

Although the framing of the findings in both studies is not identical, there is remarkable congruence among the key dimensions that affect the implementation of digital technologies. We expect that these dimensions also conditioned the implementation of digital technologies during the COVID-19 pandemic. However, we hypothesize that the malleability of these factors varies with regard to their relationship to the specific constellation brought about by the pandemic and the ability to overcome limitations within a short timeframe. In what follows, we provide a short general understanding of the effects of the COVID-19 crisis before progressing toward an empirical understanding of the drivers and obstacles in this specific context.

3. COVID-19 as a Dual Crisis

In order to identify how much and in what areas the pandemic acted as a catalyst for investment in digital technologies, we need to assess how the pandemic modified or reduced the aforementioned obstacles to the implementation of technologies. To fully grasp the impact of the pandemic, an understanding of COVID-19 as a double crisis with a health and an economic dimension is needed.

The first dimension is the health crisis, which resulted in widespread interruptions of interpersonal contacts. The immediate results of this were closures of factories, stores, and offices, disruptions in supply chains, the quick expansion of remote work, and a surge in e-commerce. COVID-19 thus demonstrated the merits of moving cognitive interactions and business transactions online wherever possible. The historic coincidence is remarkable: The sudden demand for the virtualization of social relations emerged in a situation where many of the applications that were needed to do so were already at an inflection point. An abstract possibility, therefore, became an imperative during various lockdowns and pushed the everyday use of digital applications considerably beyond the limits that had existed before. COVID-19 was also a stress test for all kinds of digital applications, including the IT infrastructure, video conferencing tools, and many other applications that facilitated the virtualization of tasks. This also convinced users that these tools could actually be applied and that they offered a broad range of options.

The second dimension of the crisis was its short-term and long-term economic impact. COVID-19 was aptly characterized as a simultaneous supply and demand shock [21]. It resulted in a deep slump in 2020 that surpassed the level of the financial crisis of 2008/2009. In Germany, the COVID-19 shock hit the economy when signs of a considerable cyclical slowdown were already widespread, and it seemed as if it could be the beginning of a major recession. These fears did not materialize, as steep recovery growth kicked in in the second half of 2020. However, the economic situation remains precarious due to the sustained fragility of supply chains, a surge in inflation, and the economic effects of the war in Ukraine with its economic and political fallout.

The macroeconomic context is of major importance for the implementation of digitalization projects. A tight macroeconomic climate can reinforce conservative investment behavior, as a return on investment is not guaranteed. Positive growth prospects, on the contrary, can lead to more confident investments with the expectation of rapid growth in a new business cycle. Historically, great economic crises and the subsequent recoveries have often resulted in shifts in the socio-technical composition of industries. They have accelerated the diffusion of base technologies in long waves of economic development [22,23], driven by new investment opportunities and Schumpeterian creative destruction. The economic effects of COVID-19 could turn out to be a cycle of economic slumps and growth of this kind. Perhaps the inflection point of "Industry 4.0" is situated in the recovery phase of COVID-19. The combination of both elements of the conjuncture—the immediate effects of the health crisis and the long-term macroeconomic effects—could result in accelerated technological adaption because there is both an awareness of new possibilities and (arguably) the economic leeway for more courageous investment behavior.

4. Research Design and Methods

The goal of our empirical study was to better understand the extent and the properties of the digitalization push during the COVID-19 pandemic. To this end, qualification and differentiation with regard to the mechanisms and the core areas of the digitalization push were needed. The rationale behind this approach was threefold. First, we found that there needs to be a more differentiated consideration of how much the pandemic acted as a driver for investments and the use of digital technologies. To what extent and in what areas did the pandemic modify the relationship between drivers of and obstacles to investment? Moreover, the theoretical reflections in Section 2 demonstrated that the obstacles to the implementation of digital technologies are manifold and include cultural, technical, organizational, economic, and regulatory dimensions. Although the pandemic certainly heightened the awareness of the importance of technological change in general, it needs to be scrutinized how much it affected the barriers with regard to each of these dimensions.

Second, the hypothesis of a universalized digitalization push leaves unspecified what technologies actually gained in weight (a byproduct of the overly vague catch-all term of 'digitalization'). Did the pandemic, for instance, lead to equal amounts of investments in technologies for the virtualization of social interaction (e.g., video conferencing tools) and in robotics? An empirical inquiry needs to paint a differentiated picture of current events and identify the rationale for investment in each case.

Third, public debates about the impact of COVID-19 on digitalization are often based on the implicit assumption of a linear progression of events, as if the issue at stake would simply be the quantity of investment, not the characteristics of the chosen approaches. Instead, we ask about possible changes in the strategic orientations of companies, i.e., the direction of the digital transformation, and new ways to combine technology, organizations, and employees under different circumstances. It might turn out in hindsight that the pandemic will have changed the way we think and act about digitalization, as different applications and different socio-technical ways of using them have become more prominent, whereas others will have lost importance. The recent trajectory is not only about more of the same, but about choices among different options.

Methods

The empirical analysis consisted of qualitative data from 36 companies in six relevant economic sectors: automotives, mechanical engineering, chemicals, logistics, financial services, and the health sector. The choice of sectors was made according to their weight in overall employment in the German economy and their exposure to the dual effects (health-related and secondary economic effects) of the COVID-19 crisis. In each sector, a sample of 4–7 companies was chosen. The case selection aimed to include companies from relevant subsectors of each industry. We deliberately included some smaller companies and suppliers in our sample in order to account for the diversity of experiences. However, roughly two-thirds of the sample consisted of larger companies of considerable economic strength and technological sophistication. This selection strategy was chosen because we suspected a particularly significant impact of the pandemic where companies had already pursued ambitious digitalization strategies that would then be impacted or modified. However, this perspective on stronger and more advanced companies in the qualitative studies was complemented by a quantitative survey consisting of a random sample of 540 companies. The sampling strategy for this survey aimed for correspondence with the actual composition of each covered sector in terms of company size and geographical distribution. In contrast to the qualitative investigation, the quantitative sample thus contained a much larger share of SMEs [14].

The interviews for the qualitative investigation were conducted between February 2021 and March 2022. The perceptions of the impact and the further development of the pandemic were in flux during this period. We counterbalanced these contingencies by aiming to reconstruct the measures undertaken since the beginning of the pandemic in each conversation. While it was not possible to eliminate differences in perceptions and attitudes related to the unfolding of the events of the pandemic, the strategy of focusing on concrete actions taken and the reasoning behind them ensures the comparability and robustness of the recorded data.

At each company, interviews of 1–1.5 h were conducted with at least two representatives. We strove to interview representatives from management and the works council in each case in order to account for diverging perspectives from the leadership and employee representatives. Where available, we also contacted managers that were responsible for the implementation of digitalization strategies or projects, who were often denominated as "chief digitalization officers (CDOs)". In each sector, we also led complementary interviews with industry experts in order to account for general trends in economic and technological development. In complementary interviews with start-ups and established technology providers, we also recorded the perspectives of firms that offered innovative technological solutions in each sector. The audio files of a total of 88 interviews were transcribed and analyzed by means of a qualitative content analysis. A deductively developed coding scheme that included the core categories of "economic situation", "digitalization measures", "relation to pandemic", "work organization", "quality of work", and "geographies" was refined through the inclusion and modification of categories that were inductively derived from the analysis of interview materials. In this way, we developed a distinction among the following key *categories of digitalization investment*:

- Remote work and virtualization of work;
- Improvement of IT infrastructure and introduction of collaboration tools;
- Virtualization of customer/supply chain interactions;
- Virtualization/digitalization/automation of administrative work/HR functions;
- Virtualization/digitalization/automation of production processes and services;
- Digital tools for training purposes;
- Business model innovation/supply of digital services.

In what follows, we present the results of the qualitative investigation of the two questions introduced in the introduction:

1. How much did the COVID-19 pandemic reduce existing obstacles to investments in digitalization measures?
2. Is there a universal digitalization push due to the COVID-19 pandemic that differs from the digitalization trajectory before the pandemic?

For this sake, we first inductively collected statements on the drivers and obstacles for digitalization during the pandemic (Section 4). Subsequently, the prevalence of the aforementioned categories of digitalization investment was identified by means of a comparative analysis of the cases (Section 5). In the presentation of the material, the findings on these questions are complemented by the interpretations derived from the interview material on why investments in a certain area took place (or did not take place) (Section 5).

5. The Effects of the Pandemic on the Motivation to and Obstacles to Investment

Our inductive analysis of the qualitative data on the motivations and obstacles for investment revealed remarkable overlaps with the conceptual frameworks of [20]. In our analysis, we categorized answers that indicated some kind of relationship between the pandemic and digitalization measures, regardless of whether they acted as drivers of or barriers to investments, in order to identify its possible effects.

As Table 2 shows, the categories from the literature are largely identical with those of the inductive analysis, with two exceptions: First, the "direct factual relationship with the COVID-19 pandemic" obviously could not have played a role before the occurrence of the pandemic. As is explained in more detail below, this category addresses whether the functions of certain digital applications helped to mitigate the impact of the pandemic, for example, by helping to sustain operations under the condition of contact limitations. In so far as this has been the case, the pandemic drove investments in such technologies, while the absence of a factual relationship mostly resulted in unchanged investment behavior or, in some cases, neglect. Second, changes related to skills and HR were not mentioned as a COVID-19-related factor affecting digitalization measures during our interviews. This is remarkable, as the pandemic involved limitations in workforce mobility and aggravated labor shortages in several sectors [24]. It is conceivable that under these circumstances, shortages of skilled labor that had acted as a barrier to digitalization could not have been removed. However, in our data, there was also no evidence that the pandemic made things worse in this respect and that this put a strain on digitalization efforts.

Table 2. Comparison of inductive categories with categories of barriers to investment in the literature.

Inductive Categories from 36 Cases	Lammers et al., 2019 [20]	Horváth/Szabó 2019 [19]
Direct factual relation to the COVID-19 pandemic	---	---
Financial	Financial	Shortage of financial resourcesRisk of fragility (uncertainty)
---	Knowledge and skills	Human resources and work circumstances
Regulatory	Regulatory	Standardization problems
Technological	Technological	Technological integration
Customer relations	Environmental	Coordination across organizational units
Organizational	Organizational	Organizational resistance
Cultural	Cultural	Concerns about cybersecurity and data ownership

The "environmental" category, by which Lammers et al. [20] understood factors related to a company's environment, was framed in a more specific way in our case studies. For this category, we summarized statements that emphasized the significance of digital technologies to address customers in order to maintain sales. As will be shown, the question of whether or not a company needed to rely on the virtualization of their customer relationship in order to maintain sales turned out to be an important factor that explained the dynamic of investment in some enterprise functions.

A closer analysis of our qualitative data on the relationship between the pandemic and digitalization measures shows that it was not linear and far from universal. In fact, the relationship can be characterized as multi-faceted and often contradictory. Table 3 displays significant reasons for or against investments in digitalization measures as a reaction to the COVID-19 crisis.

Table 3. Reasons given for investment or non-investment in digital technologies.

Categories	Drivers	Examples	Non-Investment	Examples
Direct factual relation to the COVID-19 pandemic	Investments indispensable to master the crisis	Virtualization tools and infrastructures (videoconferencing, cloud software, virtual customer interaction, digital signatures, self-service terminals, e-learning, etc.)	No relation between technologies and pandemic situation	Cyber-physical systems, robotics
Cultural	Heightened awareness and habituation	Experience of working with remote tools, new work cultures, cooperation and less researvation	Reservations against management	Concerns about deregulation, i.e., excessive working hours in a homeoffice setting;
Environmental	Mindset and habits of customers	Remote consultations, use of apps and online tools	Decline of demand for digital products	Decline of demand by corporate customers of machine vendors due to economic insecurity
Technological	Easy spontaneous implementation and scaling of digital apps	Implementation of alone standing software tools, increasing infrastructure as a service capacities	Impossibility to invest spontaneously in systemic changes	Automation of physical processes; systemic changes in healthcare information systems; implementation of remote access tools from scratch; scaling of own server infrastructures
Organizational	Additional resources for digitalization projects	Idle IT resources during lock downs, because of reduced company operaions	Lack of resources due to disruptions and crisis mode	Interruption of ongoing shopfloor-related projects during lockdown; refusal to launch additional digitalization projects in insecure context
Financial	Additional funds for investment	Political funds for strategic digitalization projects	Investment freeze	… due to insecure economic situation
Regulatory	Loosening of regulation, mode of improvisation	Acceptance of digital signatures within and between institutions; enabling innovation with regard to data transfer and collaboration; company agreements on remote work (new or extended)	Regulatory context unchangeable	Digital tools in chemical production processes

As became evident in all of our case studies, the *direct factual relation to the COVID-19 pandemic* and specific technologies were highly variable. Remote work and digital interfaces for customer interaction (websites, online shops, video calls) for many companies were indispensable means of sustaining business operations. Consequently, there was an immediate need to implement or expand technological solutions in these functions. The focus here was on means for remote communication and data exchange, which served to circumvent physical contact for health-related reasons. Conversely, in only a minority of the investigated case studies, there was investment in production-related digitalization investment. Production processes were first interrupted and then could be relaunched under safety precautions that relied on social adaptations (the modification of shift plans, hygiene and social distancing rules, etc.), not on technological solutions. Hence, it comprehensively seems that the pandemic even constituted a more difficult environment for shop-floor-related investments.

By *cultural* factors, we understand the mindset of the actors involved at the company level, i.e., the attitudes of management, works councils, and employees towards digitalization measures. In line with public discourse, we found a heightened awareness for the possibilities of digitalization. This was not only the overwhelming picture in our company case studies, but was also confirmed in the quantitative survey of 540 companies, in which over two-thirds of respondents confirmed an increased awareness of digitalization issues due to the COVID-19 crisis (45.0%: agree, 24.8%: partially agree). The results from the qualitative data illustrate that such perceptions mostly focused on the transition or expansion of remote work. This transition was mostly brought about in an improvised yet cooperative manner in the absence of major conflicts and, in some cases, with adherence to prior agreements between management and works councils. Cases in which works councils expressed profound concerns about remote work arrangements did exist but constituted a minority of our sample. Overall, our data confirmed that the experience of the pandemic strengthened positive attitudes towards digitalization in general. This may result in a more proactive and open stance towards future areas of investment that do not have any factual relationship with the pandemic.

Beyond these general observations—the *direct factual relationship* of digital applications to the COVID-19 pandemic and the *mindset of the involved* actors—the analysis of company cases highlights some additional criteria that affected the propensity of companies to invest in digital technologies.

The feasibility of investment often depended on factors related to a company's *environment*. The absence of a mindset and habits of customers that were open to digital products and services before the pandemic had often constituted a barrier to investment. Interviewees from banks, hospitals, and outpatient care providers reported a hike in the demand for such digital offers during the pandemic due to the social distancing prescriptions, which, in turn, incentivized their institutions to expand their activities in these areas. Representatives from some mechanical engineering companies also reported that there was an increased openness on the part of customers to use tools for the remote setup and maintenance of their products. Such beneficial effects related to the mindset of customers shape most service-oriented sectors, in which interactive customer relations prevail. They are less pronounced when companies provide standardized services and products in B2B supply chains, such as in the chemical industry and in logistics. In some cases, the economic effects of the crisis even resulted in a decrease in demand for digital products and, thus, constituted a barrier to digitalization: Mech.3 (the terminology refers to Table 4, which displays an overview of the company cases), for instance, reported a decline in demand for their "digital factory" products, resulting in a delay of progress by 2–3 years, because customers were holding back investments due to economic insecurity in the first phase of the pandemic.

Table 4. Results of the qualitative case studies.

		General Assessment of Relationship between COVID-19 and Digitalization	Transition to Remote Work	IT-Infrastructure, Collaboration tools	Virtualization of Customer Interaction or Supply chain	Virtualization Xautomation of Admin/HR Functions	Digital Learning and Training	Automation/ Robotics (Shopfloor)	Business Model Transformation
Auto	Auto.1	no push at all. Just infrastructure	++	+	o/-	-	o/+	-	-
	Auto.2	push for virtual communication + conference tool	+	++	x	x	+	o	-
	Auto.3	push for remote work	++	++	++	++	+	+	+
	Auto.4	no push at all	+	++	x	++	x	x	x
	Auto.5	Just remote work. No push regarding digitalization/automatization.	+	x	o	x	x	x	x
	Auto.6	push regarding remote work and conference tools	+	+	x	+	x	+	x

Table 4. *Cont.*

		General Assessment of Relationship between COVID-19 and Digitalization	Transition to Remote Work	IT-Infrastructure, Collaboration tools	Virtualization of Customer Interaction or Supply chain	Virtualization Xautomation of Admin/HR Functions	Digital Learning and Training	Automation/ Robotics (Shopfloor)	Business Model Transformation
Chemicals	Chem.1	Just remote work	+	o/+	+	-	+	o	o
	Chem.2	Push regarding remote work, greater role of IT, otherwise no accelleration	+	+	+	+	x	o	o/+
	Chem.3	New MS Office version 2o19	+	+	+	-	o/+	o	x
	Chem.4	Nothing new, but faster	++	+	x	o	o/+	o	x
	Chem5	Just remote work	++	++	+	+	x	+	x
Mechanical Engineering	Mech.1	Just remote work	+	+	x	o	-	-	-
	Mech.2	Slight push for predictive maintenance; slight increase in technology acceptance, virtualization of the sales department	++	+	+	+	o	-	-
	Mech.3	Just remote work. Much less demand for digital factory products by customers	++	++	o	o	+	x	o
	Mech.4	Retention of investments; adoption of remote infrastructure; virtualization of sales and customer introduction; introduction of RPA in HR and training; massive push for online training.	++	+	o/+	+/++	++	+	o/+
	Mech.5	Considerable acceleration at priorly little digitalized company. QR-Codes for self assembly of customers, remote work and MS Teams, acceleration of existing projects.	++	++	++	+	o/+	-	-
	Mech.6	Acceleration of certain projects, e.g., remote service assistant; introd. of HR self-service; (not only due to pandemic), plan to integrate cloud applications better.	++	+	+	+/o	o	-	o/-
Logistics	Log.1	No impact on shop floor automation, but virtualization of customs, HR self services, paperless office; more attention on digital tracking	+	+	+	+	x	o/+	-
	Log.2	No push in shop floor-related fields, but faster adoption of RPA and eSignatures.	+	+	x	++	+	-	-
	Log.3	no push; high level of digitalization already	o/+	x	x	x	x	x	x
	Log.4	Little impact of pandemic, but acceleration of some projects like eSignature	++	x	o	o	x	o	-
	Log.5	Mostly remote work, cloud and VPN infrastructures, hardware. Other projects: automation, harmonization of IT systems, RPA unaffected by pandemic	++	+	x	x	x	o	-
	Log.6	Higher demand for digitalization/automation products from clients (supply chain/ warehouse solutions). No other effects	+	x	x	x	x	+	+

Table 4. *Cont.*

		General Assessment of Relationship between COVID-19 and Digitalization	Transition to Remote Work	IT-Infrastructure, Collaboration tools	Virtualization of Customer Interaction or Supply chain	Virtualization Xautoma-tion of Admin/HR Functions	Digital Learning and Training	Automation/ Robotics (Shopfloor)	Business Model Transfor-mation
Healthcare	Health.1	Remote work in admin and some operations (radiology), indoor navigation app as cloud pilot + bed capacity management, online teaching. No impact on robotics and AGVs	+	+	++	-	+	-	-
	Health.2	Acceleration: virtual communication among employees, digitalization of admin, introduction of apps for patients,	+	o	++	+/++	o/+	o	-
	Health.3	No effect on major strategic projects, but acceleration with regard to telemedicine and automated distribution of drugs. No new digitalization projects due to difficult pandemic situation.	+/++	o	++	++	x/+	o/+	-
	Health.4	Acceleration of remote work and virtualization of communication. No effects beyond that.	++	+	x/o	+/++	x	x	-
	Health.5	No major changes. But some progress with regard to remote work, telemedicine, digitalization of admin.	+	o	x/o	x	+/++		-
	Health.6	Acceleration of virtual communication (internal and with customers). Looser regulation allows for end-to-end tracking of processes.	+	+	+	+	++	x	-
	Health.7	No changes with regard to strategic long-term projects. Acceleration of virtual communication, telemedicine. Higher acceptance by patients and employees.	x	+	++	x	+	x	-
Finance	Fin.1	No major changes. Ambitious digitalization agenda unchanged but some acceleration.	+/++	+	++	o	x	o	o
	Fin.2	Acceleration and extension with regard to tools and infrastructures (eSignature, online collaboration, cloud infrastructures)	+/++	++	++	+	x	x	o
	Fin.3	Comprehensive introduction of remote work; little changes in back office; acceleration with regard to eSignatures; slight push with regard to apps for customers.	+/++	+	+/++	+/++	+	o/+	o
	Fin.4	Little acceleration at highly-digitalized company. Remote work and cloud infrastructure, VPN, virtual training	+/++	+	+	o	+	o	o

Legend: ++: strong push through pandemic; +: medium push; o: no push; -: does not apply; x: no information provided/not mentioned.

The *technological* feasibility in the short term turned out to be a fundamental issue that determined the course of investment since the beginning of the pandemic. As many companies already had cloud infrastructures, software packages, and mobile hardware devices in place before the pandemic, the instantaneous implementation of services for remote communication could be provided with little friction. Moreover, companies could easily ramp up external cloud services and software options without buying hardware and going through arduous processes of setting up equipment and software solutions at their own premises (see Section 6). Other digital applications could not be spontaneously implemented, however. This, for instance, accounted for a company in our sample that, until the pandemic, had relied on their own server capacities and found it impossible to scale them up instantaneously. Similarly, investment in robotics mostly requires long-term preparation and an elaborate process of installation, which is hardly possible within a few weeks. At Auto.1, Mech.1, and Mech.3, it was emphasized that it would not be possible to spontaneously invest in tools for remote setup and maintenance; such software solutions and the corresponding equipment (data glasses) would need to be in place already in order to use them in the pandemic setting. Other digitalization projects constituted long-term efforts related to the creation of infrastructures and standards. This accounted for the introduction of healthcare information systems—a political target for roughly two decades—and the IoT-based interconnection of production equipment at industrial companies. Especially in healthcare, the pandemic certainly demonstrated the urgency of improved information systems. The implied complex processes of institutional innovation could not be solved on the spot, however.

Another factor that conditioned investment in digital technology was the amount of *financial* resources that a company was able to allocate to this end. In some instances, there was a considerable increase in spending. In the health sector, this was politically driven, as the government created a special fund—the so-called "hospital future fund" that was made available from the beginning of 2021. In the private sector, management mostly invested without hesitation in measures for the virtualization of communication that were considered to be low-hanging (and low-cost) fruit. Such changes also demanded *organizational* resources for ramping up capacities and readjusting work routines. In our quantitative survey, 52% of the enterprises that had intensified their investment in digital technologies stated that they complemented this investment with organizational changes. Among such measures was a flexibilization of working hours, an increase in cross-functional cooperation, and changes in leadership roles. The severity of the pandemic thus triggered a quest for more effective work organization, which required considerable effort. Sometimes, the obstruction of regular operations made it easier for companies to focus on organizational innovation. In Log.2, for instance, the IT department could continue to operate remotely, unlike the operative logistics division, which came to a halt in the spring of 2020. This freed up the capacities of the IT professionals, which were then used to intensify the automation of administrative tasks (robotic process automation). In contrast to such successful cases in which companies and other institutions could mobilize additional resources for digitalization projects, there was a minority of cases in which investment was cut. Mech.4 reported a freeze of (new) investments due to the insecure economic environment. The dominant picture, however, is that in most ongoing digitalization projects beyond those connected to remote work, there were few changes with regard to funding and schedules. Most company spokespersons emphasized that they had existed before the pandemic and, at most, experienced slight acceleration along with heightened interest.

Finally, some digitalization projects could benefit from ad hoc *changes in regulatory* circumstances. Where digital signatures had not existed before, they became accepted, thus enabling progress in paperless administration. Health insurance companies also began to accept digital signatures from customers of outpatient care units, thus facilitating the work of outpatient care providers, such as Health.6, where the need for double documentation (on paper and digitally) was eliminated. Other rules and regulations proved to be difficult to instantaneously change or could not be changed at all. In some areas of the chemical industry (such as at Chem.1), for instance, certain digital tools could not be used because of safety regulations, which is a limitation that is impossible to overcome.

The overview of the factors affecting investment or non-investment in digitalization measures during the pandemic highlights several causes for a digitalization push, but also its conditionality. Investments were most prominent where there was a strong factual relationship between the pandemic and technological change and where there was a strong conviction on behalf of the main actors in the purpose of such investments. Furthermore, the willingness of customers to accept digital offers and the ability of organizations to ramp up the available resources, as well as to loosen the regulatory context, constituted favorable circumstances for progress in digitalization. The analysis also shows that we need to distinguish between low-hanging fruits that are available with low cost and little effort (e.g., the introduction of cloud-based software packages or the broader utilization of collaboration tools) and long-term investment projects that cannot be instantaneously introduced.

6. Empirical Qualification of the Digitalization Push: Core Areas of Investment

As the prior discussion has shown, the catch-all word "digitalization" is not specific enough for a concrete analysis of recent changes in companies. "Digitalization" consists of a set of varying—not necessarily integrated—measures that affect different dimensions of an organization's activity and require different capabilities. Quantifying the changes as "more" or "less" digitalization is not satisfactory, as this omits the particular emphasis that is given to distinct aspects of digitalization. In what follows, we provide short summaries of the most important areas of investment during the pandemic at the 36 companies in our sample while highlighting the actual relationship between the pandemic and digitalization measures.

6.1. The Transition to Remote Work

The data show that the transition to remote work affected all areas of non-location-dependent white-collar work, which also implies that it amounted to different shares of the total workforce according to the core processes of a firm. It was universal in our qualitative sample with a share of between 25 and 100 percent of workers working online (with the sole exception of a logistics division that did not have its own administration at the investigated site). In the estimation of 441 respondents in our quantitative survey, the mean average from all companies with regard to the share of employees working online roughly doubled from 15 to 29.7 percent.

There was a high degree of cooperation between management and works councils in order to make this transition possible. This afforded that works councils were prepared to chart institutionally non-regulated territories by agreeing to unprecedented work arrangements, while management cultures had to instantly change to grant more leeway for independent teamwork and to loosen (or alter) performance control. The transition was smoother when agreements that specified the rules for remote work had already been negotiated before the pandemic and high independence of teams had already been a part of the company culture. However, improvisation was needed regardless of whether such formal regulations and organizational practices had existed before. In general, a well-established co-determination routine played an important role in efficiently managing the transition without major friction. At the few companies in the sample that had more conflict-ridden industrial relations, the transition towards remote work involved more friction, as the works council suspected a deterioration of standards.

6.2. Digitalization beyond Remote Work: A Polarized Picture

While the transition towards remote work absorbed much attention and was mostly associated with the digitalization push as such by practitioners, there is evidence that the pandemic had impacts on digitalization issues beyond that. As stated before, the majority of the companies in our quantitative survey reported a generally greater awareness of digitalization options (45.0%: agree, 24.8%: partially agree). A total of 64 percent of the surveyed companies also reported additional investments in digitalization measures, with considerable effects: 33 percent of the respondents stated that the level of digitalization of their enterprise before the pandemic used to be lower than at the time of the survey in the summer of 2021. Interestingly, however, the replies also suggest that some of the measures taken were rather insular, as 63 percent of the respondents denied that it amounted to a change in the overall level of digitalization at their company.

The objectives and character of the chosen measures varied considerably depending on the peculiarities of the operations in each company (characteristics of products and services, composition of the workforce, customer relations, etc.) and the extent to which the virtualization of work was imperative for maintaining operations. Table 4 provides an overview of our qualitative findings, highlighting the differentiation according to the respective dimensions of the digital transformation.

6.3. IT Infrastructure

The challenges of rapidly scaling remote work schemes were reflected in widespread *investments in IT infrastructure*. These included the purchase of additional laptops and other hardware, the ramping up of server capacities (mostly externally sourced), improvements in Wi-Fi bandwidth and virtual private network (VPN) access points, and the acquisition of a range of software tools for online collaboration. Many of these possibilities had been in place before the pandemic. However, the need to instantly rely on such tools contributed to firmly establishing them in work routines and reducing uncertainties about their usability. In this sense, the pandemic proved to be a consolidation of technological developments that were already underway but had only been partially used before. The use of video conferencing software and digital collaboration tools, such as MS Teams, experienced an especially strong push beyond the boundaries of former practices. It is notable that the scaling of remote work afforded computing capacities that could barely have been provided if there was not the option of purchasing cloud computing capacities based on the *infrastructure-as-a-service* and *software-as-a-service* options that have come to dominate the market in recent years. Several companies reported that they resorted to cloud providers in order to spontaneously ramp up capacities and bandwidth. On the contrary, Mech.5 had relied on its own server capacities on premises that it reported regarding its inability to ramp these up during the pandemic. Subsequently, it canceled its plan to acquire additional self-owned server capacities in favor of sourcing them externally. In this sense, COVID-19 surely represents a tipping point for the reach and intensity of cloud computing in the business context.

6.4. Reducing Physical Contact in Services: Customer Interaction, HR Services, and Digital Learning

Services for customer interaction proved to be vital in mechanical engineering, healthcare, and finance, which are sectors in which regular operations rely on frequent personal interactions between firms and their customers. Such findings in the qualitative study were confirmed by our quantitative survey, which also singled out the digitalization or automation of administrative processes, training, distribution, and HR as the fields with the highest level of activity, either by launching new initiatives or by ramping up existing ones (cf. Figure 1).

In mechanical engineering, remote contact with customers frequently involved the use of devices and software that could be used for interactions with the technical staff of machine providers and online manuals (Mech.1, 2, 4, 5, 6). Mech.5 introduced online services and manuals for the installment of equipment, including short videos that were distributed through the firm's own online platform. In hospitals and care facilities, tools for virtual navigation and remote contact with patients and relatives were established (Health.1, 2, 3, 7). However, our respondents emphasized that while there had been some acceleration in the introduction of such methods, they were still only partially implemented and used. Some industrial companies (Auto.3, Chem.1, Mech.2, Mech.4) also invested in additional sales channels through digital platforms and other channels. The companies that extended techniques for virtual customer interaction had already laid the basis for doing so before the pandemic. The pandemic led to the habituation and the improvement of existing approaches in these cases, which probably amounted to establishing them permanently. As mentioned before, however, virtual customer contact through data glasses and the like is difficult to spontaneously, as it requires long-term preparation and infrastructure investment.

Digitalization/automation of …

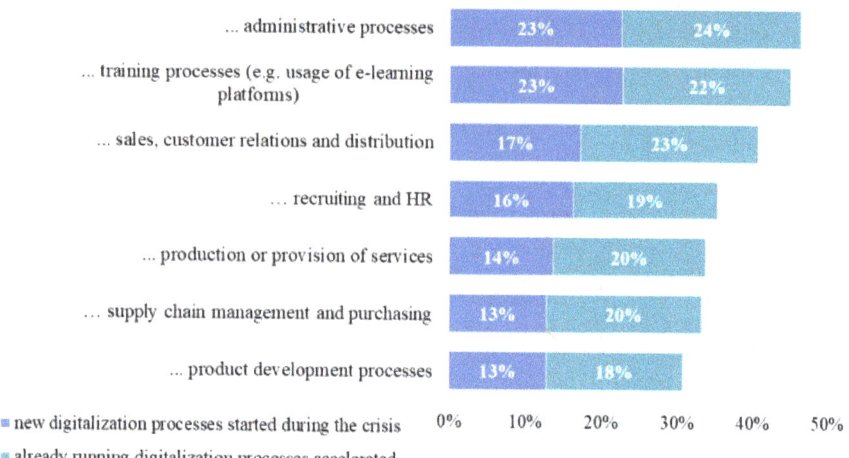

Figure 1. Core areas of investment (quantitative survey).

HR services and administrative work in general are another field that was directly related to the need to reduce physical proximity. Our quantitative survey confirmed that this constituted a major tendency, as 23 percent of the respondents said that new processes for the digitalization or automation of administrative work were newly established, and another 24 percent said that they were accelerated (46 percent said that it did not apply) (on the more specific question about the digitalization/automation of recruiting and HR, the results are as follows: newly established: 16%; accelerated: 19%; does not apply: 55%). Our qualitative investigation showed that some companies (Auto.4; Mech.6, Log.1; Health.2) established or extended the use self-service terminals for their employees, which were sometimes supported by employee apps. There was also an acceleration of initiatives for the end-to-end digitalization of administrative processes by introducing digital signatures and similar procedures (Mech.5, Log.2, 4; Health.2; Fin.2, 3). The pandemic thereby contributed to the general trend for the automation and virtualization of administrative work. This could well lead to structural changes that qualitatively transform these functions. In the long term, the changes in the media of communication could also result in a transformation of work content and the division of labor. At Mech.1, for instance, there was a geographic reshuffling of responsibilities among the HR staff. HR specialists in a particular dependency of this company were then supposed to answer requests from all employees of the entire corporation on one particular area of expertise (e.g., sickness leaves, maternity issues, etc.), instead of being generalists for employee requests from the local site only. Therefore, we suspect that the virtualization of such functions could enable structural changes ranging from adjustments of the work organization to additional possibilities for outsourcing and/or offshoring, given that no geographical co-location of such functions would be needed anymore. However, at the time of our survey, such plans were only pursued at Mech.1.

Digital technologies for supporting the training of employees were another field that, in many cases (Chem.1, 3, 4; Mech.3, 4; Health 1,2,3,6; Fin 3,4), experienced additional investment due to the need to maintain operations while avoiding direct physical contact. Some companies introduced or expanded digital learning platforms and vocational training units through video conferencing tools. Our quantitative survey confirmed the virtualization of learning processes as a major tendency. A total of 45 percent of the company representatives reported that continuous virtual training was either newly established or accelerated.

6.5. No Push for Shop-Floor-Related Investments

A striking observation at the surveyed companies concerned the absence of additional investments in the automation of physical processes. As automation can help to reduce social contacts (*'robots don't get the flu'*), it would be a likely scenario that the pandemic would also trigger investments in this field. However, this did not seem to be the case. Only five out of the 36 companies reported some additional activity with regard to shop-floor-related functions. Most companies in the automotive, chemical, mechanical engineering, and logistics industries, however, explicitly denied any correlation with the pandemic. There was a striking gap between the activities concerning remote work, IT infrastructure, and the digitalization of administrative functions and the introduction of shop-floor-related Industry 4.0 applications, which had not experienced acceleration. When companies pursued such strategic goals and projects, their pace was not greatly affected, and Auto.1 and Mech.4 even reported difficulties in pursuing them under the extraordinary circumstances of the pandemic. Industry-4.0-related applications that did experience more investments were tools for remote communication with customers for the installation of machines that were expanded at many mechanical engineering companies.

A possible explanation for the finding that there was little investment in shop-floor-related technologies is that investments in physical automation equipment need much more preparation, time and capital for their implementation than the mentioned measures for the virtualization of social interaction. They also require comprehensive adjustments of process and work organization on the shop floor, which, in many cases, had become more difficult under the conditions of the pandemic.

Moreover, the technological frontier is less permeable. As discussed in Section 2, the introduction of Industry 4.0 had progressed in a primarily incremental fashion before the pandemic. It requires long-term efforts and investment to overcome the remaining barriers to digitalization and automation, steps that often cannot be spontaneously undertaken. Most importantly, the concrete necessity to do so, apart from a general acknowledgement of the significance of digitalization strategies in general, was barely affected by the pandemic. Industry 4.0 technologies do not amount to a sweeping substitution of work to an extent that would be epidemiologically meaningful. The goal of a reduction in physical proximity out of health considerations, therefore, did not serve as a justification for automation investments in the investigated cases—especially as industrial enterprises tended to operate without major obstructions after the initial shock after the advent of the virus. Shop-floor-related digitalization projects have, thus, developed much more steadily than is the case with the dynamically developing fields of remote work, cloud computing, remote customer interaction, and the digitalization of administration and training activities.

7. Discussion

The analysis of 36 company case studies and the results from the quantitative survey in six economic sectors revealed that the pandemic affected the digitalization investments in companies in a strikingly uneven and cascaded manner. Where there was an immediate factual relationship between the pandemic and digitalization measures and where resources, technical feasibility, and the willingness of the main actors were given, the pandemic induced a push in investments. However, these conditions were not universally met, which is why there was a divergence in the digitalization patterns across our sample.

Despite this unevenness, some general tendencies could be identified as a qualification of the digitalization push in German enterprises in the six surveyed economic sectors. At its core undoubtedly lies the transition to remote work, which is not only about changes in the location of work, but also about the scaling up or addition of IT infrastructure, software applications, and new work routines. These changes seem to affect white-collar work almost universally and will continue to shape work practices after the pandemic, albeit in the form of a new synthesis between on-site and remote work. The identification of the standards and requirements that are needed in order to improve the work experience and the work–life balance of employees is of paramount importance for shaping the 'new normal'. This strong focus of digitalization activities on enabling mobile work calls into question whether it is even appropriate to interpret the effects of the COVID-19 crisis as a general push for digitalization. Accordingly, a recent survey on the matter concluded: "The so-called corona digitalization push can therefore not be called a comprehensive one. It mainly concerns processes such as interconnected work, the information exchange within companies, and digital linkages to other companies" [25].

This alludes to a second layer of process innovation that was pursued by some (but not all) enterprises in our samples: the introduction of specific applications for the virtualization of interactions with customers and between employees. The pandemic forced enterprises and organizations in which direct personal interaction with customers was frequent (in our sample, these were especially in financial services, healthcare, and mechanical engineering) to move such communication online—a measure that had already been pursued before the pandemic, but benefitted from an increased preparedness of actors to accept options for remote consultation. In most cases that concerned the virtualization of sales channels, however, this merely amounted to an ad hoc substitution of the conventional offline practices. Prospectively, however, they could support business model innovation (e.g., if a company modifies the product that it is offering) and changes in work organization (e.g., if special departments that exclusively work remotely are defined). Measures for virtualizing collaboration also characterized several areas of administrative work and activities concerning the training of employees. Goals for introducing the 'paperless office' that had been pursued in the past experienced a push as well, since organizations started to tackle remaining bottlenecks through improvisation, but also by substituting inefficient practices of the past. Just as in the general education system, e-learning experienced a surge in demand, and there has been a growing proficiency and acceptance among trainers and apprentices.

In contrast with these changes that mostly concerned cognitive work routines, the digitalization of shop-floor-related functions barely experienced progress. The dominant reply by company representatives on this issue was that the pandemic had barely affected existing digitalization projects in this area. In most cases, there was no factual relationship between the automation or digitalization of manufacturing and the pandemic, since most digitalization projects in this realm did not significantly alter the density of social interactions on the shop floor and were, therefore, irrelevant in terms of health considerations. Moreover, many applications that are subsumed under the term "Industry 4.0" do not aim at reducing the labor intensity of production, but rather at improvements in the interconnection and control of production facilities. They remain relevant for industrial companies, but the pace of investment was hardly altered through the pandemic. In this sense, the pandemic cannot be interpreted as a jump start for Industry 4.0. Investments rather concerned applications beyond the realm of manufacturing and underlined that present digitalization strategies, to a great extent, concern the automation, digitalization, and virtualization of cognitive work, especially in the administrations of companies. Exceptions to this rule were some applications for the remote setup of production equipment that were extended at some mechanical engineering companies.

However, the reasons for the lack of investment in manufacturing-related issues not only had to do with a missing *factual relationship* to the pandemic, but also with the longer time horizons of such investments. They usually cannot be instantaneously implemented because they require complex adjustments between physical and digital processes and the implementation and alignment of infrastructures (installation of sensors, construction of digital twins, and development of appropriate data analytics). This limitation is in stark contrast with the spontaneous ramping up of software and cloud capacities, which certainly constituted a challenge for the involved actors, but was feasible under the extraordinary conditions during the pandemic [2].

This correlation between pandemic-induced investment and the time horizon of digitalization projects is confirmed by other studies on recent events. An investigation of digitalization processes at mid-sized companies (the so-called "Mittelstand") observed a focus on short-term measures, while measures that require long-term preparations were often postponed [2]. A comprehensive monitoring effort initiated by the German Ministry of the Economy and Climate Changes similarly spoke of short-termism in investment. More comprehensive efforts with long-term effects, such as the innovation of products and business models, consequently received little attention during the pandemic [25].

8. Conclusions

This investigation departed from the question of whether the pandemic amounted to changes in the *quality* and the *direction* of the digital transformation in the investigated six sectors of the German economy. The results highlighted both continuity and a stronger focus on the automation, digitalization, and virtualization of white-collar work. On the one hand, the focus on short-termism and low-hanging fruits seemed to be in line with prior experiences. The narrative of a new industrial revolution belies the fact that the predominant mode of change had been incremental and focused on single applications with concrete returns, whereas systemic changes in processes and business models had been rather limited [12,26]. It could also be argued that changes that concern white-collar functions, i.e., the introduction of new software tools, collaboration through the cloud, and the partial automation of cognitive activities, had been a major axis of the contemporary wave of digitalization even before the pandemic. Consequently, the previously mentioned survey on digitalization activities by mid-sized companies during the pandemic concluded that neither the content nor the priorities of digitalization investment were significantly altered during the pandemic [2]. In this sense, the effects of the pandemic rather represent a discursive shift: It can be seen as a moment of reckoning that the virtualization of cognitive work and social interaction is at the core of many digitalization strategies. This suggests that the focus in the German "Industry 4.0" discourse, which is centered on manufacturing technologies, needs to be broadened in order to relate the transformation of manufacturing toward the mentioned changes in cognitive work in the service function of industrial (and non-industrial) companies.

On the other hand, the results of our study also include evidence that the pandemic constituted more than just a discursive shift and did bring new issues to the fore. At many companies, the introduction of schemes for mobile work and the intensified collaboration through the cloud, as well as virtual interactions with customers, constituted the entry into new modes of operation that implied more fluid work organization and more dense interactions. In most enterprises, some foundations for these steps were already present when the pandemic began. However, they often were of minor importance and did not amount to qualitative changes in the manner of collaboration. The shock of the pandemic induced not only a ramping up of capacities, but also a habitualization of cloud-based work routines, along with a set of new applications. The subsequent development is one in which new possibilities of virtual collaboration are constantly being explored and expanded. In this sense, our research highlights that the pandemic can be seen as the inflection point of cloud-based collaboration on a broader scale, which amounted to a major change in companies across our sample. This change certainly would have come without the pandemic, but this extraordinary event greatly affected the speed of this transformation and the subjective willingness to embrace it.

Our study highlights that a scientific understanding of the progression of digitalization and its effects requires one to differentiate among different dimensions and fields of application and to combine qualitative and quantitative methods in order to trace the digitalization trajectories in different economic sectors. Future research should extend such approaches to other economic sectors that were not covered by this study. Moreover, the digitalization approaches in different regions of the world should be systematically compared with sensitivity to institutional and cultural differences in order to uncover what could be called "varieties of digitalized capitalism".

Author Contributions: Conceptualization, F.B.; methodology, F.B., M.K., C.G., D.W. and J.F.; investigation, F.B., C.G., D.W., J.F., M.K., L.H. and N.D.; writing—original draft preparation, F.B.; writing—review and editing, F.B., M.K., D.W., C.G. and L.H.; visualization, F.B., L.H. and N.D.; project administration, F.B. and M.K.; funding acquisition, F.B. and M.K. All authors have read and agreed to the published version of the manuscript.

Funding: This research was funded by the German Ministry of Labour and Social Affairs.

Institutional Review Board Statement: Not applicable.

Informed Consent Statement: Informed consent was obtained from all subjects involved in the study.

Data Availability Statement: The data presented in this study are available on request from the corresponding author. The data are not publicly available due to privacy reasons.

Conflicts of Interest: The authors declare no conflict of interest.

References

1. LaBerge, L.; O'Toole, C.; Schneider, J.; Smaje, K. *How COVID-19 Has Pushed Companies over the Technology Tipping Point—And Transformed Business Forever*; McKinsey & Company: New York, NY, USA, 2020. Available online: https://www.mckinsey.com/business-functions/strategy-and-corporate-finance/our-insights/how-covid-19-has-pushed-companies-over-the-technology-tipping-point-and-transformed-business-forever (accessed on 19 May 2023).
2. KfW. KfW-Digitalisierungsbericht Mittelstand 2020. In *Rückgang der Digitalisierungsaktivitäten vor Corona, Ambivalente Entwicklung Während der Krise*; KfW Bankengruppe: Frankfurt am Main, Germany, 2021.
3. BMWK. Digitalisierung der Wirtschaft in Deutschland. In *Digitalisierungsindex 2021*; BMWK: Berlin, Germany, 2022.
4. Brynjolfsson, E.; McAfee, A. *The Second Machine Age: Work, Progress, and Prosperity in a Time of Brilliant Technologies*; W. W. Norton & Company: New York, NY, USA, 2014.
5. Sturgeon, T. *The "New" Digital Economy and Development*; UNCTAD: Geneva, Switzerland, 2017.
6. Zysman, J.; Kenney, M. The next phase in the digital revolution: Intelligent tools, platforms, growth, employment. *Commun. ACM* **2018**, *61*, 54–63. [CrossRef]
7. McAfee, A.; Brynjolfsson, E. *Machine, Platform, Crowd: Harnessing Our Digital Future*; Norton & Company: New York, NY, USA, 2017.
8. Zysman, J.; Murray, J.; Feldman, S.; Nielsen, N.C.; Kushida, K.E. Services with Everything: The ICT-Enabled Digital Transformation of Services. *SSRN Electron. J. BRIE Working Paper* **2011**. [CrossRef]
9. BMBF. Zukunftsbild "Industrie 4.0". Bundesministerium für Bildung und Forschung. Referat IT-Systeme. 2015. Available online: https://www.plattform-i40.de/IP/Redaktion/DE/Downloads/Publikation/zukunftsbild-industrie-4-0.pdf?__blob=publicationFile&v=4 (accessed on 19 May 2023).
10. Kagermann, H.; Helbig, J.; Hellinger, A.; Wahlster, W. *Recommendations for Implementing the Strategic Initiative INDUSTRIE 4.0: Securing the Future of German Manufacturing Industry*; Final Report of the Industrie 4.0 Working Group; Forschungsunion; Geschäftsstelle der Plattform Industrie 4.0: Berlin/Frankfurt a. M., Germany, 2013.
11. Sorge, A.; Streeck, W. Diversified quality production revisited: Its contribution to German socio-economic performance over time. *Socio-Econ. Rev.* **2018**, *16*, 587–612. [CrossRef]
12. Apitzsch, B.; Buss, K.-P.; Kuhlmann, M.; Weißmann, M.; Wolf, H. Arbeit in und an Digitalisierungen. Ein Resümee als Einführung. In *Digitalisierung und Arbeit: Triebkräfte—Arbeitsfolgen—Regulierung*; Buss, K.-P., Kuhlmann, M., Weißmann, M., Wolf, H., Apitzsch, B., Eds.; Campus Verlag: New York, NY, USA, 2021; pp. 9–39.
13. Hirsch-Kreinsen, H. *Industry 4.0—A Path-Dependent Innovation*; Soziologisches Arbeitspapier Nr. 56/20169; Wirtschafts-und Sozialwissenschaftliche Fakultät, TU Dortmund: Dortmund, Germany, 2019; pp. 1–25.
14. Krzywdzinski, M. Automation, digitalization, and changes in occupational structures in the automobile industry in Germany, Japan, and the United States: A brief history from the early 1990s until 2018. *Ind. Corp. Chang.* **2021**, *30*, 499–535. [CrossRef]
15. Butollo, F.; Jürgens, U.; Krzywdzinski, M. From Lean Production to Industrie 4.0: More Autonomy for Employees? In *Digitalization in Industry: Between Domination and Emancipation*; Meyer, U., Schaupp, S., Seibt, D., Eds.; Springer International Publishing: Berlin/Heidelberg, Germany, 2019; pp. 61–80. [CrossRef]
16. Marx, U. Maschinenbaugipfel. "Zehn verlorene Jahre". Frankfurter Allgemeine Zeitung. Available online: https://www.faz.net/aktuell/wirtschaft/unternehmen/gipfel-zum-maschinenbau-zehn-verlorene-jahre-18379560.html (accessed on 11 October 2022).
17. Heßler, M.; Thorade, N. Die Vierteilung der Vergangenheit. Eine Kritik des Begriffs Industrie 4.0. *TG Tech.* **2019**, *86*, 153–170. [CrossRef]
18. Mokyr, J. Chapter 17—Long-Term Economic Growth and the History of Technology. In *Handbook of Economic Growth*; Aghion, P., Durlauf, S.N., Eds.; Elsevier: Amsterdam, The Netherlands, 2005; Volume 1, pp. 1113–1180. [CrossRef]

19. Horváth, D.; Szabó, R.Z. Driving forces and barriers of Industry 4.0: Do multinational and small and medium-sized companies have equal opportunities? *Technol. Forecast. Soc. Chang.* **2019**, *146*, 119–132. [CrossRef]
20. Lammers, T.; Tomidei, L.; Trianni, A. Towards a Novel Framework of Barriers and Drivers for Digital Transformation in Industrial Supply Chains. In Proceedings of the 2019 Portland International Conference on Management of Engineering and Technology (PICMET), Portland, OR, USA, 25–29 August 2019; pp. 1–6. [CrossRef]
21. Baldwin, R.; Tomiura, E. Thinkin. ahead about the trade impact of COVID-19. In *Economics in the Time of COVID-19*; Baldwin, R., Weder di Mauro, B., Eds.; CEPR Press: London, UK, 2020; pp. 59–72. Available online: https://cepr.org/sites/default/files/news/COVID-19.pdf (accessed on 8 February 2021).
22. Archibugi, D.; Filippetti, A.; Frenz, M. Economi. crisis and innovation: Is destruction prevailing over accumulation? *Res. Policy* **2013**, *42*, 303–314. [CrossRef]
23. Louçã, F. Long Waves, the Pulsation of Modern Capitalism. In *Chapters*; Edward Elgar Publishing: Cheltenham, UK, 2007; Available online: https://ideas.repec.org/h/elg/eechap/2973_48.html (accessed on 8 February 2021).
24. Peichl, A.; Sauer, S.; Wohlrabe, K. Fachkräftemangel in Deutschland und Europa—Historie, Status quo und was getan werden muss. *Ifo Schnelld.* **2022**, *75*, 70–75.
25. DigiIndex. Digitalisierungsindex. Bundesministerium für Wirtschaft und Klimaschutz. Available online: https://www.de.digital/DIGITAL/Navigation/DE/Lagebild/Digitalisierungsindex/digitalisierungsindex.html (accessed on 11 October 2022).
26. Hirsch-Kreinsen, H. *Digitale Transformation von Arbeit. Entwicklungstrends und Gestaltungsansätze*; Kohlhammer: Stuttgart, Germany, 2020.

Review

Towards the Next Decade of Industrie 4.0 – Current State in Research and Adoption and Promising Development Paths from a German Perspective

Johannes Winter , Anna Frey and Jan Biehler *

acatech—National Academy of Science and Engineering, 80333 Munich, Germany
* Correspondence: biehler@acatech.de

Abstract: Considering the first ten years of Industrie 4.0 in Germany—the digital transformation of industry towards the goal of increased manufacturing productivity and mass customization—significant progress has been achieved. However, future efforts are required. This review first evaluates the status quo of implementation and research in Germany and finds that large-scale companies have proceeded faster than small- and middle-sized enterprises. Currently, regardless of their size, companies have in common a shortage of qualified specialists, coupled with a lack of adequate base technologies for Industrie 4.0 and an insufficient digital mindset. The creation of platform-based digital business models is particularly lagging behind, despite high research interest. This review subsequently identifies three research-driven fields of action that are particularly important for the future of Industrie 4.0: (1) resilience of value networks in the strategic area of sovereignty, (2) Open-Source as a driver for the strategic area of interoperability, and (3) the strategic combination of digitalization and sustainability as a basis for sustainable business models in the strategic area of sustainability.

Keywords: Industrie 4.0; smart manufacturing; smart factory; digital transformation; industry; sustainability; sovereignty; interoperability; mass customization

Citation: Winter, J.; Frey, A.; Biehler, J. Towards the Next Decade of Industrie 4.0 – Current State in Research and Adoption and Promising Development Paths from a German Perspective. *Sci* **2022**, *4*, 31. https://doi.org/10.3390/sci4030031

Academic Editors: Luis Norberto López De Lacalle, Sharifu Ura and Claus Jacob

Received: 25 May 2022
Accepted: 9 August 2022
Published: 11 August 2022

Publisher's Note: MDPI stays neutral with regard to jurisdictional claims in published maps and institutional affiliations.

1. Introduction

Companies have always been interested in meeting customer needs, increasing customer loyalty, improving manufacturing efficiency and flexibility, and increasing sales through improved and novel products and services [1,2]. Ten years ago, researchers from computer science, engineering, economics, and related scientific disciplines, as well as representatives from industry and the service sector, presented a concept for the future of industrial value creation and customer-specific mass production that resonated globally. Under the leadership of Prof. Henning Kagermann, Prof. Wolfgang Wahlster, and Prof. Wolf-Dieter Lukas, a conceptual framework for the fourth industrial revolution was presented at the Hannover Fair 2011 [3], which has had a lasting impact on research in the areas of product development and systems engineering, production technology, and production systems, as well as in business process management and business model innovation [4]. Companies stepped up their efforts to automate and network production systems and value chains to advance the goal of higher productivity through a smart factory approach and to progress toward mass customization through digitally enhanced, tailor-made products and services at the price of a mass-produced product [5].

From a strategic and global perspective, significant progress has been achieved in this regard [6]. However, to date, it largely remains unclear how far research and adoption has proceeded country-wise and which focus points are set in the second decade of Industrie 4.0. This also holds for Germany, where the industrial sector plays a pivotal role, and where Industrie 4.0 had been developed as a concept. This review fills this research gap by following this guiding question: What is the status quo of adoption and research in Industrie 4.0 in Germany, and which fields of action require future attention?

We find that German industry has proceeded significantly in adapting to Industrie 4.0 whereby large-scale enterprises are more mature than small- and medium-scale enterprises. Regarding research and Research and Development (R&D) across the four key themes *Value Creation Scenarios for Industrie 4.0*, *Prospective Technological Trends*, *New Methods and Tools for Industrie 4.0*, and *Work and Society*, significant insights have been gained. However, in the face of increasing volatility and rapid changes in the development of Industrie 4.0, more interdisciplinary research is required across all research fields. Regarding the three strategic areas defined by German government for the next decade of Industrie 4.0 in Germany, we find that (1) resilience of value networks in the strategic area of sovereignty, (2) Open-Source as a driver for the strategic area of interoperability, and (3) the strategic combinations of digitalization and sustainability as a basis for sustainable business models in the strategic area of sustainability deserve attention in the fields of industry and research to deepen the industrial transformation and to stay competitive from the perspective of German researchers and company representatives.

These descriptive findings add to country-specific research on Industrie 4.0 in Germany and are a valuable source for cross-country comparisons of the status quo of adoption and research. For practitioners, the findings help to identify future fields of actions and to deepen adoption of Industrie 4.0. Thereby, the review contributes to the academic and practical discussion of Industrie 4.0 not only in Germany, but also globally.

2. Methodology

By analyzing what is needed for the further dissemination and implementation of Industrie 4.0, this review firstly details the current status quo in implementation, scientific research, and R&D across the four key themes of Industrie 4.0 in German research: *Value Creation Scenarios for Industrie 4.0*, *Prospective Technological Trends*, *New Methods and Tools for Industrie 4.0*, and *Work and Society* [7].

Therefore, we rely on an efficient literature review of sources distributed across different academic disciplines and popular literature [8,9]. To cover relevant publications across engineering, production, and management, we relied on three publication databases (EBSCOhost, Emerald Insight and Google Scholar). We were searching for the terms "Industrie 4.0" and "Industry 4.0" in combination with "Germany" and "Deutschland". The title, abstracts, and keywords of the first 50 results for each search term were analyzed by two researchers independently to ensure reliability. Only those search results with a clear reference to "Industrie 4.0" were considered as relevant. For these, a forward and backward search was conducted, focusing on whether new research items tackled the implementation status of Industrie 4.0 in Germany and/or one or some of the four key themes of research, and excluding those research items that did not deliver knowledge to this research interest.

In addition, we applied a qualitative discourse analysis [10] of primary and secondary data from expert interviews originating from research projects conducted with the participation of the authors on Industrie 4.0 and in adjacent topic areas, such as Artificial Intelligence, Smart Services, or Advanced Systems Engineering, among others, to identify current and future research issues and to derive evidence from these expert interviews regarding adoption and the four aforementioned key themes of research on Industrie 4.0.

Secondly, building on the matrix method of literature review [11], we discussed one field of action each in the three strategic areas that will drive the next decade of Industrie 4.0: *sustainability* ("digitainabilized" platform-based business models), *sovereignty* (resilient value networks) and *interoperability* (Open-Source) [12].

Here, given the three strategic areas of the vision 2030 of the German Plattform Industrie 4.0, building on the efficient literature review from the first step, we searched for the terms "sustainability", "sovereignty", and "interoperability", with two researchers working independently in three publication databases (EBSCOhost, Emerald Insight, and Google Scholar) to identify fields of action requiring specific attention to drive Industrie

4.0 to success. Therefore, implications for research and practice raised in the research items received keywords and a prioritization. The two researchers then aggregated the assigned keywords and prioritization and discussed those cases with discrepancies. The three aforementioned fields of action with the highest prioritization were identified, and evidence acquired in the literature review was matched on these fields of action to round off this review on the status and outlook for Industrie 4.0 adoption and research in Germany.

However, the focused scope of the review should be taken into account when reading the review. The key themes of the research on Industrie 4.0 [7] and the strategic areas in the vision 2030 [12] were applied to structure the review. On the one hand, this approach may lead to (international) developments outside these guidelines being disregarded. On the other hand, this review gives evidence for researchers and practitioners to develop measures to strengthen Industrie 4.0 in accordance with the politically set strategy in Germany.

3. Status Quo: 10 Years of Industrie 4.0 in Practice and Research in Germany

The industry sector—predominantly comprising the automotive industry, the electrical and chemical industry, mechanical and plant engineering, and metal production—plays a pivotal role for the German economy. With a contribution to German gross value added (GVA) of 20.2% in 2021 [13], its contribution is almost twice as high as that of competitors on the world market, such as France, the United Kingdom, or the United States and is only matched among the biggest economies of the world, such as Japan, China, and South Korea [14,15]. Industry is a significant driver of innovation, with almost 60% of total German Research & Development (R&D) spending invested in this sector [16,17]. This indicates the relevance of the industrial sector for the German economy.

3.1. Development of German Industry since 2011 and Outlook on Industry Strategy until 2030

Since Industrie 4.0 was propagated as the vision for the digital transformation of German industry in 2011—aiming at connecting resources, services, and humans in real time throughout production on the basis of Cyber Physical Systems (CPS) and the Internet of Things (IoT) [3]—the gross value added of German industry increased in absolute terms, while its proportion relative to the GVA of German economy remained stable, in comparison to a decline in other countries [15,18]. German industry even experienced an internationally unmatched phase of "reindustrialization" until 2016, with increasing employee numbers, an increasing proportion of GVA, and higher productivity gains compared to other sectors, before the COVID-19 crisis and interrelated supply chain problems initiated a phase of deterioration in German industry, which was also due to its high dependency from world markets originating in its export orientation [19].

The deepening adaption towards Industrie 4.0 correlated with the continuing strength of industry as a value driver of the German economy, but could not prevent the phase of downturn during the pandemic. This phase is deemed to be temporary, due to two factors: First, extensive investments made by companies in the development of digital technologies and competencies in the first ten years of Industrie 4.0 have not amortized yet, because they are not easily quantifiable in terms of return on investment, as it is mostly intangible assets and human capital that have been targets of investment so far [19] (pp. 195–197). Second, after these significant digitalization steps in the first phase of Industrie 4.0, industry has appeared to "take a breath" regarding innovative value-driving projects in the run-up to the next wave of industrial digitalization [20], which will likely focus on industrial artificial intelligence (AI) and the creation of digital platform-based business models [6,21,22].

Put together, in the first ten years, German industry invested in shaping the breeding ground of Industrie 4.0, the base layer of digital infrastructure, architecture, products, and competences. To exploit this breeding ground—the cumulated value creation potential of Industrie 4.0 for German industry is estimated at 425 billion Euros until 2025 [23]—the German government set the ambitious goal in its Industrial Strategy 2030 of increasing the contribution of industry to GVA to 25% in 2030 (an increase of roughly five percentage points compared to 2021) [16]. After ten years of Industrie 4.0, the German government's goals thus incentivize an acceleration of research and adoption of the changes to strengthen value creation, scalation, and sustainable long-term orientation of German industry.

3.2. Status Quo: Adoption of Industrie 4.0 in German Industry

To assess whether German industry is in a position to achieve these strategic goals by deepening its digital transformation, this section provides a status quo overview on the adoption of Industrie 4.0.

In practice, Industrie 4.0 is increasingly perceived as a competitive advantage by companies in the industrial sector [7]. The call for the digital transformation of industry towards the paradigm of smart factory has received significant attention in economic circles and has also holistically incentivized research [6,7]. Capitalizing on the global market leadership of German industry, its good network with customers, high competencies in creating individual solutions, and the strength and flexibility of small and medium-scale enterprises, many companies have made significant progress regarding digitalization in the last ten years [6,24]. However, there are industry-wide internal and external barriers to the implementation of Industrie 4.0. These barriers contribute to a slanted picture regarding the current adoption of Industrie 4.0 by German industry enterprises [24,25].

One dividing line occurs between large-scale enterprises and small and medium-scale enterprises (SMEs), whose digital transformation has proceeded at a slower rate. A recent study found that some SMEs in a sound financial position are not incentivized by competitive pressure to initiate digitalization (laziness–success paradox). Correspondingly, they operate with a lack of strategic expertise and a lack of digital enterprise culture, which preempts the drive toward digitalization of core processes [24]. Cost pressure is also higher for smaller companies. It nudges them to reject digital solutions if they are not perceived as cost-effective improvements over the analogous status quo. Digital projects are among those abandoned first if SMEs face financial problems; those projects do not contribute immediately to value creation, and when they are implemented, they often serve an immediate goal in the production chain rather than strengthening basic industrial infrastructure.

As a consequence, compared with large-scale enterprises, SMEs are worse off regarding data management, which is a prerequisite for the next decade of Industrie 4.0. If SMEs apply AI solutions, they do so primarily for their products and services, and only very seldomly for improving internal processes [26]. This is due to a lack of know-how and an insufficient data base regarding internal processes, resulting from rudimental data management. Thus, in the current adoption rates, SMEs are lagging behind large-scale enterprises as industry proceeds into the second decade of Industrie 4.0.

In general, large-scale enterprises are significantly more mature regarding their progress in digitalization [25]. Only a few have not yet started digitalization, often due to a lack of enthusiasm of top management in initiating the transformation, as qualitative focus interviews indicate [24]. However, these companies are endangering their competitiveness and are exceptions. Large-scale enterprises rather face other barriers occurring in later stages in the process of digitalization; legal restrictions, data security and privacy concerns, and uncertainty regarding the expansion of digitalization activities hinder a deepening of digitalization [24].

These factors are likely to also affect SMEs in the later stages of digitalization and complement other general factors inhibiting digitalization, irrespective of company size. The insufficient digital infrastructure in Germany—Germany has the fourth-lowest IT-investment to GDP ratio within the EU [27]—and structural disadvantages in Northern and Eastern Germany are major barriers already hindering the initiation of digital projects [19].

If these internal and external barriers do not hinder digitalization ambitions, the disproportionate shortage of qualified employees in German industry does [15,28]. Without the required knowledge, successful digitalization projects that create network effects and attract follow-up investments are less likely [29,30]. Additionally, asymmetry in industrial value chains—especially in the automotive industry—hinders proactive and targeted digitalization. Companies that are highly dependent on bigger suppliers or customers in the value chain have little control over measures implemented and need to react to the digitalization requirements of those partners. This aggravates digitalization barriers, especially for SMEs who have fewer resources to facilitate such responses.

Brought together, just as industry is proceeding into the second decade of Industrie 4.0, SMEs are lagging behind large-scale enterprises (see reasons in Table 1). In general, digitalization expertise that does not meet the standards of domain expertise in German industry, insufficient external digital infrastructure, lacking creativity and courage for thinking in new business models and problems in becoming agile [24,31] (p. 22) often inhibit the adoption of Industrie 4.0 in Germany.

Table 1. Hindering factors of Industrie 4.0 in Germany per company type (Own Illustration).

Small- and medium-scale enterprises	Laziness–success paradox Lack of strategic expertise Lacking digital enterprise culture Cost pressure Insufficient digital infrastructure Shortage of qualified employees Dependency on bigger suppliers or customers in value chain
Large-scale enterprise	Lack of enthusiasm in top level management Legal restrictions (later stages) Data security (later stages) Privacy concerns (later stages) Insufficient digital infrastructure Dependency on bigger suppliers or customers in value chain

3.3. Status Quo: Industrie 4.0 in Germany from a Research Perspective

Hence, evidence-based targeted improvements in the four key themes of Industrie 4.0 [7] are required. Therefore, a more detailed assessment of the status quo setting a focus on the need for research and development in Germany is required.

3.3.1. Research Theme: Value Creation Scenarios

Regarding *Value Creation Scenarios* for Industrie 4.0, platform-based business models for industry are considered as among the most promising fields of action for Industrie 4.0 [32]. Until now, especially in the B2C sector, they have not been widely adapted [33]. Especially for SMEs, market penetration seems to be inhibited by high investment costs, challenges in determining concrete benefits, and insufficient data quality originating from the aforementioned rudimentary data management [25]. Thus, only the first steps towards digital business models have been made in many domains so far [25,32].

Correspondingly, industry-wide innovation environments that build the base layer for platform-based business models are correspondingly under-researched [24]. Developing global digital ecosystems compliant with European data protection standards that could be based on Open-Source architecture promise to drive research and implementation in this area of action [34,35]. However, to develop and further analyze value creation scenarios for Industrie 4.0, the basic technologies and infrastructure need to be put into practice.

3.3.2. Research Theme: Prospective Technological Trends

Considering *Prospective Technological Trends*, blind spots in implementation und domestic production can be identified relating to technologies with the potential to strengthen the sovereignty of German industry [24,36,37]. Considering the volatile geo-economic and geopolitical status quo, resilient ecosystems ensuring safe, secure control and self-determination for companies with regard to their data require more attention. This includes trialing certification, norming, and standardization, with the aim of creating a resilient basic infrastructure. In addition, it is important to ensure the availability of key technologies for

Industrie 4.0 (e.g., 5G) to safeguard and stabilize value chains [38]. Such a resilient infrastructure is required for the next evolutive step of Industrie 4.0 towards the smart factory, where machines, components, materials, and people exchange information seamlessly, in an automated fashion, and in real time in order to adapt production optimally to changing conditions in a self-learning manner [39,40].

In addition to the roll-out of digital twins, this requires, above all, the integration of artificial intelligence (AI) in the processes, products, and services of industry [41]. This would allow for the automation of production, the control of industrial drives (motion control), demand-driven maintenance of machinery, process optimization, and quality assurance through non-destructive testing, among other advances [42]. To implement AI industry-wide in factories, groundwork by research and in practice in the other key themes of Industrie 4.0 is required.

3.3.3. Research Theme: New Methods and Tools for Industrie 4.0

Regarding *New Methods and Tools for Industrie 4.0*, models and simulations for the AI-driven smart factory are required. Although models and simulations meeting the needs of the cyber-physical matrix systems already exist (e.g., digital twins) [36], research has not made sufficient advancements regarding these for the highest expansion stage of the AI-driven smart factory [25,39]. Here, a lack of connection of different business models with data and data analytics hinders advancements. Consequently, on a higher complexity level, coupled, hybrid, and holistic models are required [43].

Regarding implementation, even digital twins require industry-wide roll-out [44]. The lack of clarity among companies regarding the quantification of the benefits of digital projects for business operations could explain why they have not been rolled out so far [24]. KPI-driven assessments measuring indirect and direct effects of digital projects on value creation that go beyond traditional cost-benefit analysis need to be developed by academic and practice-oriented research [24].

3.3.4. Research Theme: Work and Society

Considering the fact that Industrie 4.0 is significantly interrelated with transformations in the workforce, *Work and Society* as the fourth key theme of Industrie 4.0 deserves attention as well. In the existing workforce—17% of German employees worked in the industrial sector in 2018 [19] (p. 193)—most of the employees are in wait-and-see mode and are cautious regarding Industrie 4.0. They are mostly indifferent, do not drive its implementation proactively, and are expected to react negatively toward the digital transformation of industry if its value added does not become visible, if employers do not communicate transparently about transformation steps, and if employees have the feeling that technology is being applied to control their work [29,30].

Digital transformation is accompanied by increased demand for qualified specialists with domain knowledge and information and communications technology (ICT) expertise. However, employees with the required skills to manage the transformation towards the smart factory are scarce. In 2021, more than every second company in industry sector is lacking qualified specialists in the long run [28]. Required soft skills include an interdisciplinary mindset, process know-how, leadership skills, a change and problem-solving mentality, and an orientation towards self-management. Hard skills are required in data analytics, processes, and customer-relationship-management, as well as abilities regarding IT security and IT business analytics and specific know-how about IT systems [29]. Additionally, AI-related competences are required (e.g., regarding human–machine interaction, basic knowledge about machine learning, and awareness regarding AI) [41,45].

In sum, our literature review indicates that the four key themes of Industrie 4.0 receive high attention in scientific research and in practice-oriented research, although more focus is required for interdisciplinary research in the face of accelerating technological progress [6]. The Research Council of the Plattform Industrie 4.0 (Forschungsbeirat der Plattform Industrie 4.0: advisory body of high-level experts from science and industry advising the Plattform Industrie 4.0) is continuously monitoring research progress and potentials, and defines potentials for research projects in those areas deserving more attention. This promises an evidence-driven advancement of Industrie 4.0. Complemented with academic research on Industrie 4.0, the research community leverages innovation by creating incentives for companies to fund collaborative research and by providing guidance for (public) funding programs. However, there remains future research potential across all key themes, especially in the face of the next decade of Industrie 4.0 facing a future full of ambiguities, potentials, and threats in implementing Industrie 4.0 [46].

4. Fields of Action for the Next Decade of Industrie 4.0 from a German Perspective

Against this background, German government, informed by the Research Council of the Plattform Industrie 4.0 and the Plattform Industrie 4.0 itself, has recognized the need for a strategic development of Industrie 4.0 for the next decade. It set the ambitious goal in its Industrial Strategy 2030 of increasing the share of German industry in gross value added in the economy to 25% by 2030—an increase of five percentage points [16]. Industrie 4.0 promises to be an important driver in this regard, with its target vision of the AI-driven smart factory and digital, platform-based business models. To leverage the potential of Industrie 4.0 in view of the economic, ecological, and geopolitical challenges of our time, and to follow a holistic approach for creating open digital ecosystems, strategic focus in Germany is set on three areas: sovereignty, interoperability, and sustainability [12].

Despite having been central to Industrie 4.0 already in the first ten years, German policy makers have recognized the need to intensify efforts in these areas in order to position German industry competitively in the market in the face of increasing political, ecological, and economic ambiguities [12]. Based on the matrix review assessment, one promising field of action was identified for each strategic area of the vision 2030 of the Plattform Industrie 4.0 (see descriptive Figure 1), as discussed in the following. For each field of action, an exemplary best case is presented that shows possible implementation paths.

Strategic Areas	Sovereignty	Interoperability	Sustainability
Fields of Action	Resilient Value Networks	Open Source	„Digitainable" (Digital-Sustainable) Platform-based Business Models

Figure 1. Strategic Areas [8] and Promising Fields of Action for Industrie 4.0 (Own Illustration).

4.1. Sovereignty

Globalization has caused high interdependencies in global value chains. German industry is highly dependent on raw materials, predominantly from the Far East (e.g., China). The digitalization of industry creates new dependencies, predominantly from the US, where most of the technology companies delivering services for German industry (e.g., Microsoft with Azure, Amazon with AWS, or Salesforce and Oracle with its services) are located. Under the presidency of Donald J. Trump, the U.S. gradually turned away from Europe as its primary geostrategic and economic partner [47]. Even though it might have been a temporary phenomenon, it highlighted—together with the "trade war" between China and the US—the importance of technological sovereignty for the EU and Germany in terms of economic policy [48]. This has been recognized by policymakers, e.g., in the EU White Paper on Artificial Intelligence [49], which emphasizes the importance of economic sovereignty in general and in particular for Industrie 4.0 with the vision 2030 of the German Plattform Industrie 4.0 [12]. Whereas technological sovereignty demands government action to ensure independence on a macroeconomic level, on a microeconomic level, individual companies need to operationalize this strategic directive by increasing their ability "to cope with shocks of different sorts" [50] (p. 5) by becoming resilient.

The resilience of value networks requires specific attention as the COVID-19 pandemic has unveiled the dependency of German industry on global value chains [51]. Supply chain problems have led to a widespread production standstill causing a phase of deterioration in German industry. The shortages of semiconductor electronics and cable harnesses, as well as the Russian invasion in Ukraine causing oil and electricity price explosions, are affecting companies significantly. With these factors put together, German industry companies are not resilient in facing crises and volatility on the world market [52]. To increase the resilience of its value networks, German industry requires a strategy coupling robustness, to reduce the effects of volatility on company performance, with agility, in order to respond quickly to unpredictable disruptive events. Three measures to increase resilience can be identified in this regard: (1) the adoption of value creation networks, (2) safe and secure data integration, and (3) reliance on Industrie 4.0 technologies [53].

The paradigm of single sourcing coupled with just-in-time delivery has caused a production standstill during the COVID-19 pandemic due to the strict lockdowns in divergent parts of the world over time. German industry is still dependent on global value chains, as 43% of Germany's net exports are only generated due to its embedding in global value chains [54]. This dependency needs to be reduced.

A short-term measure in this regard is a mitigation of the just-in-time paradigm by maintaining larger inventories. In the event of a catastrophe—for example, in the event of economic lockdown measures in one production country—it would be possible to quickly switch to capacities in other production countries. This mitigates the supply-side risk by diversifying production capacities, but leads to higher costs.

A medium-term approach is a local-for-local strategy, i.e., locating production and the upstream value chain in the sales market. By synchronizing supply- and demand-oriented risks in sales markets, only the respective regional clusters are affected in the event of a disaster. This concentration effect mitigates the risk of supply chain disruptions. At the same time, resources are duplicated to a much greater extent, causing efficiency losses. These solution approaches provide more resilience, but are not able to counter the increasing uncertainty and pressure for change in the market environment.

Thus, in the long run, the monolithic and highly optimized value chains of the present could be gradually replaced by resilient and adaptive value creation networks [53]: multiple sourcing based on regional local for local strategies, the creation of redundancies to allow for flexible reaction on crises, and a diversification and continuous evaluation of the supplier base as part of risk and sustainability management.

Based on the paradigm of co-creation, value creation networks require data infrastructure that allows companies to collaborate seamlessly in order to develop and strengthen resilient business models that, on a macroeconomic level, promise to increase digital sovereignty [21,55]. The German Plattform Industrie 4.0 drives such an attempt with its European Data Space Industrie 4.0, building a data infrastructure for companies from German industry and the ICT sector (see Figure 2).

These data spaces aim at not only increasing digital sovereignty, but also data sovereignty, enabling the secure, decentralized, and self-determined exchange, collection, analysis, and sharing of data. Thereby, they form the basis for new, highly scalable, data-driven, and platform-based business models—one of the focus points in the next decade of Industrie 4.0 [1–3,22].

Next to the establishment of value creation networks, data integration is a prerequisite for a resilient Industrie 4.0. Our analysis indicates that this is currently not achieved in practice, as SMEs especially face problems regarding data management [24,25,43]. They require support and innovative data ecosystem. Data integration comprises, in this regard, the use of platforms where standardized data exchange is enabled. Real-time monitoring of procurement, logistics, production, and sales processes allows for optimization of processes, and the created data could also be integrated into data exchange platforms and could train self-learning AI models for an industry-wide optimization of production.

Therefore, the industry-wide implementation of Industrie 4.0 technologies is key. Software-based risk management is required for resilient and adaptive value creation. Big data analytics and artificial intelligence are the basic technologies for reaching the paradigm of the *Smart Factory* [26]. Digital twins or digital shadows require industry-wide penetration [39,44]. Only if industry applies those technologies can it future-proof for the next decade of Industrie 4.0.

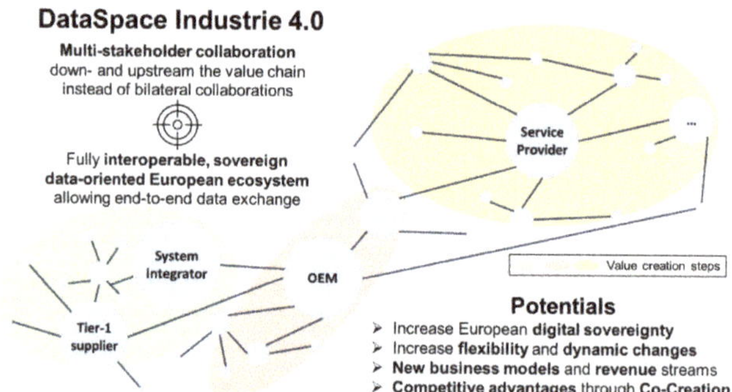

Figure 2. DataSpace Industrie 4.0—Initiative for a sovereign European industrial data ecosystem (own illustration based on Plattform Industrie 4.0 [56]).

4.2. Interoperability

Industrie 4.0, postulating the AI-driven smart factory and platform-based business models, requires the establishment of an innovative ecosystem where companies and customers interact seamlessly. To reduce transaction costs, interoperability is key. In the first ten years of Industrie 4.0, significant initiatives were started. The GAIA-X project aims at developing a safe and interconnected data infrastructure for a European digital ecosystem with industry sitting at its heart setting policy rules, guidelines, and a standard architecture framework contributing to cross-sector interoperability. The launch of RAMI 4.0 as an international industry-specific standardization and reference architecture for Industrie 4.0, the coordination of the standardization of Industrie 4.0 by the Standardization Council Industrie 4.0, the establishment of OPC UA as a standard for interoperable interfaces for AI-driven production, and the launch of the administration asset shell have improved industry-specific interoperability predominantly by means of standardization [57,58].

For the next decade of Industrie 4.0, Open-Source will play a pivotal role in implementing and exploiting interoperability for the creation of new platform-based business models [34,35]. Already at the beginning of the 2020s, Open-Source Software (OSS) contributes between €65 billion to €95 billion to the EU's GDP (equaling EU-GDP of air and water transport combined) and promises significant growth opportunities in the future, especially for Industrie 4.0 [59]. OSS could even be applied to enhance interoperability, as demonstrated by BaSyx-middleware (see Figure 3).

Open Source Industry 4.0 middleware

- **Free and collaborative** software platform
- All necessary **standards and infrastructure components** for Industry 4.0 production environments
- **Reference technology platform**: foundation for software platforms
- Built on Eclipse: **SDKs for Java, C++, C#**

Figure 3. BaSyx–Open-Source Middleware (Own Illustration).

Currently, economy-wide motivations for participating in OSS mainly lie in the need to find technical solutions, avoid vendor lock-in—a factor that is also relevant for technological sovereignty—and build knowledge in high-quality coding. At the same time, as OSS builds on voluntary contributors, its cost–benefitratio is a highly promising driver for value creation: for every Euro invested, a €12 value creation is achieved [59]. This is a significant lever, especially for SMEs in the industry, given the cost pressures they are facing.

However, to exploit this lever across the breadth of the manufacturing sector, further advancements for OSS from a technological, regulatory, and innovative perspective are required. Within companies, Open-Source needs to be integrated as an elementary step in value creation [34]. This begins with the development and the communication of an Open-Source strategy developed by an interdisciplinary team creating acceptance and support in the workforce from scratch onward. It should be embedded in a top-level vision and strategy, which could imply the application of OS to create standards, to boost business model strategies, or to gain competencies in core technologies.

Once initiated, an Open-Source Program Office (OSPO) should boost implementation. As the central business unit responsible for Open Source (OS) processes, the OSPO is responsible for quality management, checking license compatibilities and Open Source Sofware (OSS) integrity with Intellectual Property, and importantly, also for the qualification of employees regarding OSS [17]. Senior management in this phase needs to ensure the establishment of an OS mindset in the workforce alongside the selection of communities and ecosystems fitting to OS properties.

Additionally, the concrete areas of action of OS need to be crosschecked continuously. It is recommended to only develop OS for non-IP-relevant products, processes, or code, which should not be attached to the primary value proposition of a company. If the latter did occur, a company would cannibalize its market position. This indicates the need for continuous adaptation of and reflection on business models.

In industrial policy, Open-Source has played an important role. However, European governments have taken a more laissez-faire approach, and today, the EU is on the back foot when it comes to capabilities in this area. Hence, it is recommended also for the public sector to establish OSPOs, to publicly fund R&D projects related to Open-Source more intensively, to support entrepreneurial activities around OSS, and to build a European ecosystem around OSS, among other recommendations [59].

4.3. Sustainability

The need for sustainable economizing is more important than ever in view of aggravating global socioeconomic inequality and many tipping points that will be reached in the context of the climate crisis. German industry is the second biggest polluter of greenhouse gas (GHG) emissions beyond the energy sector [60]. Thus, German government has set ambitious goals to reduce the ecological footprint of the industrial sector. Greenhouse gas neutrality should be achieved by 2045 [61].

Industrie 4.0 is pivotal in this regard [62]. The smart factory standalone promises to reduce emissions and increase energy efficiency by optimizing resource need and consumption of production [63]. However, rebound effects, such as overproduction and overconsumption caused by a more efficient smart factory, could even increase the ecological footprint of German industry. The same holds for artificial intelligence as a driver for the smart factory and platform-based business models. According to a recent analysis [64], AI contributes positively to 139 of 169 indicators of the 19 UN Sustainable Development Goals (UN SDGs)–the central reference point of a holistic comprehension of sustainability–whereas it affects 59 of those 169 indicators negatively at the same time.

Thus, to exploit the positive effects of ICT while mitigating its negative externalities on sustainability, a strategic combination of digitalization and sustainability is required to create win–win–win scenarios for economy, ecology, and society [65] (pp. 64–70); 'Digitainability: [65,66], as a strategic perspective on Industrie 4.0 based on circular economizing and sustainable business models, is the vision in this regard [67].

Building on the approach of an integrated sustainability strategy for AI [65,68], an integrated strategy framework of Industrie 4.0 and sustainability could support companies in becoming future proof for the next decade of Industrie 4.0. Similarly to the integrated sustainability strategy for AI [68] and the adoption guide for SMEs to AI [69], an integrated sustainability strategy for Industrie 4.0 could comprise the five stages of Status & Stakeholder Assessment, Sustainability Target Setting, Impact Check, Adaption of Technological strategy (AI and other ICTs), and Implementation and Continuation [68].

The *Status & Stakeholder Assessment*, as the first step in entrepreneurial transformation towards future-proofing and sustainable economization demands, the evaluation of the current value chain [70], stakeholder analyses, and assessments of the impact of ICTs on a company's sustainability, identifies those critical aspects that require attention in sustainability management.

Derived from this, *Sustainability Target Setting* includes operationalizing a company's sustainability vision in measurable dimensions. These dimensions include the three aspects of economic, ecological, and social sustainability in addition to ESG criteria [68].

This is followed by a concrete *Impact Check* to assess where action is needed in the creation and design of products, services, or processes in the company, and how measures need to be designed so that they contribute to these sustainability targets. The UN SDGs are a productive reference point in this regard.

What follows is the merging of the sustainability strategy with the digitalization strategy of a company in the stage of *Adaption*. Here, companies need to apply ICTs only in those areas where they are really creating benefits based on the current adaption status of ICTs. They need to assess how their products, processes or services can be optimized, or (and) how ne sustainable business models based on AI and other ICTs could be developed.

In the stage of *Implementation and Continuation*, it is important to create a 'digitainabilized' mindset within an industrial company. This includes employee-related measures regarding the use of digital technologies in the company and the permanent integration of the company's individual sustainability goals into the company organization, as well as the propagation of a change-oriented and agile orientation in implementation [68].

Strategically merging sustainability and AI-driven Industrie 4.0 offers potential for companies to create new and resilient value creation networks or to even design sustainable platform-based business models that promise to increase and solidify revenue streams for industrial companies and in affiliated sectors (e.g., agricultural machinery technology; see Figure 4) [62]. Smart Services are important drivers in this regard [71].

AI-driven Smart Farming Services Platform

- Optimization of agricultural processes with platform-based Smart Services
- Sustainability increase: optimized resource efficiency
- Profitability increase: optimized food production

... Embedded in value creation network

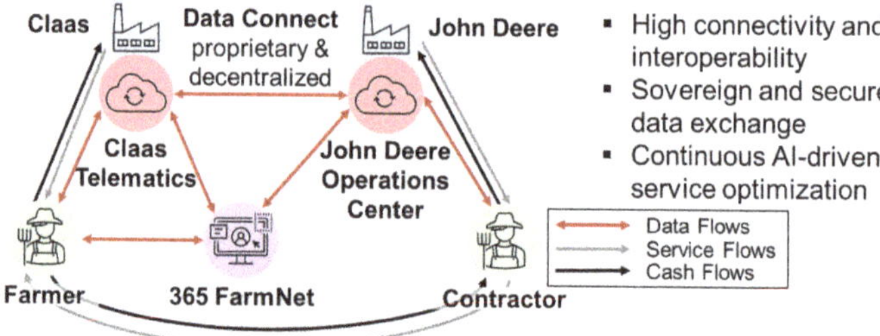

Figure 4. Smart farming services as an example of 'digitainabilized' platform-based business models (Own Illustration based on Plattform Lernende Systeme [58] (p. 16)).

If companies become that innovative, this creates market opportunities for German and European industry to position itself as a market leader regarding sustainable services, products, and service-product system offerings on B2B platforms. Such an ecosystem—given a sufficient level of interoperability—would promise to strengthen the sovereignty of German and European industry. This indicates that the focus areas discussed in this paper are mutually dependent and offer symbiotic potential. Therefore, they deserve future attention in economic implementation and research.

5. Conclusions

Based on an efficient literature review, this review details the status quo regarding the digital transformation of German industry towards Industrie 4.0 ten years after it had been postulated as the new paradigm. On the path towards higher productivity and mass customization, large-scale enterprises proceed faster than small- and middle-sized enterprises, which face competitive pressure and internal barriers towards digitalization to a greater degree. Lack of know-how and strategic courage is problematic in this regard. Industry-wide, the shortage of qualified specialists with both high domain expertise and ICT know-how significantly inhibits the adoption of Industrie 4.0. Coupled with a lack of adequate base technologies for Industrie 4.0 and the lack of data exchange platforms, the creation of platform-based digital business models in industry is especially lagging. In general, it has become evident that academic and practice-oriented research drives and supports the implementation of Industrie 4.0 across all relevant key themes. Its insights require accelerated application into practice.

Furthermore, relying on the matrix method of literature review, this contribution identifies three fields of action in the three strategic areas of the next stage of Industrie 4.0. To strengthen *sovereignty* of the industrial sector, resilience needs to be increased by complementing efficient but failure-prone value chains with flexible and robust value creation networks based on a high level of data integration built on Industrie 4.0 technologies. Interoperability is key in this regard, as it allows seamless cooperation within value creation networks. Open-Source Software promises to strengthen technical *interoperability* and is a significant lever for value creation and particularly attractive for SME facing cost and time pressure to adapt faster towards Industrie 4.0. Interoperable value creation networks also allow for the creation of sustainable platform-based business models as a core element for the next stage of Industrie 4.0. *Sustainability*—given rising socioeconomic inequalities and the climate crisis, and the corresponding policy incentives for sustainability—already impacts industrial economizing in the 21st century [72]. If it is strategically coupled with digitalization it promises to become a success factor, shaping new business opportunities for German industry; 'Digitainability made in Germany' could couple industrial excellence with digital skills on platforms for the good of the economy, environment, and society [65,66]. An intertwinement of these strategic areas and fields of action leveraging synergetic potentials promises to cure current weaknesses in industry regarding adoption of Industrie 4.0, to overcome the obstacles in the industry sector, and to exploit the potentials of Industrie 4.0 technologies in a productive manner in the next decade of Industrie 4.0.

For the practice-oriented research community, this review details the status quo of adoption and research on R&D from a German perspective based in the strategic objectives set by the federal government. This creates an evidence base for future cross-country comparisons of the status quo of Industrie 4.0. The identified differences between SMEs and large-scale enterprises regarding its adoption are of significance for interdisciplinary research to identify their causes and interrelations. Intensified research efforts are required in this regard. The identified future fields of action in the strategic areas of Industrie 4.0 necessitate further research attention, as they might allow acceleration of the digital transformation of the industrial sector. Here, a holistic perspective comprising the four key themes applied on these development paths is of importance.

The implications of this review are not restricted to research, but are also directed to industrial companies and political stakeholders. The review on the status quo of Industrie 4.0 adoption focusing on the causes of different degrees of maturity serves to inform corporate management about current weaknesses and opportunities to increase competitiveness. All industrial companies must anchor digitalization within core company strategy, comprising the strengthening of the digitalization culture, introducing clear accountability among top-level management, establishing a program for lifelong learning, and taking a long-term perspective in assessing the added value of digitalization measures to company revenues. The German government needs to upgrade the basic infrastructure for Industrie 4.0, especially for Industrial AI; must develop an ICT skills strategy for the industrial sector; and is advised to emphasize the strategic areas of Industrie 4.0 to maintain and increase the competitiveness of German industries. Concurrently, the future fields of action raised by this review could inspire strategic middle-term orientation of industrial companies, as well as providing evidence for political stakeholders for the orientation of future Industrie 4.0 funding programs. In this way, this review extensively supports the accumulation of knowledge for science, application, and politics with regard to the current and future orientation of Industrie 4.0 in Germany.

Author Contributions: J.W., A.F. and J.B. have developed all sections of the paper jointly. All authors have read and agreed to the published version of the manuscript.

Funding: This research received no external funding.

Institutional Review Board Statement: Not applicable.

Informed Consent Statement: Not applicable.

Data Availability Statement: Not applicable.

Acknowledgments: The authors are delighted to thank Joachim Sedlmeir and an anonymous referee for many helpful suggestions on a draft of this entry, and to Victoria Neubert for valuable proof-reading.

Conflicts of Interest: The authors declare no conflict of interest. The funders had no role in the design of the study; in the collection, analyses, or interpretation of data; in the writing of the manuscript, or in the decision to publish the results.

References

1. Autio, E.; Thomas, L.D.W. Value co-creation in ecosystems: Insights and research promise from three disciplinary perspectives. In *Handbook of Digital Innovation*; Nambisan, S., Lyytinen, K., Yoo, Y., Eds.; Edward Elgar Publishing: Cheltenham, UK, 2020; pp. 107–132. [CrossRef]
2. Lingens, B.; Miehé, L.; Gassmann, O. The ecosystem blueprint: How firms shape the design of an ecosystem according to the surrounding conditions. *Long Range Plan.* **2021**, *54*, 102043. [CrossRef]
3. Kagermann, H.; Wahlster, W.; Helbig, J. *Recommendations for Implementing the Strategic Initiative Industrie 4.0: Final Report of the Industrie 4.0 Working Group*; Research Union of the German Government: Berlin, Germany, 2012.
4. Spath, D.; Gausemeier, J.; Dumitrescu, R.; Winter, J.; Steglich, S.; Drewel, M. Digitalisation of Society. In *Handbook of Engineering Systems Design*; Maier, A., Oehmen, J., Vermaas, P.E., Eds.; Springer: Cham, Germany, 2021; pp. 1–27. [CrossRef]
5. Salvador, F.; Piller, F.; Aggarwal, S. Surviving on the long tail: An empirical investigation of business model elements for mass customization. *Long Range Plan.* **2020**, *53*, 101886. [CrossRef]
6. Kagermann, H.; Wahlster, W. Ten Years of Industrie 4.0. *Sci* **2022**, *4*, 26. [CrossRef]
7. Forschungsbeirat der Plattform Industrie 4.0 (Ed.) Key Themes of Industrie 4.0. Research and Development Needs for Successful Implementation of Industrie 4.0. 2019. Available online: https://en.acatech.de/publication/key-themes-of-industrie-4-0/download-pdf/?lang=en (accessed on 19 April 2022).
8. vom Brocke, J.; Simons, A.; Niehaves, B.; Riemer, K.; Plattfaut, R.; Cleven, A. Reconstructing the Giant: On the Importance of Rigour in Documenting the Literature Search Process. In Proceedings of the 17th European Conference on Information Systems, Verona, Italy, 26 June 2009; pp. 2206–2217.
9. Denyer, D.; Tranfield, D. Producing a systematic review. In *The Sage Handbook of Organizational Research Methods*; Buchanan, D.A., Bryman, A., Eds.; Sage Publications Ltd.: Thousand Oaks, CA, USA, 2009; pp. 671–689,. ISBN 978-144-620-064-3.
10. Mayring, P. Qualitative Inhaltsanalyse. In *Handbuch Qualitative Forschung: Grundlagen, Konzepte, Methoden und Anwendungen*; Flick, U., von Kardoff, E., Keupp, H., Rosenstiel, L.V., Wolff, S., Eds.; Beltz: München, Germany, 1991; pp. 209–213, ISBN 362-127-229-1.
11. Klopper, R.; Sam, L.; Hemduth, R. The matrix method of literature review. *Alternation* **2007**, *14*, 262–279. [CrossRef]

12. Plattform Industrie 4.0 (Ed.) 2030 Vision for Industrie 4.0. Shaping Digital Ecosystems Globally. 2019. Available online: https://www.plattform-i40.de/IP/Redaktion/EN/Downloads/Publikation/Vision-2030-for-Industrie-4.0.pdf?__blob=publicationFile&v=9 (accessed on 21 April 2022).
13. Statistisches Bundesamt (Ed.) Volkswirtschaftliche Gesamtrechnungen. Inlandsproduktberechung 2021. Fachserie 18 Reihe 1.1. 2022. Available online: https://www.destatis.de/DE/Themen/Wirtschaft/Volkswirtschaftliche-Gesamtrechnungen-Inlandsprodukt/Publikationen/Downloads-Inlandsprodukt/inlandsprodukt-erste-ergebnisse-pdf-2180110.pdf?__blob=publicationFile (accessed on 15 April 2022).
14. WTO. World Trade Report 2021. Economic Resilience and Trade. 2021. Available online: https://www.wto.org/english/res_e/booksp_e/wtr21_e/00_wtr21_e.pdf (accessed on 13 May 2022).
15. OECD. Main Economic Indicators. Volume 2022 Issue 5. 2022. Available online: https://www.oecd-ilibrary.org/deliver/bb256505-en.pdf?itemId=%2Fcontent%2Fpublication%2Fbb256505-en&mimeType=pdf (accessed on 13 May 2022).
16. Bundesministerium für Wirtschaft und Energie. Industrial Strategy 2030. Guidelines for a German and European Industrial Policy. 2019. Available online: https://www.bmwi.de/Redaktion/EN/Publikationen/Industry/industrial-strategy-2030.pdf?__blob=publicationFile&v=7 (accessed on 9 May 2022).
17. Destatis. Germany. Statistical Country Profile, Edition 02/2022. 2022. Available online: https://www.destatis.de/EN/Themes/Countries-Regions/International-Statistics/Country-Profiles/germany.pdf?__blob=publicationFile (accessed on 13 May 2022).
18. Kinkel, S. Industrie in Deutschland: Kern Wirtschaftlichen Wachstums und inländischer Wertschöpfung. In *Die Modernität der Industrie*; Priddat, B.P., West, K.-W., Eds.; Metropolis-Verlag: Marburg, Germany, 2012; pp. 193–214, ISBN 978-3-89518-936-4.
19. Institut für Weltwirtschaft. Analyse der industrierelevanten wirtschaftlichen Rahmenbedingungen in Deutschland im Internationalen Vergleich. Endbericht and das Bundesministerium für Wirtschaft und Energie, Referat I C 4. 2020. Available online: https://www.bmwi.de/Redaktion/DE/Publikationen/Studien/industriestudie.pdf?__blob=publicationFile&v=4 (accessed on 13 May 2022).
20. Kagermann, H.; Winter, J. The second wave of digitalization. In *Germany and the World 2030: What Will Change. How we Must Act*; Mair, S., Messner, D., Meyer, L., Eds.; Econ: Berlin, Germany, 2018; pp. 216–227. ISBN 343-021-010-0.
21. Kagermann, H. Change Through Digitization—Value Creation in the Age of Industry 4.0. In *Management of Permanent Change*; Albach, H., Meffert, H., Pinkwart, A., Reichwald, R., Eds.; Springer: Wiesbaden, Germany, 2015; pp. 23–45. ISBN 978-3-658-05014-6.
22. Kagermann, H.; Nonaka, Y. Revitalizing Human-Machine Interaction for the Advancement of Society—Perspectives from Germany and Japan. Acatech DISKUSSION. 2019. Available online: https://en.acatech.de/publication/revitalizing-human-machine-interaction-for-the-advancement-of-society-perspectives-from-germany-and-japan/ (accessed on 3 May 2022).
23. Bundesministerium für Wirtschaft und Energie. Digitale Strategie 2025. 2016. Available online: https://www.bmwi.de/Redaktion/DE/Publikationen/Digitale-Welt/digitale-strategie-2025.pdf?__blob=publicationFile&v=8 (accessed on 3 May 2022).
24. Forschungsbeirat der Plattform Industrie 4.0 (Ed.) Blinde Flecken in der Umsetzung von Industrie 4.0—Identifizieren und Verstehen. 2022. Available online: https://www.acatech.de/publikation/blinde-flecken-i40/download-pdf/?lang=wildcard (accessed on 3 May 2022).
25. Schuh, G.; Anderl, R.; Dumitrescu, R.; Krüger, A.; Ten Hompel, M. (Eds.) Using the Industrie 4.0 Maturity Index in Industry. Current Challenges, Case Studies and Trends. Acatech COOPERATION. 2020. Available online: https://www.acatech.de/publikation/der-industrie-4-0-maturity-index-in-der-betrieblichen-anwendung/download-pdf/?lang=en (accessed on 9 May 2022).
26. Forschungsbeirat der Plattform Industrie 4.0 (Ed.) Künstliche Intelligenz zur Umsetzung von Industrie 4.0 im Mittelstand. Expertise des Forschungsbeirats der Plattform Industrie 4.0. 2020. Available online: https://www.acatech.de/publikation/fb4-0-ki-in-kmu/download-pdf/?lang=de (accessed on 3 May 2022).
27. European Commission. The Digital Economy and Society Index (DESI) 2020. 2020. Available online: https://ec.europa.eu/newsroom/dae/document.cfm?doc_id=67086 (accessed on 3 May 2022).
28. Deutscher Industrie- und Handelskammertag. Fachkräfteengpässe schon über Vorkrisenniveau DIHK-Report Fachkräfte 2021. Available online: https://www.dihk.de/resource/blob/61638/9bde58258a88d4fce8cda7e2ef300b9c/dihk-report-fachkraeftesicherung-2021-data.pdf (accessed on 5 May 2022).
29. Acatech (Ed.) Kompetenzentwicklungsstudie Industrie 4.0. Erste Ergebnisse und Schlussfolgerungen. 2016. Available online: https://www.acatech.de/publikation/kompetenzentwicklungsstudie-industrie-4-0-erste-ergebnisse-und-schlussfolgerungen/download-pdf/?lang=de (accessed on 5 May 2022).
30. Acatech (Ed.) Kompetenzen für Industrie 4.0. Qualifizierungsbedarfe und Lösungsansätze. Acatech POSITION. 2016. Available online: https://www.acatech.de/publikation/kompetenzen-fuer-industrie-4-0-qualifizierungsbedarfe-und-loesungsansaetze/download-pdf/?lang=de (accessed on 5 May 2022).
31. Mrugalska, B.; Ahmed, J. Organizational Agility in Industry 4.0: A Systematic Literature Review. *Sustainability* **2021**, *13*, 8272. [CrossRef]
32. Gausemeier, J.; Dumitrescu, R.; Echterfeld, J.; Pfänder, T.; Steffen, D.; Thielemann, F. *Innovationen für die Märkte von Morgen: Strategische Planung von Produkten, Dienstleistungen und Geschäftsmodellen*; Carl Hanser Verlag: München, Germany, 2019; ISBN 978-3-446-42824-9.

33. Hermann, M.; Pentek, T.; Otto, B. Design Principles for Industrie 4.0 Scenarios. In Proceedings of the 49th Hawaii International Conference on System Sciences (HICSS), Honolulu, HI, USA, 5–8 January 2016; pp. 3928–3937. [CrossRef]
34. Forschungsbeirat der Plattform Industrie 4.0 (Ed.) Open Source als Innovationstreiber für Industrie 4.0. Expertise des Forschungsbeirats der Plattform Industrie 4.0. 2022. Available online: https://www.acatech.de/publikation/open-source-i40-innovationstreiber/download-pdf/?lang=de (accessed on 5 May 2022).
35. Zeid, A.; Sundaram, S.; Moghaddam, M.; Kamarthi, S.; Marion, T. Interoperability in Smart Manufacturing: Research Challenges. *Machines* **2019**, *7*, 21. [CrossRef]
36. Thoben, K.-D.; Wiesner, S.; Wuest, T. "Industrie 4.0" and Smart Manufacturing—A Review of Research Issues and Application Examples. *Int. J. Autom. Technol.* **2017**, *11*, 4–16. [CrossRef]
37. Glass, R.; Meissner, A.; Gebauer, C.; Stürmer, S.; Metternich, J. Identifying the barriers to Industrie 4.0. *Procedia CIRP* **2018**, *72*, 985–988. [CrossRef]
38. Trappey, A.J.C.; Trappey, C.V.; Govindarajan, U.H.; Sun, J.J.; Chuang, A.C. A Review of Technology Standards and Patent Portfolios for Enabling Cyber-Physical Systems in Advanced Manufacturing. *IEEE Access* **2016**, *4*, 7356–7382. [CrossRef]
39. Forschungsbeirat der Plattform Industrie 4.0 (Ed.) Modellierungs- und Simulationsbedarfe der Intelligenten Fabrik. Expertise des Forschungsbeirats der Plattform Industrie 4.0. 2021. Available online: https://www.acatech.de/publikation/modellierungs-und-simulationsbedarfe-der-intelligenten-fabrik-expertise/download-pdf/?lang=de (accessed on 3 May 2022).
40. Drossel, W.-G.; Ihlenfeldt, S.; Langer, T.; Dumitrescu, R. Cyber-Physische Systeme. Forschen für die digitale Fabrik. In *Digitalisierung*; Neugebauer, R., Ed.; Springer Vieweg: Berlin, Germany, 2018; pp. 197–222, ISBN 978-3-662-55889-8.
41. Lee, J.; Davari, H.; Singh, J.; Pandhare, V. Industrial Artificial Intelligence for industry 4.0-based manufacturing systems. *Manuf. Lett.* **2018**, *18*, 20–23. [CrossRef]
42. Chen, B.; Wan, J.; Shu, L.; Li, P.; Mukherjee, M.; Yin, B. Smart Factory of Industry 4.0: Key Technologies, Application Case, and Challenges. *IEEE Access* **2018**, *6*, 6505–6519. [CrossRef]
43. Sansana, J.; Joswiak, M.N.; Castillo, I.; Wang, Z.; Rendall, R.; Chiang, L.H.; Reis, M.S. Recent trends on hybrid modelling for Industry 4.0. *Comput. Chem. Eng.* **2021**, *151*, 107365. [CrossRef]
44. Kennet, R.S.; Bortman, J. The digital twin in Industry 4.0: A wide-angle perspective. *Qual. Reliab. Eng. Int.* **2021**, *38*, 1357–1366. [CrossRef]
45. André, E.; Bauer, W.; Aurich, J.C.; Bullinger-Hoffmann, A.; Heister, M.; Huchler, N.; Neuburger, R.; Peissner, M.; Stich, A.; Suchy, O. Kompetenzentwicklung für KI. Veränderungen, Bedarfe und Handlungsoptionen. Plattform Lernende Systeme, Ed. 2022. Available online: https://www.plattform-lernende-systeme.de/files/Downloads/Publikationen/AG2_WP_Kompetenzentwicklung_KI.pdf (accessed on 21 April 2022).
46. Forschungsbeirat der Plattform Industrie 4.0 (Ed.) Forschungs- und Entwicklungsbedarfe zur Erfolgreichen Umsetzung von Industrie 4.0. 2. Überarbeite Fassung. Available online: https://www.acatech.de/publikation/themenfelder-i40-akt/download-pdf?lang=de (accessed on 7 August 2022).
47. Nielsen, K.L.; Dimitrova, A. Trump, trust and the transatlantic relationship. *Policy Stud.* **2021**, *42*, 699–719. [CrossRef]
48. Leonard, M.; Pisani-Ferry, J.; Ribakova, E.; Shapiro, J. Redefining Europe's Economic Sovereignty. 2019. Available online: https://www.jstor.org/stable/pdf/resrep28498.pdf (accessed on 21 April 2022).
49. European Commission. On Artificial Intelligence—A European Approach to Excellence and Trust. White Paper. 2020. Available online: https://ec.europa.eu/info/sites/default/files/commission-white-paper-artificial-intelligence-feb2020_en.pdf (accessed on 21 April 2022).
50. Neumann, K.; van Erp, T.; Steinhöfel, E.; Sieckmann, F.; Kohl, H. Patterns for Resilient Value Creation: Perspective of the German Electrical Industry during the COVID-19 Pandemic. *Sustainability* **2021**, *13*, 6090. [CrossRef]
51. Cai, M.; Luo, J. Influence of COVID-19 on Manufacturing Industry and Corresponding Countermeasures from Supply Chain Perspective. *J. Shanghai Jiaotong Univ. (Sci.)* **2020**, *25*, 409–416. [CrossRef]
52. Guan, D.; Wang, D.; Hallegatte, S.; Davis, S.J.; Huo, J.; Li, S.; Bai, Y.; Lei, T.; Xue, Q.; Coffman, D. Global supply-chain effects of COVID-19 control measures. *Nat. Hum. Behav.* **2020**, *4*, 577–587. [CrossRef]
53. Forschungsbeirat der Plattform Industrie 4.0 (Ed.) Wertschöpfungsnetzwerke in Zeiten von Infektionskrisen. Expertise des Forschungsbeirats der Plattform Industrie 4.0. 2021. Available online: https://www.acatech.de/publikation/wertschoepfungsnetzwerke-in-zeiten-von-infektionskrisen-expertise/download-pdf/?lang=de (accessed on 21 April 2022).
54. World Trade Organization. Beyond Production. Global Value Chain Development Report 2021. 2021. Available online: https://www.wto.org/english/res_e/booksp_e/00_gvc_dev_report_2021_e.pdf (accessed on 21 April 2022).
55. Kagermann, H.; Streibich, K.-H.; Suder, K. (Eds.) Digital Sovereignty. Status Quo and Perspectives. Acatech IMPULSE. Available online: https://www.acatech.de/publikation/digitale-souveraenitaet-status-quo-und-handlungsfelder/download-pdf/?lang=en (accessed on 10 July 2022).
56. Plattform Industrie 4.0 (Ed.) Creating the DataSpace Industrie 4.0. Position Paper. 2021. Available online: https://www.plattform-i40.de/IP/Redaktion/EN/Downloads/Publikation/PositionPaper-DataSpace.html (accessed on 21 April 2022).
57. Tantik, E.; Anderl, R. Integrated Data Model and Structure for the Asset Administration Shell in Industrie 4.0. *Procedia CIRP* **2017**, *60*, 86–91. [CrossRef]

58. Bundesministerium für Wirtschaft und Energie. Von der Vision in Die Praxis. Industrie 4.0-Umsetzungsprojekte. 2020. Available online: https://www.plattform-i40.de/IP/Redaktion/DE/Downloads/Publikation/Umsetzungsprojekte.pdf?__blob=publicationFile&v=10 (accessed on 21 April 2022).
59. Blind, K.; Böhm, M.; Grzegorzewska, P.; Katz, A.; Muto, S.; Pätsch, S.; Schubert, T. The Impact of Open Source Software and Hardware on Technological Independence, Competitiveness and Innovation in the EU Economy. European Commission, Ed. 2021. Available online: https://digital-strategy.ec.europa.eu/en/library/study-about-impact-open-source-software-and-hardware-technological-independence-competitiveness-and (accessed on 19 April 2022).
60. Bundesministerium für Umwelt, Naturschutz und Nukleare Sicherheit. Klimaschutzplan 2050. Klimaschutzpolitische Grundsätze und Ziele der Bundesregierung. 2019. Available online: https://www.bmuv.de/fileadmin/Daten_BMU/Download_PDF/Klimaschutz/klimaschutzplan_2050_bf.pdf (accessed on 19 April 2022).
61. Bundesregierung. Klimaschutzgesetz 2021. Generationenvertrag für das Klima. 2021. Available online: https://www.bundesregierung.de/breg-de/themen/klimaschutz/klimaschutzgesetz-2021-1913672 (accessed on 19 April 2022).
62. Sautter, B. Shaping Digital Ecosystems for Sustainable Production: Assessing the Policy Impact of the 2030 Vision for Industrie 4.0. *Sustainability* **2021**, *13*, 12596. [CrossRef]
63. Chen, M.; Sinha, A.; Hu, K.; Shah, M.I. Impact of technological innovation on energy efficiency in industry 4.0 era: Moderation of shadow economy in sustainable development. *Technol. Forecast. Soc. Change* **2021**, *164*, 120521. [CrossRef]
64. Vinuesa, R.; Azizpour, H.; Leite, I.; Balaam, M.; Dignum, V.; Domisch, S.; Felländer, A.; Langhans, S.D.; Tegmark, M.; Nerini, F.F. The role of Artificial Intelligence in achieving the Sustainable Development Goals. *Nat. Commun.* **2020**, *11*, 233. [CrossRef] [PubMed]
65. Lichtenthaler, U.C. Digitainability: The Combined Effects of the Megatrends Digitalization and Sustainability. *J. Innov. Manag.* **2021**, *9*, 64–80. [CrossRef]
66. Gupta, S.; Motlagh, M.; Rhyner, J. The Digitalization Sustainability Matrix: A Participatory Research Tool for Investigating Digitainability. *Sustainability* **2020**, *12*, 9283. [CrossRef]
67. Plattform Industrie 4.0 (Ed.) Zehn Thesen, Wie Digitale Geschäftsmodelle Nachhaltigkeit in der Industrie 4.0 fördern. 2019. Available online: https://www.plattform-i40.de/IP/Redaktion/DE/Downloads/Publikation/Thesen-Nachhaltigkeit-Geschaeftsmodelle.pdf?__blob=publicationFile&v=6 (accessed on 19 April 2022).
68. Plattform Lernende Systeme (Ed.) Mit KI den Nachhaltigen Wandel gestalten. Zur Strategischen Verknüpfung von KI und Nachhaltigkeitsmanagement in Wirtschaft, Wissenschaft und Gesellschaft. 2022. Available online: https://www.plattform-lernende-systeme.de/files/Downloads/Publikationen/PLS_Booklet_Mit_KI_den_nachhaltigen_Wandel_gestalten.pdf (accessed on 7 August 2022).
69. Plattform Lernende Systeme (Ed.) KI im Mittelstand. Potenziale Erkennen, Voraussetzungen Schaffen, Transformation Meistern. 2022. Available online: https://www.plattform-lernende-systeme.de/files/Downloads/Publikationen/PLS_Booklet_KMU.pdf (accessed on 10 July 2022).
70. Gawellek, M. Digitale Vernetzung für nachhaltige Geschäftsmodelle. In *Klimawandel in der Wirtschaft. Warum Wir ein Bewusstsein für Dringlichkeit Brauchen*; Hildebrandt, A., Ed.; Springer Gabler: Wiesbaden, Germany, 2020; pp. 189–208. [CrossRef]
71. Frank, M.; Koldewey, C.; Rabe, M.; Dumitrescu, R.; Gausemeier, J.; Kühn, A. Smart Services—Concept of a New Market Offering. *Z. Wirtschatftlichen Fabr.* **2018**, *113*, 306–312. [CrossRef]
72. Stock, T.; Obenaus, M.; Kunz, S.; Kohl, H. Industry 4.0 as enabler for a sustainable development: A qualitative assessment of its ecological and social potential. *Process Saf. Environ. Prot.* **2018**, *118*, 254–267. [CrossRef]

MDPI

St. Alban-Anlage 66

4052 Basel

Switzerland

www.mdpi.com

Sci Editorial Office

E-mail: sci@mdpi.com

www.mdpi.com/journal/sci